Energy Demand and Efficient Use

ETTORE MAJORANA INTERNATIONAL SCIENCE SERIES

Series Editor:
Antonino Zichichi
European Physical Society
Geneva, Switzerland

(PHYSICAL SCIENCES)

Recent volumes in the series:

Energy Demand and Efficient Use

Edited by

Fernando Amman

University of Pavia
Pavia, Italy

and

Richard Wilson

Harvard University
Cambridge, Massachusetts

Plenum Press · New York and London

Library of Congress Cataloging in Publication Data

International School on Energetics (4th: 1980: Erice, Italy)
   Energy demand and efficient use.

   (Ettore Majorana international science series. Physical sciences; v. 9)
   "Proceedings of the Fourth International School on Energetics held July 15-24,
1980, in Erice, Sicily"–Verso of t.p.
   Bibliography: p.
   Includes index.
   1. Power resources–Congresses. 2. Energy conservation–Congresses. I. Amman,
Fernando Maria. II. Wilson, Richard, 1926-   . III. Title. IV. Series.
TJ163.15.I5673 1980   333.79   81-2803
ISBN 978-1-4684-3961-8   ISBN 978-1-4684-3959-5 (eBook)   AACR2
DOI 10.1007/978-1-4684-3959-5

Proceedings of the Fourth International School on Energetics,
held July 15–24, 1980, in Erice, Sicily

© 1981 Plenum Press, New York
Softcover reprint of the hardcover 1st edition 1981
A Division of Plenum Publishing Corporation
233 Spring Street, New York, N.Y. 10013

PREFACE

The Fourth International School on Energetics was held in July 1980 at Erice, Sicily and was devoted to the subject of Energy Demand and Efficient Use. In contrast to the Third School, we chose to concentrate on the demand side of the energy equation.

The initial emphasis was on the methodology for determining demand; but it soon became clear that it is necessary to control demand also.

All too often energy policy is set by the large industrial nations, or those nations blessed with a plentiful supply of fossil fuels. It seemed to us important to have the views of some representatives from the less developed countries.

The manuscripts were collected and ordered by Ms. Diane Rolinski of Harvard University whose work was invaluable.

Fernando Amman
Richard Wilson
Directors of the School

The course would not have been possible without the financial support of the Italian Ministry of Public Education, Italian Ministry of Scientific and Technological Research, Italian National Research Council, National Electric Energy Council, National Nuclear Energy Council and Sicilian Regional Government.

CONTENTS

ENERGY DEMAND CONTROL IN ENERGY POLICY

J.M. MARTIN

Institut Economique et Juridique de l'Energie

Grenoble, France

INTRODUCTION

Why try to control the energy demand, whose growth is often presented as an expansion factor of economic activity? This is the question still asked by certain energy-policy makers, who hesitate to take into account the effect on demand in the same way as the effect on supply control, especially the effects which aim to increase production capacity.

We are going to try to answer this question by showing how, in the course of the last ten years, the idea has grown up that it is not only possible, but to be hoped for, that the energy demand can be controlled, and made into an endogeneous energy-policy variable (chap. 1).

Later on, we will consider what the limits of this control are, that is to say whether or not it is compatible with the continuation of economic growth, which leads us to bring up the subject of the debate between those who support the theory that economic growth and increased energy-consumption are strictly related, and those who oppose it (chap. 2). In reality, as we shall see, the situation is not as harsh as it seems: beyond the structural rigidity, choice of technique allows a certain margin for flexibility, which could be put to use in energy policies aiming to control the demand (chap. 3).

But to what extent is it desirable to limit energy consumption? Is it necessary, as certain technicians suggest, to exploit the entire technical potential of energy saving? Are they per-

1

haps forgetting that energy-saving investments must be treated in
the same way as all other investments if one wants to avoid wasting economic resources? In order to eliminate this risk, it is
necessary to seek the criteria which will allow energy policy makers to work towards specific aims and to fix norms. These norms
allow energy supply and demand to be treated coherently. (chap. 4)

CHAPTER 1 - ENERGY DEMAND, AN ENDOGENEOUS OR EXOGENEOUS VARIABLE?

Regardless of the characteristics of their energy policy,
no country had considered controlling the energy demand before
the mid 1970s. Energy policies have always been supply-policies,
in that they are only concerned with adapting the supply volume
and structure (of domestic or exterior origin) to an evolution
of the overall demand, (often called energy needs), considered
as an exogeneous variable. How is this conception of energy
policy recognisable? What is it based on? Why was it modified
during the course of the 1970s?

When a list of the measures taken by public authorities
concerning energy is drawn up, it is very rare to find authorities whose object is to limit the quantities of energy required
in order to obtain a given service (see Graph n°1). The French
legislation of 1948 making obligatory an initial technical test
before the installation of boilers of a certain size, destined
to guarantee its thermal efficiency, is explained by the energy
shortages immediately following the war.Most often, public authorities esteem that they do not need to take a special interest
in the techniques employed or in the specific equipment used to
convert the various energy-sources (coal, fuel-oil, petrol, natural gas, electricity), with the intention of using it as efficiently as possible (lighting and heating for housing, production
of driving force, drying out of materials ...). This domain is
implicity considered to be uniquelly the concern of the design
engineers, and the inference is that they have access to the
best available technology. This is in fact true for the most
part, since specific energy consumption has been in steady
decrease.

The limitations in the field of energy policy are shown
even more clearly when its conception is taken into consideration. For example, Michael Posner (1) considers that the working
-out of an energy policy consists of:
        - specifying cost functions for the different accessible
sources of primary energy (coal, oil, natural gas, hydraulics,
nuclear energy ...) a function which we can eventually modify
if we think that the market prices do not correctly reflect
the social costs;
        - specifying demand functions, that is to say to go from
a deduced agregate demand of the expected GNP growth to a specific demand obtained by the elasticity-substitution;

Graph n° – From Primary Sources to Energy Services

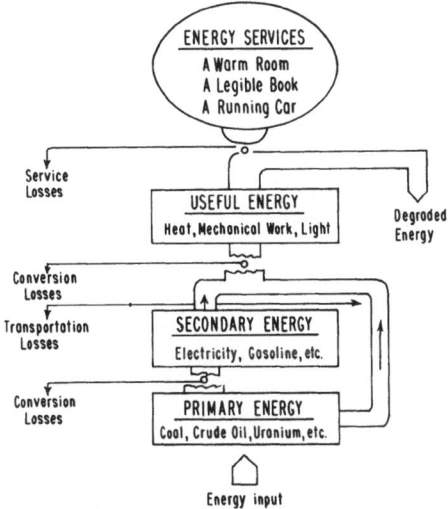

Source : HAFELE– Energy Demand, IIASA

- combining these two functions groups by iterative
methods untils we get a sectoral optimum, that is to say "a set
of investment, output and pricing decisions which subject to the
uncertainties and inflexibilities to which we have drawn attention, will minimise the (social) cost of producing a given output of fuel and simultaneously determine what the output should
be" (p. 3).

It is worthwhile considering the conception of energy-
demand. In actual fact it is noticeable that:
- If the various specific demands, (coal, fuel-oil, natural gas, L.T. electricity, H.T. electricity) are closely linked
up, over a projected time-span, to the expected change in respective energy prices (for example, $\underline{\text{Heavy fuel-oil price}}$);
$$\text{Bituminous coal price}$$
- the resulting substitutions, which change the structure of the final energy demand, are included in an overall
energy demand which is totally exogeneous, in that it is independent of the change in relative energy price (for example,
$\underline{\text{Average energy price}}$).
Average price of equipment

So, whether a country's energy is supplied at an increasing real cost, or at a decreasing real cost, the expected
overall evolution of energy demand, supposed to be price
inelastic, does not change.

For the energy policy makers, this demand is therefore
an exogeneous factor; a result of GNP growth to which it is
linked by output-elasticity.

It is precisely, this whole conception of energy policy
which changes during the course of the 1970s when (as we shall
see in the next chapter), the relationship between economic
growth and energy consumption is questioned. If the latter is
not just a specific by-product of economic growth, it might be
sensible to consider its possible evolution. If the same supply
can be obtained with variable quantities of primary sources, it
could be less expensive to save one ton of crude oil than to
produce an extra ton. The overall energy demand goes from being
an exogeneous variable to being an endogeneous one, which can
then affected by energy policy markers.

Before examining the conditions of this action, it is
suitable to discuss the domain it can be used in, with regard
to economic growth.

CHAPTER 2 - ECONOMIC GROWTH AND ENERGY CONSUMPTION

Is it possible to control the energy demand, with the

Inset n° 1 – Elasticity coefficients

"Elasticity is the dimensionless slope of a curve plotted on
double – log – paper"

T.C. KOOPMANS – Economics among the science, p. 13

We can use 3 kinds of elasticity coefficients

1/ Income or output elasticity

$$\lambda E = \frac{y}{E} \times \frac{\delta E}{\delta y}$$

Relation of the same sign between the relative variations of E (energy demand) and y (GNP)

$$= \frac{d \log (E)}{d \log (y)}$$

2/ Price elasticity

$$\gamma E = \frac{p}{E} \times \frac{\delta E}{\delta p}$$

Relation (sign changed) between the relative variations of E (energy demand) and p (energy price)

$$= \frac{d \log (E)}{d \log (p)}$$

3/ Substitution elasticity

$$\varepsilon \, h/k = \frac{ph/pk}{qh/qk} \times \frac{\delta (qh/qk)}{\delta (ph/pk)}$$

q and p being the quantities and the prices of two energy sources h and k which may be substituted, the elasticity measures the change qh/qk which indicated a variation of relative prices ph/pk.

The energy demand E is more or less elastic to the GNP (y) or
to the price (p) according to whether the absolute value of
the elasticity coefficient is more or less greater within a
variation field which, in practice, goes from 0 to 1 in the
first case end from 0 to −1 in the second. It is said to be very
elastic when a small change of y or p leads to an important modification of E. On a logarithmic graph a demand curve, with a
constant elasticity, is represented by a straight line.

intention of limiting its growth, without interrupting economic
growth ipso facto? Or, which comes to the same thing, can an elasticity decrease of the total energy consumption be included in
the aims of an energy policy, in relation to the GNP? If so, to
what extent is it possible?

At the centre of this debate are two opposing theses,
which we are going to try and sum up in order to understand the
discussion fully.

Let us begin with the theory of the so-called "Neo-
classical" economists, although it is not chronologically the
first, because, in the past, economists were primarily interested in substitutions among energy-sources, that is to say that
they tried to understand the structure changes rather than the
level changes of the energy-demand. But for several years now,
they have been intervening more often in the debate about the
possibility of reducing the growth of energy demand. Generally-
speaking, they can be analysed in the following way: in the neo-
classical thesis there is not one single production factor
(energy, work, capital) whose rareness would dominate the others,
and which would become a unity of natural value for all. Let us
hear T.C. KOOPMANS on this particular point: (2)

> "The model of production is such that —not by logical
> necessity, but as an empirical fact— any primary input to
> production can be substituted to some extent for any other.
> If such substitutions do not take place within one-and-
> the-same production process, then it can still eome about
> through suitable changes in the levels of several proces-
> ses and in the inputs of these. In this view, "the energy
> problem" is not one of "just saving energy", regardless of
> the cost in other ressources. It is rather one of seeing
> to it that the increased market power of OPEC, are —over
> time— reflected in the real prices of primary energy-forms
> relative to other primary inputs, and thereby in different
> degrees in the prices of all other goods and services"(p.7)

It ought to be possible therefore over a long period of
time, to reduce the growth in energy-consumption quite drastically, on two conditions:
- that the relative price of energy (that is to say in relation to that of other production factors) rises in proportion to its relative scarcity;
- that nothing hinders factor substitutions (at all levels
and in all their forms), which ought, logically, to follow
on from the changes of the price system.

But to what extent can one envisage this reduction? With
the help of models, several people have proposed answers to this.
The "Modeling Resource Group" of the "Committe on Nuclear and

Alternative Energy System" (CONAES), for example, has come to
the conclusion that reductions of 10 to 20% in energy-availabi-
lity, in relation to the base case, would only provoke a con-
traction of 1 to 2% of the discounted GNP value in the USA
between 1975 and 2010! Greater reductions (up to 60% of the estimated requirements for 2010) would lead to contractions of the
same GNP between 2 and 20% according to the hypotheses of price
elasticity adopted (from −0.5 to −0.25) which gives an elasticity
coefficient which is far removed from 1. There is not point in
insisting on the limits and fragility of this type of exercice,
which is based on unchecked hypotheses concerning the possibility
of substituting production factors (3).

It is precisely this possibility of substitution between
energy and other production which is much disputed by those who
support the second theory. Malcolm SLESSER, for example, has
just written a book to show (4):
"Why energy cannot be treated as just another input"(p.IX).
According to him, the hypothesis of wide substitution between
factors prevents economists from understanding that:
"There is a real possibility that energy, which we have
come to depend upon for our life styles and wealth, may
become the limiting factor in economic growth" (p. 5).

Where does this specific feature of energy amongst all
the production factors come from? From a physical condition of
all material production, recognised, moreover, by certain economists. Thus for Hollis CHENERY mentioned by M. SLESSER (p. 42):
" To the economist, production means anything that happens
to an object or set of objects which increases in value.
Usually, this results in a form, but it may be merely a
change in space or time. The basic physical condition
necessary to effect any of these changes (except the last)
is that energy must be applied to the material. Application of energy in some form is one element, common to the
economist's and the engineer's concept of production".

A very similar argument is developed by K. BOULDING, another economist, when he seeks the physical basis in the role
played by energy in the growth of material production (affluence)
(5). If there exists a very close connection between the latter
and energy consumption, it is because the relationship between
affluence and entropia is a negative one. The commodities which we
use most often constitute an improbable arrangement of material,
whose propension disintegrating and disappearing is very widespread. The only thing which maintains their original state is the
introduction of a certain amount of energy. Therefore no quantitative growth of commodities can be brought about without an increase in the amount of energy available.

Up until the mid 1970s, those who considered there to be
a firm relationship between energy growth and energy consumption
put forward the results of correlations between primary energy
consumption (explained variable) and the GNP (explanatory variable).

Sometimes, research into a correlation concerns a given
country, over a given period of time. Most often, in order to give
the results the universal nature associated with all scientific
laws, the correlation is sought taking into account a large number of countries over one given year (Cross-section). One then
assumes considered to be a universal, homogeneous phenomenon, each
country being obliged to follow the same path. "The laggards try
to catch the leaders" wrote P. PUTNAM on this subject.

Y. MAINGUY uses the findings of E.S. MASON in 1955 (concerning the year 1952), and makes a rearrangement by a line which
"in real quantities, corresponds to a parabola (in a wide sense):
$y = bx^a$"(6). He concludes: "one notices, in the event, that "a"
is hardly different from 1, which means that energy-consumption
is roughly proportional to the national income" (p. 2). This proportional coefficient is even valued at 4000 tce per million US
dollars, that is, a similar value to the one, on finds in the
drawing up of a report for France between 1920 and 1956. This
coincidence cannot be due to chance alone and Y. MAINGUY deduces:
"that at first sight, one sees that the relationship between a
country's energy-consumption and its national income, is presu-
med invariable" (p. 4).

Since then, this procedure (quoted above), has been rewor-
ked several times, notably by J. DARMSTADTER, who calculates the
following equations:
-   on the findings of 1965:
    Log E/p = - 0,34 + 1,21 Log G/p
    with r^2 = 0,87 or 0,89, according to the equivalence
    coefficient chosen for hydroelectricity (graph. n°2)

-   on the findings of 1972:
    Log E/p = 0,082121 + 1,0048 Log G/p
    with r^2 = 0,85.

Although the authors of this correlation are well aware of
the limits of the exercise (J. DARMSTADTER, for example, calls it
a "snapshot"), the salient idea is that a statistically accurate
"Law" exists, according to which all GNP increases imply a growth
proportional to the primary energy consumption (output elasticity
= 1).

However, the attention of certain researchers is drawn to

Graph n°3 Relationship Between Useful Energy Consumed Per Capita
GNP Per Capita

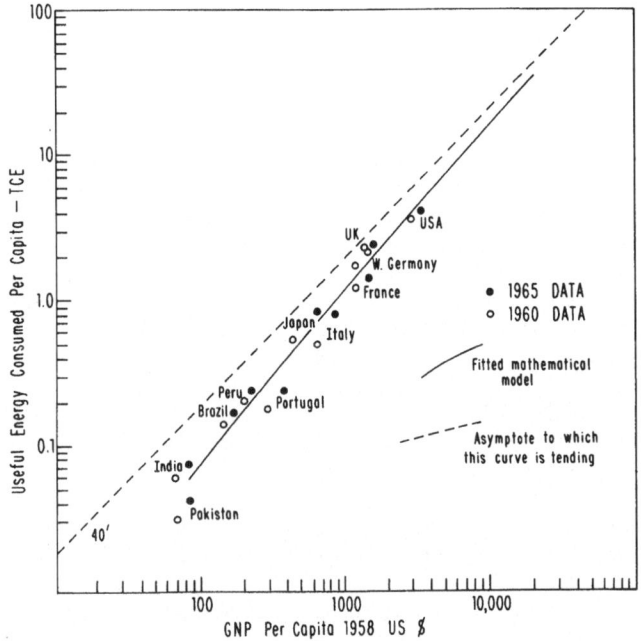

the very wide field of variation for the output elasticity coefficients, when these have been calculated country by country.
P.E. JANOSI and L.E. GRAYSON (7), who deal with 30 countries
during the period 1953-1965, observe values which range from
            - 2,07 (Philippines); 1,93 (Thaïlande); 1,89 (India);
            - to 0,96 (United States); 0,88 (France); 0,82 (Belgium);
              0,66 (Federal Germany); 0,48 (United Kingdom).
Even if it is not the only one, the level of industrialisation
(value added by industry/GNP) seems to be the most significant
explanatory factor, unless the gaps may be first of all attributable to the data, since only the so-called commercial energy-
ressources are taken into account, and one doesn't take into
account the efficiency of energy-use.

        This last way of considering the question had already
been brought to light by F.G. ADAMS and P. MIOVIC, who wrote,
as early as 1968 (8),
            "It is likely that the energy-elasticity is being under-
            estimated" (p. 42),
because (in order to calculate the global primary energy consumption) variable quantities (varying from one year to the next),
are put together, of products whose effective usefulness is
different from those expressed by the usual equivalence-coef-
ficients. This does not hinder a constant energy-consumption
structure (by sources); but it is more of a hindrance to a
variable structure. In other words, in the countries where
fuel-oil is rapidly taking the place of coal, it is probable
that the increase of useful energy consumption is larger than
the one mesured in terms of primary energy.

        L.G. BROOKES (4) finds in the thesis of ADAMS & MIOVIC
(for the most part, confirmed), the explanation for the extremely low output-elasticity observed in his country (the United
Kingdom). But, using this as a basis, he aims to reconstruct a
new "Law" linking GNP increases and useful energy consumption,
in a direct way. He sets out his hypothesis in the following
way:
            "It is postulated that, given universal energy units (ie
            incorporating coefficients of usefulness of various fuels)
            and ignoring, for the time being, changes in efficiency
            over time, the energy coefficient should gradually fall
            as output per capita rises, tending asymptotically to
            unity" (p. 4).
This hypothesis is based on the representation used by L.G.
BROOKES of the economic development: a sector I (Industry) grows
at the same rate as the commercial energy sources and gradually
takes over a sector A (Agriculture), which only uses non-commercial energy-sources. From then on, as the GNP increases, the energy consumption output elasticity decreases between:

Graph n°2 Relationships Between Per Capita GNP and Per
Capita Energy Consumption

Source — J. DARMSTADTER. World energy...op-cit, p. 66

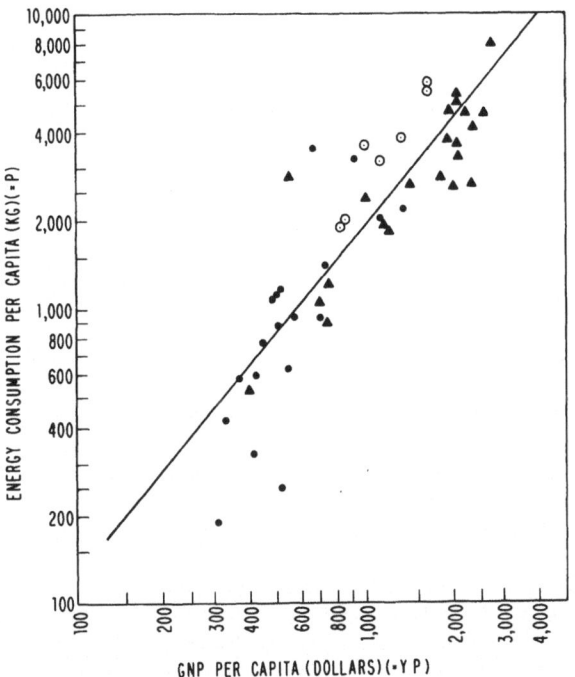

COUNTRY LEGEND : ▲ North America, Western Europe, Oceania, South Africa, Japan
⊙ East European Communist countries and U.S.S.R.
● Latin America, Other Africa and Non Communist Asia

- plus the infinite (+∞) represented by the first infini-
tessimal increase in commercial energy demand,
- and 1, because A is totally eliminated, so any GNP increase comes from an increase in sector I which, by definition, demands a proportional increase in the consumption
of useful energy.

In this way, the effective energy consumption would be
linked to the economic growth by the following function:
$$Eu = G. K f (G)$$
- f (G) is a saturating function, tending towards 1 when
G tends towards infinity;
- K is a parameter giving the asymptotic value of energy
consumption per unit of GNP, which was 1,85 Kce per US $
in 1958 (graph. n°3).

But to prove this "Law", L.G. BROOKES, who doesn't know
the useful energy consumption and its development in each country
any better than anyone else, has had recourse to sleight of hand,
in supposing that in every country in the world, the consumption
structure is the same, and that the technical efficiencies are
same too.

If one joins this second simplification with the first
(representation of economic development), it can be seen that
under this sophisticated guise, the "Law" is no more satisfactory
than the others. The two increases are not, at all times and in
all places, linked to each other by an elasticity coefficient
equal to 1, anymore than they are by a unique and known function.

Does it follow that it is those who defend the first thesis who are right, and that there is wide scope for substitution
between energy and the other production factors?

We do not think so, at least not when the reasoning is
done on a finite time plane, (several decades), and when countries whose economic and technical structures are the results
of a given historical evolution, are considered.

In this respect, the study of the economic activity's
energy content is particularly informative.

CHAPTER 3 - THE ENERGY CONTENT OF ECONOMIC ACTIVITY

Numerous researchers have been interested in this question since 1973 as J. DARMSTADTER, J. DUNKERLEY and J. ALTERMAN
underline, not without a certain irony, in one of the best works
devoted to the subject (10):

Table n°1 - Energy content of the Gross National Product of the main industrialized countries

	Evaluation of the CEREN (MEDINA)		Evaluation of Resources for the Future (DARMSTADTER)	
	in TOE per million of EUR 1971(a)	as an index in relation the U.S.	in TOE per million US 1972	as an index of in relation the U.S.
United States	1544	100	1480	100
Canada			1772	120
France	960	62	795	54
West Germany	1174	76	1031	70
Italy	1148	74	915	62
Netherlands	1448	93	1272	86
United Kingdom	1776	115	1121	76
Sweden			1072	72
Japan	1260	81	849	57
Belgium	1436	93		
Luxemburg	4400(b)	280		

(a) This refers to the European monetary unit which in 1971 was work approximately 5,50 FF.

(b) This very high content is largely due to the consumption of an important steel industry in a small country.

Table n°2 - Explanation of the differences of the energy content of Industry's Value Added in some european countries in relation to France

	Structural Factor (VA_i/IVA)	Technological Factor (C_i/VA_i)
West Germany	+ 46	+ 54
Italy	- 50	+ 150
Netherlands	- 29	+ 129
Belgium	- 23	+ 123

Source: C.E.R.E.N.

"A few years ago the question with this study deals
might have engaged the attention of a contributor to an
obscure technical journal; today it has become an issue
vigorously debated by economists, environmentalists, and
policy-markers. Why is per capita consumption of primary
energy resources so much higher in the United States than
it is in other advanced industrial countries -such as
Sweden, Germany, and France- whose per capita income does
not differ appreciably from that of the United States?"

E. MEDINA (CEREN, France) was the first to try to answer
this question, which he had asked in slightly different terms:
amongst all the industrialised countries, why is it France which
has the lowest energy-content per unit of GNP? (11) J. DARMSTADTER'S
evaluations confirm this observation, although his findings are
slightly different from MEDINA'S, having aimed to correct certain
evaluation errors of the GNP and of primary energy consumption
(see Table n°1).

Numerous factors come into the explanation of these
differences:
      - The average efficiency of the transformation of combus-
tibles into electricity is higher in Italy and France than in
Luxemburg and the United Kingdom;
      - The importance of non-energy-producing uses (petro-
chemical, gaso-chemical ...) which is particularly great in the
Netherlands and in Italy, which are rich in natural gas;
      - The climatic differences which play a very important
role in the difference between household consumption in Italy
and Germany, for example.

But beyond these easily-identified, easily-comprehensible
factors, one also discovers that the difference in final energy
consumption per unit (value) of industrial production, between
one country and another, are sizeable. This could originate from
two factors:
      - Differences in production-structure = VA_i/IVA (IVA =
        Industrial value added)
      - Differences in production techniques = C_i/VA_i (C_i =
        Energy consumption in branche i).

It can be seen (Table n°2) that it is always the second
of these two factors which proves the more important. The differences in industrial structure only intervene in a positive
sense in the difference between Federal Germany and the others,
whereas the technological factor plays an important role in all
the observed differences. This is confirmed by a section-by-
section examination: with the exception of the metallurgy in-

dustry in the Netherlands, the C_i/VA_i are higher in every country
for metallurgy, chemicals, non-metallic mineral products ...

This conclusion is expanded and confirmed by J. DARMSTADTER
who aims to explain the difference of 490 toe per million US $ of
GNP between the United States on one hand and the 6 most industrialised european countries on the other. Energy intensity, expressed by the technological factor, comes into the differences
observed for all uses, and appears to be a determining influence
in industrial uses.

Whenever this is possible, ie when the sectors of activity have a homogeneous production, it is interesting to compare
not C_i/VA_i but he specific consumption C_i/P_i (P expressing a
physical quantity), and this permits the elimination of price-
influence. Where do these differences, which sometimes go from
1 to 2, come from? All the characteristics of industrial equip-
ments (age, size, technical process, energy sources used ...)
play a role. We lack precise studies which would enable us to
identify the exact incidence of each of these factors, but
J. DARMSTADTER notices that:
  - In the cement industry, the wet process, which uses
on average 50% more than the dry process, represents 59% and 60%
of the production capacity of the United States and of the United
Kingdom, as against 5% and 13% in Germany and Italy;
  - In the steel industry, the open furnace which consumes
more than the oxygen furnace, prevails in the United States, and
in the United Kingdom, whereas it is being phased out in the
Netherlands and in Japan; inversely, factories which work 24
hours a day and which diminish the specific energy consumption
are much less widespread in the Unites Kingdom (3%) and the
United States (6%) than in Germany (17%) and in Japan (21%).

The main interest of empirical studies on economic activity's energy-content lies in the bringing to light of two
intervening factors in the explanation of a country's energy-
consumption level. Effectively, one can admit that the part of
energy consumption depending on the structural factor far esca-
pes the control of energy policy, whereas the consumption depending on technical factors can be changed to a certain point by
the same control.

In the studies made by E. MEDINA and J. DARMSTADTER, the
structural effect is reduced to the relationships and proportions
of the various industrial branches (metallurgy, chemicals, textiles, ... etc). This analysis can be extended to consider that
in every country, the economic activity's energy content depends
on the structures' condition from 3 points of view:

- the sharing of economic activity amongst the different branches;
- the distribution of the national revenue amongst the different social groups;
- the space distribution of population, especially between town and country.

These 3 basic structural characteristics command the secondary structural characteristics to a great extent. These are types of habitat (individual or collective); forms of transport (automobile or railway) ... Energy demand control cannot modify these structural elements which, on the one hand change very slowly and which, on the other hand are modified as the result of many factors, amongst which energy supply is not necessarily the most important. So within that there is a rigid energy consumption limiting all voluntary measures aiming to slow down its growth.

But, as we have seen, the structural factor only partly commands the energy consumption. The energy consumption also depends on technical choices, adding the following to structural choice:
- a single industry can have recourse to various processes;
- a single form of transport (aviation, automobile, railway) can use various machines;
- a single type of dwelling can be heated in different ways ...

To conclude, if we come back to the two theses previously outlined, we can see that they are both true to a certain extent, in that:
- All economic growth implying structural change of the same kind known in industrialised countries demands ready supplies of energy;
- within this structural evolution the matter of technical choice is quite open, and makes various energy-consumption growth-rates possible.

CHAPTER 4 - FROM TECHNICAL POTENTIAL TO ECONOMIC OBJECTIVES

If, instead of tackling the problem in a general way, as we have just done, we consider it analytically, we obtain similar results. Specialists in each branch of industry, of every form of transport (aviation, railway, automobile), and of individual and collective housing ... have evaluated the energy gains obtainable by a certain number of technical improvements. By comparing the effective energy consumption obtainable by the most efficient known techniques, each specialist evaluates a potential method of saving energy.

It is sometimes thought that this potential ought to be
completely transformed into effective savings, that is, all the
measures likely to bring about a technically minimal energy consumption, ought to be put to use. Generally-speaking, this view of
things is based on distrust of economic criteria. But it is not
without certain risks, as can be seen when considering the simple
case of a production process where K (value of the machines) and
E (value of the energy they use) can be substituted, and their
relationship represented by a linear homogeneous production function of the following type:

$$Y = \alpha K.E^{(1-\alpha)}$$

For a given output and constant Y, it is possible to turn
to several input combinations between $e_1 \, k_1$ and $e_2 \, k_2$.

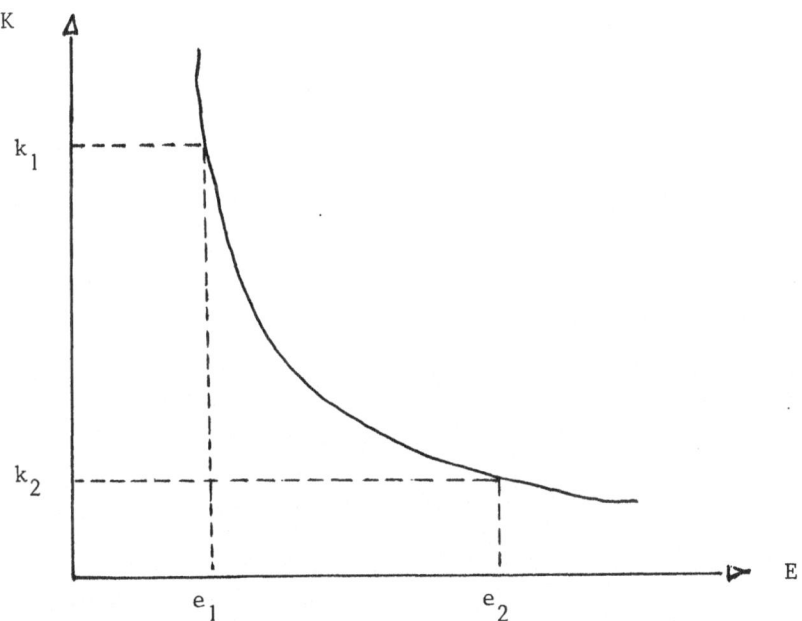

If the one which minimises the energy-consumption is systematically chosen, without taking into account energy price in relation to the machine-price, there is a risk of making investments

which could have bad results, for the company concerned and the
overall economy alike.

For companies, the choice of a productive combination which
does not take into account the relative input price is recognisable, other things being equal, by higher production costs than
competitors' (national or foreign), therefore by a loss of compe-
titivity and perhaps even disappearance.

In a closed economy, this risk can be eliminated if all
companies are persuaded to adopt the same techniques. But if these
techniques, other things being equal, are too far removed from
those which would minimise production cost for a given relative
input price, it is the entire national economy which will suffer.
The latter has to invest far more than necessary with the aim of
saving energy. If the annual share of GNP for investment is considered as a matter of fact, other investments could not be made
in the sectors where they would be much more effective. The national economy would therefore use its production factors badly, and
inhibit its long term development.

The situation just described does not correspond at all
to what can be covered up nowadays in most countries. Usually
the reverse situation is observed: the production schemes conti-
nually chosen are more closely related to $e_2 k_2$ than to $e_1 k_1$.
However, our aim was to show that systematic adoption of techniques which minimise energy consumption, would not be a reasonable
solution. The only correct solution is to choose the technique
capable of minimising the discounted production cost of the equipment and the energy it uses. How would we manage this?

At the micro-economic level, the problem is, in its principle, simple. It is a question of calculating what is the energy saving investment which is justified by the growth of the
energy price. Under the name of cost-effectiveness, this measure
consists in seeking "an investment in conservation (which) will
reduce energy consumption over some period of time such that,
suitably discounted, the present value of the savings minus
the capital cost, is positive" (13).

$$(1) \left\{ \left[\sum_{t=0}^{t=T} \left\{ S_t P_e \right\} \quad dt - K_0 \right] > 0 \right.$$

S_t = energy saved in year t

P_e = price of that energy

K_0 = assumed capital cost in
           year 0

$$T = \text{time horizon}$$
$$d_t = \text{discount factor given by}$$
$$\frac{1}{(1+r)}$$
were r is the rate of
discount

If the price of energy is supposed to increase by x% per year,
one may claim that:

$$(2) \left[\left\{ \sum_{t=0}^{t=T} \left\{ S_t P_e \ (1 + x)^t \ dt \right\} - K_0 \right\} > 0 \right]$$

"Inequalities (1) and (2) are expressed in terms of discrete time.
They could equally well be presented in continuous time units"
(p. 346).

If we suppose:
- that all energy users behave in a perfectly rational
way in an economic world devoid of inertia
- that all correctly anticipate the future evolution of
the level price of energy
- that all included in their calculations a discount rate
which balances the offer and the demand of capital
- and finally that all may obtain without difficulty the
means of financement necessary to investment,
the problem is solved. When the anticipations relative to the
energy price are modified (during the rise) the volume of energy
saving investments increases and the energy consumption (final
and primary) decreases. But all the "ifs" above mentioned are
never all present at the same time.

In as much as it is the concern of the authorities:
- on the one hand to publish a certain amount of information to help decisions makers in their choice;
- on the other hand to eliminate a certain number of
obstacles making this choice more difficult.
So we move from micro-economics to macro-economics.

It is the public authorities who ought to provide two
pieces of information, which the market does not spontaneously
provide, and which are nonetheless indispensable to the calculation of the investments to be made in energy economy. In the
"cost effectiveness" formula, the two necessary pieces of information concern:
- the discount rate,
- the energy price in t
that is its value in the years to come.

At present, any governement wishing to allow decision makers
(company managers in particular) to calculate correctly the energy-
saving investments to be made ought to provide these two details:
the first is a norm resulting from the economic policy adopted;
the second is both an estimation of perceptible tendencies on the
energy market and a norm resulting from the energy policy adopted.

Still on the subject of publicising information, energy
policy makers can extend their work further by working out the
various values K_0 can take in overall energy use, themselves. As
an example, here is the approach taken by the "Agence pour les
Economies de l'Energie", in France:
  - In various sectors and uses, the investment value allowing a saving of 1 toe was calculated;
  - Investments were filed in increasing order to obtain a
general relationship between the amount of money invested and the
volume of energy saved.

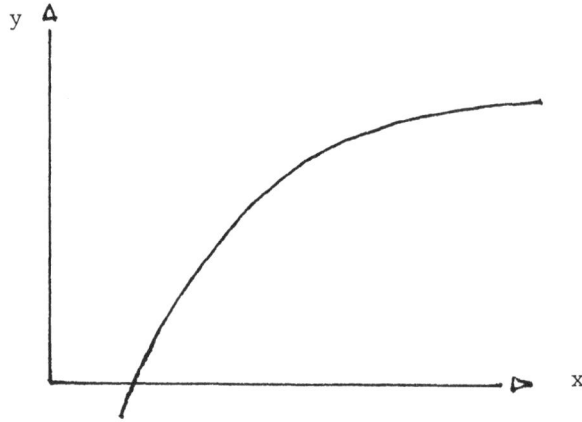

where y = % of the total consumption saved ,
      x = the average cost of investment saving 1 toe

The curve established and interpreted is
    log y = 0.768 Log (x-700) - 3.846

That is 5% saved at a cost of 1905 F/toe
       10%   "   "   3668 F/toe
       15%   "   "   5728 F/toe
       20%   "   "   8009 F/toe
       25%   "   "   10470 F/toe
       30%   "   "   13084 F/toe

Up to what point should energy-saving investements be encouraged?
It all depends on the anticipated energy price. In France, for
example, this   threshold was fixed in 1976, at 2500 Frs per toe
saved per annum. Any investment below this price was considered
worth encouraging as it was inferior to the least costly investment for obtaining a supplementary toe. This threshold was, in
fact, much too low as shown by the "Agence pour les Economies
d'Energie" which recalculated it in 1978. This is the conclusion
of this study (4).

"We may assert that the energy saving investment costing
less than 2500 Frs/toe must be awarded the greatest priority in
respect of the nuclear program and that this priority should be
extended to the industrial investments, especially as regards
heavy fuel consumers, way beyond this threshold, up to a value
included between 4000 and 6000 Frs/toe. The case of investments
in other sectors should be considered with when taking into account the longevity of the installation and the cost, for the
residential and tertiary sector, the threshold would probably
be higher than that established in this note".

By taking the approach outlined, a country can provide
itself with:
    - an aim = for example, to economise 20% of its energy
    consumption for 1990;
    - a norm = to make the energy-saving investments, whose
    cost does not excede 8000 F/ toe.

But it is not enough to fix aims and norms to bring about
effective investment by all energy users. Many obstacles appear
along the way:
    - the norm can seem, to certain energy users, to be too
    high, if they do not anticipate the same results as the
    energy policy makers;
    - the sectors encouraged to invest because they are in
    favour of an energy price increase are not those who
    decide the investment rate (the case of tenants and
    appartment owners);
    - industrialists and individuals who would like to invest
    are not able to find the necessary sums of money ...

All these situations ought to be studied by the energy
policy makers who would, according to the circunstances:
    - set out rules to be followed if and when the economic
    calculation only intervenes to a slight extent (public
    administration, for example);
    - make the norm more attractive by subsidies in the form
    of tax deductions;
    - give decision makers easier access to means of financing.

BIBLIOGRAPHY

(1) POSNER Michael.- Fuel policy: a study in applied economics,
    Mac Millan, 1973, 355 p.

(2) KOPMANS T.C.- Economics among the sciences. The Am. Econ.
    Rev. march 1979, vol. 69 n°1, pp. 1-13.

(3) Modeling Energy Group. Energy Modeling for an uncertain future.
    Report prepared for the Committee on Nuclear and Alternative
    Energy Systems, National Academy of Science,1978.

(4) SLESSER M.- Energy in the economy. Mac Millan, 1978, 200 p.

(5) BOULDING K.- The economics of energy. The Annals of American
    Academy of Political Science, november 1973, pp. 120-126.

(6) MAINGUY Y.- L'économie de l'énergie, DUNOD, Paris, 1966, 532 p.

(7) JANOSI P.E. and GRAYSON L.E.- Patterns of energy consumption
    and economic growth and structure, Journal of Development
    Studies, n°2, January 1972, pp. 241-249.

(8) ADAMS F.G. and MIOVIC P.- On relative fuel efficiency and
    the output elasticity of energy consumption in Western Europ,
    The Journal of Industrial Economics, vol. XVII, 1968, pp. 41-
    56.

(9) BROOKES L.G.- Energy and economic growth, Industrial Marke-
    ting Management, 1971, n°1, pp. 3-10.

(10) DARMSTADTER J., DUNKERLEY J., ALTERMAN J.- How industrial
     societies use energy. Resources for the future, J. Hopkins,
     1977, 282 p.

(11) MEDINA E.- Consommation d'énergie, essais de comparaisons
     internationales, Economie et Statistique, n°66, avril 1975,
     pp. 3-21.

(12) CHATEAU B. et LAPILLONNE B.- La prévision à long terme de
     la demande d'énergie. CNRS, Paris, 1977, 180 p.

(13) DOYLE Guy and PEARLE David.- Low energy strategies for the
     UK, ENergy Policy, vol. 7, n°4, december 1979, pp. 346-350.

(14) INSEE.- Présentation du modèle MINI DMS ENERGIE, Ronéoté,
     Paris, mars 1980, 44 p.

(15) Agence pour les Economies d'Energie. Economie d'énergie et
     investissement de production d'électricité, note interne,
     avril 1978, 6 p.

ENERGY IN EUROPE: DEMAND, FORECAST,

CONTROL AND SUPPLY

H.-F. Wagner

Federal Ministry Re-
search and Technology
Heinemannstr., 53 Bonn

0. Introduction

Adequate and reasonably-priced energy supplies are fundamental to the functioning of the economy and to the stability of the
society of all countries. Energy questions, therefore, have become
of steadily increasing importance during the last 20 years and have
been charged, in particular in times of actual and threatened
difficulty, with the highest political significance both nationally
and in the field of foreign relations. More fundamentally, the
standards of public services and the level of business activity
and hence employment can decline or remain at an unacceptably low
level if the energy situation is such as to jeopardise world trade
and economic growth.

In this context energy demand forecasts and control are of
great importance for actual politics both in the field of national
economics and foreign relations. However, forecasting of energy
demands in particular for the longer term and even in the frame
of a single nation cannot be done with a high degree of confidence.
The problem is very complex, as energy demand and the mix of fuels
to meet this demand are dependent on a large number of factors
which include population growth, economic growth, technological
development in supply and utilisation of fuels, resource availability and national and international policies.

The situation in Europe is characterized by the fact that a
variety of nations with different domestic situations and thus
differing energy policy goals is trying to come to a common
approach in solving the energy problems of this continent both
now and for the future. Most of all, two organizations form the

framework for this endeavour: the European Economic Community
(EEC) in Brussels and, in a broader frame, the International Energy
Agency (IEA) within the OECD in Paris. The European Community is
comprising 9 countries: Belgium, Denmark, France, Federal Republic
of Germany, Italy, Ireland, Luxembourg, Netherlands and United King-
dom. The european countries participating in the IEA are eight
EEC-countries (France is not member in IEA) and Austria, Greece,
Norway, Spain, Sweden and Switzerland.

Of course, energy policy in Europe is linked to the respective
policies in other countries, in particular to major industrialized
countries, the oil supplying countries and the third world.

Cooperation and Coordination with the major industrialized
nations in the western world is achieved by EEC, the IEA and to some
extend by the economic summits annually held by the heads of seven
governments: Canada, France, Federal Republic of Germany. Italy.
Japan. United Kingdom and USA.

Current and near term European energy problems are most of all
directly related to the oil problem: the instable and critical
supply situation as well as the attempt to reduce the dependence
on oil imports as fast as possible. In addition, the uncertain
situation in the development of nuclear power is another destabilizing
factor for the future energy supply of Europe.

The EEC may serve as an example to illustrate the situation
in the oil sector for Europe during the last 18 months. During the
first half of 1979 tremendous difficulties arose due to the un-
foreseen interruption in oil supplies from Iran, coupled with a
severe winter and a high underlying energy demand due to a relatively high economic growth. This led to a shortage, which occured
despite of total oil supplies to the Community in the first half
of 1979 about 6 % higher than in the same period in 1978. Net imports of crude and products together increased by about 1.5 %
(net imports of crude oil alone increased by 4.0 %) and domestic
crude production increased by 40 %. Nevertheless, after allowing
for climatic, seasonal and cyclical factors, total community
supplies in the first half of 1979 fell 3 % short of theoretical
potential consumption. The consequences of this relatively small
shortfall were tremendous and the people in Europe will have them
in painfull memory for a long time. In particular the oil price
exploded dramatically: by the end of August 1979, consumer prices
(net of tax and duty) for oil products had on average increased by
52 % as compared to the situation at the end of 1978. Prices on
the spot market had even increased by 90 % during the first eight
month in 1979.

The entire year 1979 has brought an oil consumption for
the community of 525 million tonnes and has, therefore, surpassed

the original goal of 500 million tonnes by 5 %. The forecast for
1980, as of April this year, given by the Commission of the European Communities (CEC) is 517 million tonnes. Although the absolute figures of oil consumption have increased the share of oil
in the total energy supply of the EEC is reduced during the last
years: It was 54.5 % in 1978, 53.3 % in 1979 and is expected to
decline further to 52.4 % in 1980.

Energy forecasts in Europe are mainly based on national plans
and views. They are mostly focused on the short and medium term,
i.e. 10 or 20 years ahead. There are almost no national long term
forecasts covering 40 or 50 years from now. This situation is
slightly different for the international scene: long term forecasts
exist for instance by the World Energy Conference (1977), Nuclear
Energy Agency of OECD (1978), International Institute for Applied
Systems Analysis (1978), CEC and IEA (in preparation).

Control of energy consumption in times of peace is a relatively
young undertaking both on the national and the international level.
National controls, if they exist, are varying from country to
country. Mostly in Europe they consist of comparing energy consumption goals with actual comsumption and in giving incentives for
energy conservation either by increasing energy prices or by
granting tax advantages (or similar measures) for energy saving
measures. In principle, all nations rely on free market forces and
economic incentives. International controls are performed by the
CEC and the IEA. They conduct regular reviews on the national
situation of their member countries and give appropriate recommandations. In particular IEA is performing so called in depth reviews
in member countries with extensive visitation of a group of international experts and. afterwards, comprehensive discussions on the
special situation of the country reviewed. This results in comprehensive recommandations for that country and in a certain annual
control of the execution of these recommandations.

Since, as mentioned before, the development of future energy
demand in Europe is primarily the composition of national developments, in the following chapters I will explain in more detail
what could be concluded for the future. My scope of work, however,
is limited to the non-communist european countries, with exception
of Finnland and Portugal, since for both these countries I was not
in a position to gather sufficient material on this topic.

I. Austria

A. Overview

1. Overall situation

Austria is heavily dependent on imports to meet its energy

needs. In 1978, energy imports represented about 60 % of its TPE
requirements and the share is expected to rise to 70 % in 1990.
Although gas and coal imports should rise, oil imports will continue to dominate total energy imports. Due to this situation
Austria decided early in the seventies to introduce nuclear power
for the generation of a major share of the electricity demand.
However, in a public referendum on nuclear power in November 1978
the majority in essence opted for a stop of the Austrian nuclear
energy programme. As a consequence of this decision a shortfall of
4.2 Twh has to be compensated in 1985 and 1990 by other energy
sources. In the short term this should be achieved by electricity
imports and by running old thermal conventional plants previously
planned to be set out of operation.

Indigenous energy sources of Austria are hydropower and some
rather modest resources of oil, gas and coal.

Hydropower is with almost 70 % the major energy source for
electricity generation. Capacities are planned to be increased
in the next years as can be seen from the analysis of the Austrian
electricity supply structure (table I.1). However, to fill the gap
created by the stop of the nuclear programme in the long term, a
very firm coal policy is required in order to get in due time the
quantities necessary to supply future plants.

Table I.1: Electricity generation in Austria (capacities in GW)

	Hydro/ Geothermal	Coal	Oil	Gas	Total
Operating	7.8	1.0	1.5	1.9	12.2
under construction	1.4	0.3		0.1	1.8
authorized and site confirmed	0.4				0.4
planned until 1995	3.4	1.1	0.8	1.2	6.5
Total	13.0	2.4	2.3	3.2	20.9

2. Energy policy goals

The outcome of the November 1978 referendum on nuclear power has led to a decisive limitation of the possible options to meet future Austrian energy requirements. As a consequence the goals for the national energy policy had to be changed. According to the latest publications they are:

- to increase the share of TPE covered by indigenous energy resources as far as possible,
- to achieve a wiedespread diversification of the energy forms
  to the actual needs in end-use,
- to encourage energy conservation measures,
- to make non-conventional forms of energy competitive,
- to reduce the demand of imported oil,
- to keep the financial burden of energy imports on the balance
  of payment as low as possible.

B. Energy demand: forecast and control

1. Forecast

Energy demand figures up to 1990 were published in April 1979 by the "Energy Report 1979". Figures beyond that time frame are presently not available. The projections for 1985 and 1990 as given on table I.2 were prepared by the Austrian Institute for Economic Research. They seem, however, to be somewhat out of date since they are nearly the same as of 1978.

The figures show the continuously growing dependence from energy imports. In particular, the steadily increasing dependence from imported oil and gas causes great concern for the energy policy decision makers in Austria. The analysis shows for oil that despite of all policy efforts net oil imports will increase by a growth rate of 3.7 % and 4.9 % for the periods 1978-85 and 1985-90, respectively. This increase will mainly flow into the residential/commercial sector, which will increase its share of the total final consumption from 37.6 % in 1978 to 42.9 % in 1990.

2. Control

Austria is a Federal Republic with some countries called Länder having their own governments and the Federal Government in Vienna. This gives some constitutional difficulties for the attemps of the Federal Government to control the future deve-lopments of the energy demand in Austria. Main item of the energy policy for reducing the growth rates of the energy consumption is the conservation sector.

Table I.2

AUSTRIA
Key Energy Indicators and Data
(Mtoe)

	1960	1973	1978	1985	1990
General					
Energy Demand (A)	12.4	23.5	24.7	32.3	37.6
Energy Production (B)	8.7	9.7	10.3	10.4	11.3
B/A (%)	70.2	41.3	41.7	32.2	30.1
Net Oil Imports	0.6	9.7	10.4	13.4	17.0
Total Oil Consumption	3.1	12.3	12.0	15.0	18.4
TPE/GDP Ratio	1.37	1.38	1.26	1.27	1.24
TPE/GDP Elasticity	1.02	1.33	1.04	0.90	
Per Capita TPE	1.76	3.13	3.29	4.35	5.09
Total Final Consumption	8.1	18.0	17.8	23.4	27.3
Industry (incl. non energy)	4.1	8.7	6.9	8.9	10.1
Transport	1.5	3.7	4.2	5.0	5.6
Commercial	2.5	5.6	6.7	9.4	11.7
Production					
Solids	1.3	0.7	0.6	0.4	0.4
Oil	2.5	2.6	1.9	1.7	1.6
Gas	1.3	2.0	2.1	1.0	0.9
Nuclear	–	–	–	–	–
Hydro/Geothermal	3.6	4.3	5.6	7.3	8.4
Electricity (Twh)	16.0	31.3	38.1	52.4	62.8
Trade					
Coal Exports	–	0.1	0.1	–	–
Imports	3.5	3.0	2.4	3.3	3.3
Oil Exports	1.1	0.1	0.1	0.1	0.1
Imports	1.7	9.8	10.5	13.5	17.1
Gas Exports	–	–	0.1	0.1	–
Imports	–	1.4	2.5	5.4	6.1

References:	1960	1973	1978	1985	1990
GDP (1970 $US billion)	9.0	17.0	19.7	25.5	30.2
Population (millions)	7.04	7.52	7.51	7.42	7.39

Policy is focussed on new and more stringend regulations on thermal insulation standards in buildings, size and control of large central heating systems, and energy labeling.

These efforts are supported by some legal measures set into force with the beginning of 1980. They provide by law for fiscal incentives for investments in heat pumps, solar, total energy module systems and facilitaties for heat recovery and the use of biomass as well as for additional insulation measures. In addition advanced depreciation of investments for switching from oil to coal burner in power stations, for small hydro power plants (up to 10 MW), for cogeneration plants, for district heating and for the production, distribution and storage of gas.

However, all these endeavous are aiming only at special problems and fields. Efforts have to be increased for planned conservation measures on sectoral consumptions and total energy demand. One of those measures is the increase of prices for enduse energy. There exists a price control mechanism for gasoline and heating oil, but these prices were raised only once in 1979 with the exception of premium oil where the price has been stept up twice.

C. Supply situation

Major supplier is oil which will maintain its present share of about 50 % through 1990. It is followed by hydro, gas and coal.

1. Oil

There are some indigenous oil resources in the country, however by far too small to be a significant supplier. Current estimates of annual supplies show the following figures (million tons):

1980 : 1.625,   1985 : 1.360,   1990 : 0.970

Based on the geological knowledge as of beginning of 1979 the following oil reserves are estimated:

proven reserves: 17.81 million tons
probable reserves: 3.28 million tons
possible reserves: 1.60 million tons
speculative reserves: 25.3-26.3 million tons

2. Gas

There are also some modest gas resources in Austria. Nearterm production is estimated to be (billion cubic meters):

1980 : 1.994, 1985 : 1.380, 1990 : 1.038

Reserves (billion cubic meters) are estimated to be:

proven reserves   8.99
probable reserves   3.49
possible reserves   1.34
speculative reserves 36.4-29.4
as of January 1979.
In addition, gas has to be imported. On the long term imports will be
based on supply contracts with Algeria, USSR and the North-Sea
suppliers.

3. Coal

    Austria owns deposits of hard and brown coal, which should be
used to replace to some extend heavy fuel oil.
Brown coal production on the near term is estimated to be
1980 : 2.568 million tons, 1990 : 2.278 million tons.

The reserves (million tons) as of the beginning of 1978 are estimated to be:

	hard coal	brown coal
proved reserves	1.0	145.8
additional reserves	3.0	35.2
hypothetical reserves	6.0	85.0

    Although proved reserves in brown coal are rather high,
economically recoverable are only 58 million tons, based on present
knowledge. Additional coal will be imported most of all from
Eastern European Countries like Hungary and Poland.

4. Nuclear

    The nuclear programme has been canceled as a consequence of the
referendum in fall 1978.

5. Hydropower

    At the moment the economically usable potential of hydropower
(without plants already under construction) amounts to 21 Twh/a. It
will be under operation to a great extent in the beginning of the
1990's. In addition to the big plants Austrian government intends
to support the construction of small-scale plants for local supplies.

6. Other energy sources

    There are some efforts to support the development and utilisation
of solar devices for warm water preparation. In addition R+D in bio-

mass and geothermal techniques is support by the public authorities.

II. Belgium

A. Overview

1. Overall situation

Belgium which has one of the highest energy consumptions per capita in Europe is very strongly dependent on energy imports. Only 20 % of the demand can be supplied by domestic resources. This situation is not expected to be changed during the next 10 years. The only indigenous energy source of importance is coal presently supplying almost 5 million tons oil equivalent. Dependence on net oil imports was 58 % of total energy consumption in 1978. This dependence is planned to be reduced by a strong nuclear programme.

For 1985-1986 50 % of electricity will be generated by nuclear power plants. Further expansion of nuclear capacities is dependent on the new energy plan of Belgium commencing in 1985. This plan however, has not yet been approved by Parliament.

2. Energy policy goals

Major energy policy goal is to reduce energy consumption by at least 10 % in 1985 and 15 % in 1990 from the levels of the 1978-1977 historical trends with an averaged annual increase of 3.5 %.

The new energy plan calls for: a strong energy conservation policy; optimum diversification of non-nuclear energy sources; security of supplies including direct state contracts with producers and the holding of emergency stocks; increased non-nuclear R+D; energy pricing covering total costs.

B. Energy demand: forecast and control

1. Forecast

The energy demand forecast by the government through 1990 is displayed on table II.1. The figures show a continuous increase in net oil imports from 26.7 Mtoe in 1978 to 33.0 Mtoe in 1990. In relative figures, however, dependence on imports is planned to be reduced from 58 % to 52 % in 1978 and 1990, respectively.

The analysis of the consumption structure reveals the dominant position of the Belgium industry. In 1978 energy consumption in this sector was 48.6 % of the total final consumption and this figure is scheduled to be increased to 51.2 % in 1990. According

to the energy plan oil consumption in industry should be substituted by coal as fast as possible. It is assumed that the utilisation of coal is increased to 7.0 Mtoe in 1980 as compared to
5.2 Mtoe in 1978.

In the residential and commercial sector only a modest growth
is expected through 1990. This should mainly be achieved by strong
conservation measures. Energy supply in this sector will be based
most of all on electricity and gas imports.

2. Control

Belgium is planning to carry out a strong conservation programme for the coming years. However, due to several changes in
governments firm decisions are pending.

The energy plan puts emphasis on all three consumption sectors.
For industry grants are foreseen for energy saving investments
as well as audit systems and the development of more efficient
techniques. In the residential and commercial sector grants will
be provided for building insulation and heating systems improvements.
The new programme for the transport sector will promote greater use
of public and freight transport, impose controls on engine fuel
efficiencies and encourage the development of electric engines.

C. Supply situation

As in many other countries oil is the major energy supplier
followed by gas, coal and nuclear energy.

1. Oil

There are no major oil resources in the country so that in
practice all the fuel has to be imported.

2. Gas

The same holds for gas, which currently has a share of 20 % of
TPE. This is expected to remain constant through 1990. This will
be mainly achieved by imports from Algeria, Norway and the Netherlands.

3. Coal

Domestic coal production at present is about 7 million tons
(5 Mtoe). However, this is not sufficient to cope with the
requirements. Therefore steam coal (for electricity generation) is
imported and it is expected that the present level of about 2 Mtoe
is increased to over 4 Mtoe in 1990.

Table V.1

BELGIUM
Key Energy Indicators and Data
(Mtoe)

	1960	1973	1978	1985	1990
General					
Energy Demand (A)	25.4	46.7	46.1	57.0	64.0
Energy Production (B)	15.9	6.4	7.4	11.4	13.1
B/A (%)	62.6	13.7	16.1	20.0	20.5
Net Oil Imports	7.7	30.9	26.7	31.0	33.0
Total Oil Consumption	7.2	27.6	24.4	28.0	30.0
TPE/GDP Ratio	1.60	1.51	1.36	1.27	1.17
TPE/GDP Elasticity	0.95		0.78	0.60	
Per Capita TPE	2.8	4.8	4.7	5.8	6.4
Total Final Consumption	18.1	37.5	36.2	44.7	50.0
Industry (incl. non-energy uses)	9.1	19.2	17.6	22.7	25.6
Transport	2.6	5.1	5.4	7.1	8.6
Residential/Commercial	6.4	13.2	13.1	14.9	15.8
Production					
Solid fuels	15.7	6.2	4.6	5.0	5.0
Oil	-	-	-	-	
Gas	0.1	0	0	0	0
Nuclear	-	-	2.7	6.2	7.9
Hydro/Geothermal	0.1	0.2	0.1	0.2	0.2
Electricity (Twh)	15.2	41.1	48.4	68.6	83.4
Trade					
Coal Exports	2.4	0.6	0.4	0.4	0.4
Imports	3.0	5.9	5.1	6.4	8.7
Oil Exports	2.9	15.1	15.1	19.0	19.0
Bunkers	0.7	3.1	2.8	3.0	3.0
Imports	10.6	46.0	41.8	50.0	52.0
Gas Exports	-	0.2	0.4	0.5	0.5
Imports	-	7.5	9.9	12.5	13.5
Reference:					
GDP (1970 US$ billion)	15.9	30.1	34.0	44.9	54.7
Population (million)	9.2	9.7	9.8	9.9	10.0

4. Nuclear

Belgium has a strong nuclear programme as compared to the
size of this country. According to the power plants under construction 12 % of TPE will be produced by nuclear in 1985 as
compared with 6 % in 1979. This means that capacities are increased
from 1.7 GW in 1978 to 5.4 GW in 1985 allowing for a 50 % share
of the entire electricity production in 1985-1986. Further
expansion of the nuclear programme depends on further decisions
about the energy plan mentioned before. However, according to the
figures submitted to the International Nuclear Fuel Cycle Evaluation (INFCE) 1977-1980, nuclear capacities might range between
6.8 MW and 9.4 MW in 1990 with a potential further increase to
a range between 9.4 MW and 14.6 MW in 2000.

5. Hydropower

Hydropower is negligible in Belgium's energy supply and there
is no significant potential for this energy source for the future.

6. Other energy sources

Efforts to develop other energy sources are small but will be
increased for the future. They comprise solar power for warm water
production, geothermal energy and nuclear fusion.

III. Denmark

A. Overview

1. Overall situation

Until very recently Denmark was almost 100 % dependent on
energy imports. Oil was predominant with a share of 79 % of the
energy demand in 1978. This situation will be changed after discovery and exploitation of oil and gas fields in the North-Sea.
According to the present plans indigenous resources will contribute
25.6 % to the energy supply in 1990.

In addition uranium deposits in Greenland could also improve
the supply situation from indigenous sources in the years to come.
However, despite governments intention to introduce nuclear power
in Denmark a firm decision has not been taken due to public interventions. In the case of abandoning the nuclear programme entirely
the gap in electricity supply should be closed by coal, though
Danish government is fully aware of all the problems associated
with such a decision. In 1990 electricity will constitute one-third
(one-quarter today) of Denmark's total energy consumption.

2. Energy policy goals

     Denmark has a comprehensive energy plan as outlined by the
Energy Report in March 1979. This plan is presently due to revision
and the revised form will be published at the end of 1980. The new
plan will operate with two sets of assumptions: one with nuclear
power and one without. The main objectives of the present plan,
however, will in principle not be changed, They are:

- reduction of the dependence on imported oil by
  . increasing the production of indigenous oil and natural gas;
  . increasing the imports of coal;
- increase of energy conservation;
- development of alternative energy sources;
- comprehensive R+D efforts.

B. Energy demand: forecast and control

1. Forecast

     The energy demand forecast by the government through 1990 is
displayed on table III.1. Due to the increasing production from
North-Sea sources absolute figures of net oil imports will decrease
by about 40 % during the next ten years. In parallel, the share of
oil of the total energy demand will also decrease from 79 % in 1978
to 51.2 % in 1990. Within the same time interval the share of gas
and coal is considerably increased, the latter by 88 %.

     The analysis of the consumption structure reveals the strong
role of the residential and commercial sector, which is to a large
extend due to the climatic circumstances. In 1978 energy consumption
in this sector was 56.7 %. This rather high figure should be brought
down to 49.5 % in 1990 if the strong conservation programme can be
successfully conducted. Small increases in energy consumption are
foreseen for the industry and transportation sector. They will both
expand from currently about 22 % to about 25 % in 1990.

2. Control

     As mentioned before Denmark is planning to perform a strong
conservation programme, as part of the comprehensive energy programme. The importance attached by the government to energy problems
may be indicated by the creation of a separate Ministry for Energy
in late 1979.

     Goal number 1 is the residential and commercial sector. Major
tool for the government are incentives and legal measures as for
instance the Heat Supply Act of 1979, by which a comprehensive heat
planning for the country should be achieved. To act in the heating
sector is of major importance for 90 % of the Danish heat consump-

tion is currently covered by oil.

The present conservation plan calls for stricter requirements for the insulation of new buildings, energy saving investments in public buildings, and grants to private property owners for insulation.

District heating systems are well advanced in Denmark, presently covering 30 % of the demand for residential heating. However, two thirds of the distric heating production is currently based on fuel oil. This should be substituted in the medium and longer term by fluidised bed coal, straw waste and gas or by the introduction of heat pumps. For the comming years the distric heating system will be enlarged and technically improved.

In 1978 oil use made up 69 % of the industrial energy. This should be declined to 49 % by 1990 by the introduction of new energy conserving technologies. In addition, this reduction process will be supported by an extensive scheme for energy saving investments consisting of grants of up to 40 % of eligible investments and supplemented by tax credits.

Great importance is attached to improvements of the public transportation system. It is planned to electrify 1065 km of railways in order to save oil by the substitution of coal-fired electricity generation plants.

C. Supply situation

As outlined before energy supply by domestic resources has been markedly improved during the last years due to the beginning exploition   of the North-Sea oil and gas fields. It expected that by the mid eighties Danish-produced oil and natural gas will be able to cover 30-40 % of the total energy requirements, the mix of these two fuel will be roughly 50 : 50.

1. Oil

Present indigenous oil production is stemming from the Dan Field which is in operation since 1977. Oil production 1979 has amounted to almost 500 000 tons. Nearterm development of production is shown on table III.2.

As can be seen production may decline to the end of this decade. However, here are good prospects for the discovery of new fields. Reserves as of end of 1979 were estimated to be 50 million toe. In 1981 when Gorm Field is expected to come on-stream, domestic crude oil production should cover about 20 % of Danish oil consumption.

Table III.1

DENMARK
Key Energy Indicators and Data
(Mtoe)

	1960	1973	1978	1985	1990
General					
Energy Demand (A)	9.0	19.7	20.5	21.5	24.2
Energy Production (B)	0.6	0.1	0.7	5.4	6.2
B/A (%)	0.6	0.1	3.4	25.1	25.6
Net Oil Imports	5.0	17.9	16.8	10.0	10.7
Total Oil Consumption	5.1	17.5	16.2	12.5	12.4
TPE/GDP Ratio	0.93	1.14	1.05	0.83	0.77
TPE/GDP Elasticity		1.38	0.31	0.16	0.64
Per Capita TPE	1.96	3.94	4.02	4.13	4.65
Total Final Consumption	7.1	15.9	16.4	17.3	19.8
Industry (incl. non-energy use)	2.1	4.1	3.6	4.4	5.1
Transportation	1.3	3.5	3.5	4.2	4.9
Residential/Commercial	3.7	8.3	9.3	8.7	9.8
Production					
Oil	0.1	0.1	0.4	2.8	2.3
Gas	-	-	-	2.0	3.0
Nuclear	-	-	-	-	-
Solids	-	-	0.2	0.3	0.3
Hydro/Geothermal	-	-	-	-	-
Renewables	-	-	-	0.3	0.6
Electricity (TWh)	5.5	19.1	22.9	31.0	35.8
	_1960_	_1973_	_1978_	_1985_	_1990_
Trade					
Coal Exports	-	-	-	-	-
Coal Imports	3.8	2.0	4.3	6.7	8.1
Oil Exports (incl. bunkers)	-	3.6	2.6	3.0	3.0
Oil Imports	5.3	21.5	18.9	15.2	12.5
Gas Exports	-	-	-	-	-
Gas Imports	-	-	-	-	-
Reference					
GDP (1970 US $ billion)	9.7	17.3	19.6	25.8	31.4
Population (million)	4.6	5.0	5.1	5.2	5.2

Table III.2: Oil production (million toe) from Danish oil fields
         including oil condensates.

	1980	1985	1990
Dan Field	0.5-1.0	0.5-1.5	0-0.5
Fields under development (Gorm, Skjold, Cora, Bent)	0	1.5-3.0	0.5-3.0

2. Gas

Current domestic gas production from the Dan Field is in the
order of some 0.1 million toe. With the beginning of the operation
of Gorm Field production from this source will increase the total
production of natural gas to 1.5-3.5 million toe in 1985 and 3.0-
5.5 million toe in 1990. Estimated reserves as of end 1979 are
80 million toe. They could be sufficient to cover Denmark's needs
for a period of 25-30 years. Main consumption sector for gas will
be the heat market since aim of the Danish policy is that 30-40 %
of heat supply should be provided from natural gas.

3. Coal

Since the first oil crisis in 1973/74 Denmark's imports of
coal have been heavily increased from 2.2 million tons in 1972 to
7.2 million tons in 1979, of which the electricity supply under-
takings accounted for 6.4 million tons corresponding to almost
90 % of the total figure. This reflects also the rapid switching
of Danish utilities from oil to coal. In 1978 60 % of the electricity was generated by coal, in 1990 this share is expected to
increase to 80 %, depending, however, on the decision about the
future of the unclear programme. Major current coal suppliers are
Poland (40 %) and South Africa (33 %). This supply pattern is
intended to be changed in the medium and long term by negotiations
with Australia, Canada and Colombia.

4. Nuclear

Nuclear power is an intensively debatted issue in Danish
energy policy. The government is strongly supporting the use of
nuclear energy and on February 13, 1980 the new energy minister
has declared in Parliament, that the governments's view that
supply safety considerations speak clearly in favour of including

nuclear power in Denmark's multiple energy supply system is un-
changed. However, due to public opposition the execution of the
nuclear energy programme is postponed for the beginning of at least
1981. Main focus of the debate is the final disposal of radioactive
waste. Now, the Danish Government is considering an information
campaign concerning energy problems in general prior to a referendum to be held on the nuclear issue. The origianl plan of the
government calls for the first nuclear power plant of 1.3 GW in
operation in 1988, then adding one 1.3 GW plant every two years
achieving 6.5 GW total capacity in the 2000.

Denmark is one of the few countries in Europe having own
uranium deposits located in Greenland. Proved reserves are 27 000
tons and additional reserves are estimated at 15 000 tons. Production costs are 80-130 Dollar US per field. This uranium would
be sufficient for 8 of the planned reactors for their entire life-
time.

5. Hydropower

Hydropower is negligible in Denmark.

6. Other energy sources

Denmark has launched a strong R+D programme on solar heating,
wind power, geothermal and biomass technology. In particular wind
power is expected to contribute in the order of 1-3 % to the total
final energy consumption in the nineties.

IV. France

A. Overview

The most striking feature of the French energy programme is
the vigorous nuclear component. It is the French answer to the
oil crisis of 73/74. According to the present planning nuclear
energy will increase its current share of total energy consumption
of about 7 % to about 26 % in 1990, and to about 33 % in 2000.

As many other countries in the world France has increased the
oil consumption by a factor of 10 from 1950 to 1980, almost all of
it had to be imported. In 1974 oil imports amounted to 64 % of
total primary energy consumption. Since there are no major
indigenous energy resources in the country (despite of some hydropower sources) France is trying to substitute oil by nuclear power.
In order to avoid new dependences in the long term the nuclear
programme is consequently aiming at a large scale introduction of
fast breeders as quickly as possible. Therefore, the current nuclear
programme calls for 15-21 GW of fast breeders in 2000 and for
58-112 GW in 2025.

2. Energy policy goals

    Major energy policy goal is assurance of energy supply which
means reduction of energy demand by conservation and reduction of
import dependence. Presently, import dependence is more than 70 %.
Due to current plans import dependence should be reduced to 55-60 %
in as early as 1985. In addition, it should be achieved that in
the future the import share of one country should not exceed 15 %.
This diversification should help to stabalize the energy supply in
the long term.

B. Energy demand: forecast and control

1. Forecast

    The energy demand forecast as given by the Direction Générale
à l'Energie et aux Matières Premières is displayed on table IV.1.

Table IV.1

France
Key Energy Indicators and Data

Mtoe

	1960	1973	1978	1985	1990	2000
Demand						
Energy Demand	85.6	175.4	182.2	215-230	240-270	300-355
Total Oil Consumption	26.8	116.9	108.8	96-104	15-109	103-120
Industry			66.1	76.3-83.7	85.7-100	
Transportation			42.4	50.3-52.7	55.7-62	
Residential			73.7	88.4-93.6	98.6-108	
Supply						
Oil	26.8	116.9	108.8	96-104	95-109	103-120
Gas	3.0	15.0	20.8	34-36	40-43	45-50
Nuclear	0.03	3.0	6.5	44.46	63-72	96-123
Solids	46.8	30.5	32.4	25-28	24-27	33-36
Hydro	9.0	10.0	16.0	14-14	15-15	16-16
Renervables	-	-	-	2-2	3-4	7-10

    The uncertainties in the figures for 1985 and 1990 reflect
differences in GDP-growth rates: They vary between 3 % and 4.5 %.

    The supply structure shows clearly that by the year 2000 oil

and nuclear power will be equivalent, both contributing with about
33 % to the primary energy demand. In addition to nuclear supplies
from gas will also be increased considerably during the next 20
years, whereas coal and hydropower will in absolut figures remain
constant for the years to come.

No major changes in the demand structure can be seen. Within
small margins the residential sector will maintain its leading
position with about 40 % followed by industry with 37 % and the
transport sector with 23 %. This seems to indicate that slightly
more emphasis is given by French energy planners to securing the
supply side than to conservation measures while most other
countries in Europe are aiming at reducing energy demand in the
residential sector. This, of course, might be changed since France
has also an energy conservation programme. However, for coping with
the energy problem one could argue what it is more efficient in the
short and medium term: to strengthen more vigorously the supply
side first and than, slightly later on, to change the consumption
structure or vizeversa.

2. Control

In principle the French conservation programme is similar to
that of other European countries, i.e. increasing the insulation
of buildings, reduction of energy consumption in the transport
sector by more fuel efficient cars and enlarged public transportation systems and by gradually decreasing industrial demand by
improved technologies. To support this programme tax reduction
incentives and grants for the investment in conservation measures
are given.

C. Supply situation

Unlike some other European countries France has only rather
limited domestic energy sources, most of it are some coal deposits
and hydropower. Therefore nuclear power will be enlarged since
some uranium resources are available in the country and, most of
all, in the former colony of Niger. These resources, however, would
not help so much in achieving some energy import independence, if
used only in light water reactors (LWR). Therefore, the French
nuclear programme is vigorously aiming at introducing fast breeders
(FBR) on a large scale as quickly as possible. French nuclear power
programme is by far the strongest in Europe. It is comparable in its
strength and consequence to some extent only to that of Japan and
perhaps the USSR. The programme comprises the construction of an
LWR capacity of 53-60 GWe in 1990 and of 71-86 GWe in 2000; an FBR
capacity of 4.4-5.9 GWe in 1990 and of 15-20 GWe in 2000. In addition
large scale uranium enrichment capacities are set up in Tricastin,
a large scale reprocessing facility is under construction at La
Hague. With regard to all these comprehensive efforts there is

little doubt that France will be one of the first countries in the
world having already switched its energy supply to a major share
away from oil in the medium term.

V. Federal Republic of Germany

A. Overview

1. Overall situation

Germany is like many other European countries heavily dependent
on foreign supplies of oil and,to a less extend natural gas. In
1979 51 % of the total energy consumption was supplied by oil, which
had almost completely to be imported. Of these imports 81 % have
been purchased from OPEC-countries, the rest was supplied from the
North-Sea.

However, energy dependence from abroad with 44 % in 1978 is not
as high as in other countries. This is mainly due to large indigenous
coal resources consisting of both hard and brown coal. Unfortunately,
the price in particular for hard coal, is rather high so that competitiveness of this energy source can only be achieved by government supports. Coal resources are also the main competitor to
nuclear power, which was assigned a major role in future energy
supplies still in the middle of the seventies. However, due to
growing public acceptance problems, the share of nuclear power in
the supply system has been markedly decreased and the future is
still uncertain.

The longterm perspective for coal, however, will be in the
area of liquefaction and gas supply and not in electric power generation, therefore leaving an important sector open for a potential
contribution by nuclear power.

2. Energy policy goals

Germany has a comprehensive energy programme launched for the
first time in 1973. Its second revision was published in Dezember
1977 and confirmed once more by the cabinet in May 1979.

The main objective of the programme is to reduce the growth
in energy demand over the long term and to provide a wider and
secure range of supplies to meet this demand. To achieve this
objective a package of measures is aiming at:

- reducing the growth of energy consumption by efficient and economical use of energy,
- cutting down the share of oil in energy supply,
- attaching priority to the use of domestic hard and brown coal,

- developing nuclear energy to the extent absolutely necessary to
  secure electricity supply, with priority being given to the
  safety of the population,
- limiting the threat of import disruptions by diversifying sources
  and entering into international agreements and cooperation,
- continuing systematically energy research in order to make use
  of all new technologies and removable energy forms available to
  Germany given its particular geographical position.

   This energy programme is regularly reviewed by the goernment
and additional measures, if required, are concluded on this
occasion.

B. Energy demand: forecast and control

1. Forecast

   The latest forecast available is that of 1977/78. It is displayed on table V.1. This forecast was not established by the
government but by independent economic research institutes. It is
somewhat out of date in sofar as it does not take into account the
severe slipage of the nuclear programme and the effects of all the
conservation measures initiated during the last 3 years. In addition,
total energy consumption was estimated by use of growth assumptions of 4 % per year between 1975 and 1985 and 3.5 % per year
between 1985 and 1990. These figures appear to be rather high in
the light of recent developments in particular in the oil sector.
Nevertheless the table gives some indications on the supply and
demand structure as well as on the intended future development in
these structures.

   The development of the total energy demand shows that the
dependence from foreign supplies will prevail through the next
decade. Apparently, there is no decline and, therefore, one can
conclude that this dependence will not be considerably lowered
even within this century.

   It is very difficult to immagine how this could be achieved
in the long term, if the nuclear programme and in particular the
deployment of fast breeder reactors would be halted; for the
German coal deposits could not be exploited to a much larger extend due to economic and severe environmental constrains. Thus
considering the situation on a long term basis it could happen
that Germany is becomming an importer of growing amounts of coal
and natural gas while mainting its present share of oil consumption. It is rather difficult, however, to reconcile such a likely
development with the long term objectives of the German energy
programme.

Table V.1

GERMANY
Key Energy Indicators and Data
(Mtoe)

	1960	1973	1978	1985	1990
General					
Energy Demand (A)	145.8	265.8	272.3	337.8	371.0
Energy Production	129.1	120.5	118.7	147.7	171.1
B/A (%)	88.5	45.3	43.6	43.7	46.1
Net Oil Imports	29.4	146.5	140.0	154.0	156.9
Total Oil Consumption	31.3	147.8	142.3	156.0	157.9
Consumption					
TPE/GDP Ratio	1.25	1.27	1.19	1.11	1.03
TPE/GDP Elasticity	1.04		0.25	0.77	0.54
Per Capita TPE	2.63	4.29	4.44	5.70	6.37
Total Final Consumption	102.1	203.8	204.0	243.4	259.4
Industry	51.8	92.2	84.4	105.4	114.5
Transport	15.8	32.9	37.9	39.6	42.0
Residential/Commercial	34.5	78.7	81.7	98.4	102.9
Production					
Solids	118.3	92.4	85.1	88.9	91.2
Oil	6.4	6.7	5.2	5.0	4.0
Gas	0.7	15.4	16.2	14.4	12.5
Nuclear	-	2.6	8.3	34.9	58.2
Hydro/Geothermal	3.7	3.4	3.9	4.5	5.2
Electricity (Twh)	119.0	299.0	353.4	534.0	635.0
Trade					
Coal Exports	21.1	17.4	20.1	15.7	15.7
Imports	7.8	7.0	6.7	5.5	8.9
Oil Exports	2.9	8.9	6.2	7.0	7.0
Imports	32.3	155.4	146.2	161.0	163.9
Bunkers	2.3	3.6	2.8	3.0	3.0
Gas Exports	-	0.1	0.2	-	-
Imports	-	12.3	26.4	47.1	50.2
Reference:					
GDP (1970 $US billion)	116.7	208.9	229.6	303.5	360.6
Population (million)	55.4	62.0	61.3	59.3	58.2

The consumption structure shows an equivalence of industry
and residential/commercial sector, both having a share of about
40 %.Whereas the latter sector will maintain its share it is expected that the industry sector can be enlarged to 44 % in 1990,
indicating the importance of the industry in German economy.

2. Control

Major tool in controlling the development of the energy
demand are reliance on the free market system, regulation of coal
and electricity prices and a vigorous conservation programme.
Main idea in the free market policy is to not shield the consumer
from energy price increases but allowing the market to determine
energy demand. Regulation of coal and electricity prices are used
in this context to maintain the market share of the only domestic
energy source of abundance. This measure is understood to increase
assurance of supply.

Energy conservation is most of all necessary in the residen-
tial/commercial and the transport sector. This can be well demonstrated by a comparison between 1977 and 1978 where the total final
energy consumption was increased by 4.3 %. This increase was lead
by the above mentioned sectors averaging almost to 6 % while the
industrial sector grew only by 1.9 %.

To indicate the nature and structure of the German conservation programme the following measures should be mentioned:

- Introduction of a retrofit tax credit and grant programme for
  houseowners under the Housing Modernisation Act. This programme
  began in July 1978 and will run 5 years at a cost of 2.35 billion
  US Dollars (1 Dollar = 1.85 DM).
- Requirement of a regular service of heating systems. In the case
  of new installations, devices to regulate heat supply are compulsory.
- Voluntary programme by automobile manufacturers to improve auto
  fuel economy by at least 10 % until 1985.
- Amendment to the Investment Allowance Act of 1975 to provide a
  7.5 % allowance for investment in combined heat and power plants,
  industrial waste heat systems and district heating systems.
- Grants to accelerate marketing of new energy saving technologies.
- Raising of the level of thermal insulation and heating system
  standards.
- Increase in R+D funding (1978 : 0.85 billion Dollar, 1979 : 1.04
  billion Dollar, 1980 : 1.04 billion Dollar).

It is expected by the German government that all the efforts
of the energy conservation programme will result in energy savings
of about 35 Mtoe annually by 1985.

C. Supply situation

In 1978 over 95 % of the German energy supply was stemming from fossile sources: oil 51 %, coal 27.7 %, and natural gas 16.5 %. Coal resources are big enough to cover domestic needs, however oil had to be imported by 96 % and gas by 62 %.

1. Oil

According to the statistics as of January 1979 the recoverable proven and probable oil reserves in the Federal Republic of Germany are about 65 million tonnes.

2. Natural gas

On January 1979 177.3 billion cubicmeters proven and 86.1 billion cubicmeters probable were estimated to be the natural gas reserves of the Federal Republic of Germany, both together account for a calorific value of 9.77 kWh per cubicmeter. Future dis-coveries are estimated at a level of about 200 billion cubicmeters. Due to these indigenous gas resources, Germany is in a position to supply in 1980 27 % of the consumption (68 billion cubicmeters) from own sources.

However, this production rate cannot be markedly increased so that (consumption expected to be 82 billion cubicmeter) this share will be decreased to only 23 % in 1985. Within the total supply structure natural gas will play a continously growing role. Ex-pansion of its share will continue to the mid nineties and will remain constant from then on.

3. Coal

German coal resources are very well investigated since many years. The Federal Institution for Geoscience and Raw Materials has catagorized these resources into the following groups:

I. Identified resources (proven reserves down to 1 500 m)

  A. Technically recoverable reserves (only worth opening condi-
     tionally)

  B. Economically recoverable reserves (worth opening)

II. Probable resources (forecast reserves deeper than 1 500 m)

The situation in the Federal Republic of Germany is as follows:

	billion tonnes hardcoal	liquite
IB. Economically recoverable reserves	6.0	10.0
IA. Technically recoverable reserves	24.0	35.0
IA. and B. identified and II. probable resources (total resources)	230.0	62.0

As indicated by the table economically recoverable reserves of hard coal are only very small in comparison to the total resources. This is due to geological (very deep deposits) as well as to price problems. The latter are caused by the very high income of miners and by rather comprehensive and expensive environmental requirements. Therefore, domestic hard coal can be only maintained competitive to other energy sources by a strict government legal subsidy policy. In addition, government interventions on the electricity market has forced the utilities to use 33 million tonnes of hard coal per year for electricity generation and this amount will be steadily increased to 50 million tonnes per year in 1995.

In contrast to hard coal lignite is produced by open air mining and is, therefore, the cheapest energy carrier in the Federal Republic. German open air mines are among the largest in the world. However, production cannot be increased due to the severe environmental problems associated with this kind of coal mining.

Since a major role was assigned to coal on the way to reduce oil dependence and since economic as well as environmental reasons are impeding future expansion of domestic coal production, it is expected that Germany will become a major coal importer within the next decades. This trend will even be amplified if the German nuclear programme is continued to be serverely delayed. For the time being, however, the import level is limited to 6.6 million tonnes annually, but it was indicated recently that this quota could be increased by a factor of 10 within 10 years. In the long term it is planned to use coal for the production of synthetic fuels and gases. A strong R+D programme to develop the respective techniques is underway and the construction of larger demonstration plants is expected within the next years.

4. Nuclear

Like France the German answer to the enrgy crisis of 1973/74 was a vigorous nuclear programme. However, due to some public opposition this programme was heavily delayed. For the time being

all forecasts are drastically revised downwards and it is expected
that nuclear capacities will be in the order of only 30 GWe in
1990 in contrast to about 50 GW originally planned for that year.
Major problems in the public debate are nuclear safety and the
backend of the fuel cycle. The Federal Government has opted for
recycling of plutonium in light water reactors (LWR) in the medium
term and is developing the option to deploy fast breeder reactors
(FBR) in the long run. Radioactive wastes should be ultimately
stored in salt domes. Both, the necessary reprocessing plant and
the final waste disposal facility, are the crucial points in the
current nuclear programme, since both these plants encounter strong
public opposition, mainly from the local population concerned.

5. Hydropower

Development of further hydropower sources will require great
efforts since the economically recoverable resources are almost
completely exhausted in the Federal Republic. It might, however,
be that the present share of about 2 % could be maintained for the
next 20 years.

6. Other energy sources

There are good prospects that in the long term solar, wind and
geothermal energy could contribute in the order of 5-10 % to
primary energy requirements of the Federal Republic of Germany.
Comprehensive R+D programmes were launched in the past, both
nationally and internationally. First measurable effects by these
energy sources are expected to be seen by the mid nineties.

In addition, fusion energy is developed with great efforts,
however mostly in the frame of the joint European fusion programme.

VI. Greece

A. Overview

1. Overall situation

Energy demand in Greece has grown rapidly from 12.2 Mtoe to
15.2 Mtoe between 1973 and 1978 (4.5 % per year). It is projected
to grow even faster at an annual rate of 6.3 % through 1990. This
growth reflects the Greek situation as a developing economy with
a very low energy consumption per capita. Thus, growing energy consumption will be mainly due to an increase in the residential/
commercial area and the industrial sector in order to attain a higher
economic level within the next ten years. This development shall be
achieved by an increase in oil consumption, utmost exploitation of
indigenous resources (lignite, hydro, oil, gas) and increasing
amounts of imported steam coal. It will be interesting to observe

in how far the goals set for the coming year will be attained, in
particular in view of the near term attendance of Greece to the
European Economic Community.

2. Energy policy goals

The main energy policy goals are outlined in the "Report on the
Energy Policy of Greece 1977". They are to:

- decrease the growth rate of energy consumption relative to the
  national income growth rate, and
- minimise the total social cost of energy.

To support the first goal the Greek government has set targets
for the TPE/GDP elasticity ratio. These targets are 1.4 between
1985 and 1990. They have to be compared with the historical ratio
of 1.8 between 1960 and 1973. Prospects for such a reduction are
good since in particular the efficiency of the current Greek industrial energy systems is low and can, therefore, be considerably
improved in the years to come.

To support the second goal an "integrated pricing policy" will
be introduced gradually. Such a policy is aiming at a rational
system of interdependent prices for all energy carriers. First step
will be the introduction of more flexible oil product pricing systems
which can more closely reflect conditions in the world market.

B. Energy Demand: forecast and control

1. Forecast

The energy forecast for Greece for the next ten years is dis-
plyed on table VI.1. As can be seen Greece is heavily dependent on
foreign energy imports. However, all efforts will be undertaken to
reduce the almost 100 % dependence of today to 86 % in 1990. Major
energy source is oil which had a level of 78 % of total energy demand in 1978. But this percentage should be reduced to 68.2 % in
1990. However, it should not be overseen that despite this reduction
in relative figures the absolute amount of oil imports will be increased by 70 %.

Major sector for energy consumption is the Greek industry with
41.5 % in 1978 followed by transport with 33.0 and residential/
commercial with 25.5 %. The first and the latter one should be enlarged to 44.0 % and 29.7 %, respectively, by 1990. This change of
the consumption structure is reflecting the endeavor of the Greek
government the raise the living standard in this country in order
to achieve the levels of other European countries within the next
10-15 years. The decrease of the consumption share of the transportation sectors indicates that the increase of Greek living standards

Table VI.1

GREECE
Key Energy Indicators and Data
(Mtoe)

	1960	1973	1978	1985	1990
General					
Energy Demand (A)	2.8	12.2	15.2	24.2	31.6
Energy Production (B)	0.6	2.4	3.7	10.0	13.6
B/A (%)	21.4	19.7	24.3	41.7	43.0
Net Oil Consumption	2.1	9.4	11.2	14.7	18.2
TPE/GDP Ratio	0.59	1.11	1.16	1.32	1.38
TPE/GDP Elasticity	1.82	1.28	1.38	1.23	
Per Capita TPE	0.34	1.39	1.63	2.49	3.16
Total Final Consumption	2.0	8.8	10.6	16.2	20.2
Industry (including non energy uses)	0.7	3.7	4.4	7.2	8.9
Transport	0.9	2.7	3.5	4.5	5.3
Residential/Commercial	0.5	2.4	2.7	4.5	6.0
Production					
Solid Fuels	0.4	1.8	2.9	7.5	9.7
Oil	-	-	-	1.1	1.1
Gas	-	-	-	0.4	0.2
Nuclear	-	-	-	-	1.0
Hydro/Geothermal	0.2	0.6	0.8	1.0	1.6
Electricity (Twh)	2.3	14.8	21.0	38.6	54.2
Trade					
Coal Exports	-	-	-	-	-
Imports	0.1	0.5	0.3	0.6	0.9
Oil Exports	-	5.0	4.5	5.0	5.0
Bunkers	0.5	0.9	1.0	1.6	2.0
Imports	2.6	16.3	16.5	21.5	25.4
Gas Exports	-	-	-	-	-
Imports	-	-	-	-	-
References:					
GDP (1970 $ US billion)	4.8	11.0	13.1	18.4	22.9
Population (millions)	8.3	8.8	9.3	9.7	10.0

might not go in parallel with an increase of individual car traffic.

2. Control

    Greece is just embarking on an energy conservation programme.
For the time being it is not as comprehensive as in other countries
but it is expected that it will be considerably improved within the
next 5 years. Most effective instrument for the moment are price increases of oil products and electricity. As an example, in 1979
prices increased as follows: Gasoline: 37 %, Diesel-Oil: 83 %,
Heavy Fuel Oil: 35 %.

    In the industrial sector mandatory auditing and reporting for
energy intensive industries were introduced. They were also required
to reduce their consumption of fuel during the second half of 1979
by 5 % with reference to the corresponding consumption in 1978.
Interest-free loans to cover up to 60 % of investment for energy-
saving equipment and retrofitting are granted. Of particular interest
are co-generation and recuperation of waste heat which are also
supported by the government.

    In the transport sector great importance is attached to an en-
largement of public transport systems. Privat car traffic should be
reduced in growth by heavily increased fuel prices, speed limits,
traffic restrictions on weekends and road and purchase tax increases
for larger cars.

    In the residential/commercial sector main measures to conserve
energy are new insulation standards for houses and tax reductions
for solar heating devices.

C. Supply situation

    There are only modest indigenous energy source in Greece as
hydropower, some lignite and very soon oil and gas.

1. Oil

    Oil production in Greece will most probably start in 1981 at
the Prinos oil field and the South Cavala oil and gas field in the
Aegean Sea. It is projected to reach a peak rate of 30-35 000 barrel/
day which would represent about 10 % of the probable oil consumption
in 1985.

    The recoverable reserves of the Prinos field are 53 million
barrels. In the long term Greece is trying to diversify its oil
supply basis as far as possible. Presently the main suppliers are
the USSR, Saudi Arabia, Iraq and Libya.

2. Gas

Natural gas production in Greece will start in the early 1980s in the South Cavala gas fields with a maximum production rate of 350 000 cubicmeters/day. The total recoverable reserves of this field are estimated to be 800 million cubic meters. It might be that additional resources from the Western Sea could be tapped by the end of the 1980s.

In the long run quantities in the order of 2 billion cubic-meters/year are scheduled to be imported. Major source might be the USSR. In addition plans are underway for a long term co-operation with Algeria.

3. Coal

Greece has some domestic lignite resources which are most of all used for electricity generation. Lignite and hydropower will in the long term dominate the electricity sector. Due to the current planning it is anticipated that in 1985 lignite and hydro will account for 82 % in electricity generation as compared to oil with 18 %.

Proved reserves are about 900 million tons; 400 of which can be recovered economically, given today's international energy prices. Greek lignite, however, has unfortunately a low calorific value of 1500-2000 Kcal/kg. Annual lignite production is expected to be 26.8 million tons in 1980, 56.8 million tons in 1985 and 72.9 million tons in 1990.

For the medium and long term domestic lignite production will no be sufficient to meet the demand. After 1988 no new lignite fired plant can go on line. Greece government, herefore is conside-ring long term coal supplies from abroad.

4. Nuclear

For the time being no nuclear power plant is established in Greece. However, for the medium and long term it is planed to intro-duce nuclear power in order to replace oil and the decreasing lignite resources for electricity generation. Nuclear energy and hard coal are anticipated to be the partners of hydropower for electricy pro-duction towards the end of this century and beyond.

The present nuclear programme calls for a first power plant of 0.6 GWe in 1987, for 1.2 GWe in 1990 and 3.2 GWe in 2000.

VII. Ireland
A. Overview
1. Overall situation

Ireland is to almost 85 % dependent on energy imports. More than 75 % of its energy needs are met by imported oil. Although the relative dependence from oil should be decreased by the next ten years to 63 % the absolut figures will continue to go up from about 8.5 Mtoe of today to 15.0 Mtoe in 1990.

According to present plans Ireland is on the way to strengthen its industrial capacity so that the current share of final energy consumption in this sector should double through 1990. This is one of the reasons for the doubling of the total energy demand of this country from 7.7 Mtoe in 1978 to 15.0 Mtoe in 1990. In parallel to this increase in energy demand supply from domestic resources will also be increased by more than 100 %, so that the import dependence situation will remain the same despite the drastically raised energy consumption.

Presently, the only indigenous energy source is peat. Its potential, however, is rather limited and its contribution to the total energy supply will even decrease within the next ten years from currently 14.3 to 10.7 % in 1990. There are good prospects that in the next years natural gas resources will be tapped. The potential is so that one can expect a contribution of 7.3 % in 1990 to the total energy demand.

2. Energy policy goals

The general policy in energy for the next ten years is to switch to more reliable (coal, peat) and less costly energy sources and in addition to conserve energy as far as possible. The main features of this energy policy as outlined in the 1978 discussion paper "Energy Ireland" are:

- to promote economy in the use of fuels, particularly oil;
- to encourage greater use of coal;
- to further development of indigenous fuels;
- to progress towards nuclear energy;
- to encourage energy conservation measures;
- to promote the exploration for hydrocarbons and uranium; and
- to participate in research aimed at discovering and developing alternative energy sources.

Underlying to this policy is the general policy as to let reflect energy prices at home the continously raising levels on the world market. By this policy the prices of petroleum products, gas, coal and electricity have been increased steadily over the past years.

Table VII.1

IRELAND
Key Energy Indicators and Data
(Mtoe)

	1960	1973	1978	1985	1990
General					
Energy Demand (A)	3.9	7.0	7.7	12.0	15.0
Energy Production	1.5	1.3	1.3	2.9	2.9
BA (%)	38.5	18.6	16.7	21.6	17.1
Net Oil Imports	1.3	5.3	5.9	8.0	9.5
Total Oil Consumption	1.2	5.2	5.8	7.9	9.4
TPE/GDP Ratio	1.49	1.59	1.46	1.61	1.58
TPE/GDP Elasticity	1.14	0.51	1.31	0.91	
Per Capita TPE	1.39	2.26	2.33	3.22	3.84
Total Final Consumption	3.0	5.1	5.9	9.4	11.5
Industry (Inc. non-energy uses)	0,8	1.5	1.8	3.7*	4.8*
Transport	0.5	1.3	1.7	2.6	3.2
Commercial/Residential	1.7	2.3	2.4	3.1	3.5
Production					
Solids	1.3	1.1	1.1	1.6	1.6
Oil	–	–	–	–	–
Gas	–	–	–	1.1	1.1
Nuclear	–	–	–	–	–
Hydro/Geothermal	0.2	0.2	0.2	0.2	0.2
Electricity (Twh)	2.3	7.5	9.8	15.8	22.2
Trade					
Coal Exports	–	0.1	–	–	–
Imports	1.2	0.6	0.6	1.2	2.7
Oil Exports	0.6	0.5	–	–	–
Imports	1.9	5.8	5.9	8.0	9.5
Bunkers	0.1	0.1	0.1	0.1	0.1
Gas Exports	–	–	–	–	–
Imports	–	–	–	–	–
Reference	1960	1973	1978	1985	1990
GDP ($ US billion)	2.62	4.39	5.29	7.44	9.50
Population (million)	2.8	3.1	3.3	3.6	3.8

* Includes 0.4 Mtoe of natural gas used as feedstock

B. Energy demand: forecast and control

1. Forecast

The energy forecast for Ireland for the next ten years is displayed on table VII.1. This forecast presents the best projection
available to the government as of mid-August 1979. They are, however, subject to revision, which possibly might take place during
fall of this year.

The future development of the consumption structure shows as
indicated before an increase of energy consumption in industry from
30.5 % in 1978 to 41.7 % in 1990. This is one of the largest efforts
in promoting industry in whole Europe. As the share of the transportation sector remains constant the increase in industry will be
achieved by a reduction in the commercial/residential sector.

2. Control

Control on future energy increase is performed by setting energy
prices at world market levels and by a strong and comprehensive conservation programme. The latter has two components: mandatory and
voluntary measures.

Mandatory measures as introduced so far are:

- Regulation made prohibiting the sale of gas/diesel for central
  heating in homes.
- regulation made providing for minimum sales of motor fuel.
- regulation made reducing overall speed limit from 60 to 55 m.p.h.
- Standards for thermal insulation of houses extended from 1st
  July, 1979.

The voluntary measures are comparable to those in many other
countries in Europe: grants for switching from oil to solid fuel
fired central heating devices, grants for R+D and introduction of
new energy conserving techniques, public information campaigns
stressing the importance of energy conservation in private households and industry, incentives given by the government for
strengthening public transportation systems, reduction of home
temperature to 18° C (compulsory for government departments), etc.

In addition, strong efforts are made to discourage the use of
oil. In particular, in the electricity generation sector there are
good prospects for the future to decrease the share of oil: at
present 37 % of the electricity is made from imported oil, 17 %
from hydro, coal and peat, 13 % from natural gas and 33 % from
imported coal. For the future no new oil fired plants are allowed
and new plants should use coal, peat and natural gas. There are
also plans for the introduction of nuclear power for the 1990s.

C. Supply situation

Faced with limited indigenous energy resource, Ireland plans to meet its anticipated energy needs over the next decade by increasing its oil imports by more than 62 % (4.1 % per year) and by more than quadrupling its coal imports. The bulk of the increased oil supplies will be used in industry (1.7 Mtoe) and transportation (1.5 Mtoe). The growth in coal utilisation will be consumed as indicated before in electricity generation (1.4 Mtoe).

1. Oil

There are practically no oil reserves in Ireland.

2. Natural gas

As of July 1979 natural gas reserves are estimated to be 25.4 Mtoe.

3. Coal

Coal, predominantely peat, ist the only indigenous energy source for the time being. Reserves are estimated to be about 40 million tons. The average annual output is 55 000 tons mainly from the Arigua and Rossmore. Collieries at Arigua, Co.Roscommon and Rossmore, Co.Loois, respectively. Output is expected to increase somewhat as the Kealy Mines operation in Co.Tipperary builds up to full production which is estimated at 40 000 tons maximum.

4. Nuclear

Originally Irish government has planned to set up a first nuclear power plant by 1990. However, this plan was postponed in 1979 and is now subject to a special legislation and a public enquiry. There might be some uranium deposits in Ireland, however exact figures are presently not available.

5. Hydro

Exploitation of hydro resources is practically exhausted in Ireland. Contribution to the total energy demand was 2.6 % in 1978 and is anticipated to be only 1.3 % in 1990.

6. Other energy sources

A policy for the commercialisation of other energy sources, in particular renewables, has not yet been developed. However, R+D efforts are promoted by the government.

VIII. Italy

A. Overview

1. Overall situation

According to the actual energy plan energy demand in Italy is anticipated to grow from 138.9 Mtoe in 1978 to 224.2 Mtoe in 1990. More than three quarters of the demand had to be imported, mostly crude oil. This dependence should continue through the next 10 years. However, even this high degree of dependence could only be maintained by conducting a vigorous nuclear programme: 1.0 Mtoe nuclear in 1978, 16.5 Mtoe nuclear in 1990. For the time being one can doubt whether such a programme could really be carried out and thus it might happen that for the years to come Italy's dependence from foreign energy imports will either grow further or the energy plan with its GDP growth of 4 % annually will be abandoned. The dilema with the Italian nuclear programme becomes even more apparent if one compares the figures of today with those of two years ago where a nuclear contribution was expected of 7.0 Mtoe in 1985 and of 45.0 Mtoe in 1990. That corresponded to 7.4 GW and 33.4 GW in 1985 and 1990, respectively. The figures to date are 2.4 GW and 13.4 GW.

Italy's main indigenous energy sources are natural gas and hydro/geothermal. Whereas gas production will decline there is some potential for an expansion of energy generation from hydro/geothermal.

2. Energy policy goals

Italy's energy policy is laid down in the National Energy Plan which was approved by the Inter-Ministerial Committee for Economic Planing (CIPE) on December 23, 1977. It is still valid. Main items of the plan are reduction of oil dependence, enhanced exploitation of domestic resources, energy conservation, increased use of nuclear power, comprehensive R+D efforts for new energy sources. Implementation of the Energy Plan was assigned mainly to the Ministry of Industry. However, due to the complexity of the Energy Plan, which contains numerous technical, financial, tax and legislative measures, many other institutions are involved. Therefore, an institutional reform is currently debated for creating a single and transparent administrative structure required for setting priorities and for conducting all the measures needed to implement the Energy Plan as quickly as possible.

Table VII.1

ITALY
Key Energy Indicators and Data
(Mtoe)

	1960	1973	1978	1985	1990
		General			
Energy Demand (A)	49.8	132.6	138.9	188.7	224.2
Energy Production (B)	23.8	25.7	25.8	29.9	46.9
B/A (%)	49.8	19.4	18.6	15.9	20.9
Net Oil Imports	22.4	104.7	95.2	128.4	131.0
Total Oil Consumption	20.5	97.9	93.9	124.9	127.0
TPE/GDP Ratio	0,92	1.28	1.11	1.14	1.12
TPE/GDP Elasticity	1.53	n.a	1.13	0.88	
Per Capity TPE	1.0	2.42	2.45	3.25	3.77
Total Final Consumption	36.2	101.0	103.7	138.1	155.5
Industry (incl. non-					
energy uses)	19.3	49.3	46.6	64.1	73.4
Transport	6.8	19.6	22.9	28.6	32.1
Residential/Commercial	10.1	32.1	34.2	45.4	50.0
		Production			
Solids	3.8	1.8	1.1	1.5	1.5
Oil	2.1	1.1	1.5	3.0	3.0
Gas	5.4	12.9	11.3	10.0	10.0
Nuclear	-	0.7	1.0	1.9	16.5
Hydro/Geothermal	13.1	9.2	10.9	12.0	12.6
Other	-	-	-	1.5	3.3
Electricity (TWh)	55.3	139.1	167.3	253.2	348.2
		Trade			
Coal Exports	0.1	0.4	0.2	0.8	-
Imports	7.7	8.8	9.3	15.5	24.5
Oil Exports	8.6	29.0	23.2	23.0	23.0
Bunkers	3.1	7.2	5.6	6.5	7.0
Imports	31.0	133.8	118.4	151.4	154.0
Gas Exports	-	-	-	-	-
Imports	-	1.7	11.7	22.2	28.8
Reference					
GDP (1970 $ US billion)	53.9	103.5	125.6*)	165.2	201.0
Population (million)	50.2	54.9	56.7	58.0	59.4

*) Due to changes in statistical basis for GDP, this figure as well
   as TPE/GDP ratio and elasticity cannot be compared to 1960, 1973
   (and 1977) figures. An estimate based on historical figures so
   far used would give:

	1973	1978	1985	1990
GDP (1970 $ US billion)	103.5	114.8	151.1	183.8
TPE/GDP ratio	1.28	1.21	1.25	1.22
TPE/GDP elasticity		0.48	1.12	0.88

B. Energy demand: forecast and control

1. Forecast

The energy forecast for Italy for the next ten years is displayed
on table VIII.1. It shows that imported oil will remain the main
supply source, but coal and gas imports are stept up together with
nuclear and hydro sources. Main sector in energy increase will be industry
with 3.9 % per year, reflecting the importance attached to this area in Italy.

The energy forecast has been revised downwards during recent
years mainly due to a slipage in GDP growth as compared to the
planning. Due to the development of oil prices during the last year
it is anticipated that even the 4 % GDP growth underlying to the
actual forecast might have to be reduced, at least for the immediate
future.

2. Control

Most energy prices in Italy are controlled and do in general
not reflect the international level. In particular, electricity
tariffs have decreased in real terms over the past few years and
about two-thirds of total residential consumption of electricity
is priced at "social tariffs" which do not even cover average costs
of production and distribution. But it is intended to reduce this
subsidized consumption considerably by limiting it to households
with an annual consumption of less than 900 KWh/year with beginning of August 1980.

However, in general it seems that price instruments for controlling future development of energy consumption in Italy are not
very effective, at least for the near term. Conservation, although
intended to play a major role in the future, is for the time being
more on a rather modest level.

A coherent programme is in the preparation phase and so far
only strong information campaigns and a programme designed to cut
oil demand for space heating and transport, by mandatory measures,
are in place.

For industry funds are provided (Restructuring and Reconversion
of Industry Law 675/77) which can be allocated to energy conservation. In addition, a plan to provide loans on favourable terms
has been prepared. Government is also attempting to improve Co-
operation between various industries in order to cave energy. In
this respect combined energy-heat production is of particular interest.

In transport sector one is aiming at making public transportation systems more efficient and if possible, at enlarging it in
appropriate areas.

The residential/commercial conservation programme is mainly based on mandatory measures. Room temperatures in winter are not allowed to exceed 20°C. Codes with minimum thermal efficiency standards for new and renovated buildings and for heating systems have been implemented. It is expected that effective incentive programmes for retroffitting existing buildings will be implemented for the comming years.

C. Supply situation

1. Oil

Italy has some modest oil resources. Recoverable reserves as of end of 1978 were about 50 million tons. Major efforts are being made for exploration on further oil resources in the country.

2. Natural gas

Natural gas resources as of end of 1978 were about 200 billion cubic meters. For the time being production from own fields cover about 45 % of the domestic needs. The rest is imported from Libya, the Netherlands and the USSR. The most important factor for the future is the new gas pipeline from Algeria which is under construction. Supply from this new sources will be mainly for the Mezzogiorno area.

3. Coal

Coal will be one of the major long term energy sources in Italy. Almost all of it will have to be imported since the domestic resources are rather small. By switching thermal power plants to coal, consumption is expected to increase to 6.5 million tons by 1985 and by building new coal-fired base load plants consumption could increase still further to 15-27 million tons in 1990. Environmental problems, however, might constrain those enormous extension plans.

4. Nuclear

The originally very ambitious nuclear power programme has been drastically delayed in recent years. Since nuclear power was assigned a key role in further energy supply, it may turn out, that this delay will be a bottle neck for the expension of the Italian economy. At the moment CIPE and Parliament have accepted the immediate construction of 8 000 MW which should go on line between 1985 and 1990. However, even this plan may not be realized due to servere siting problems. In fact, there is only one site today (Montalto di Castro) where construction is underway.

5. Hydro/Geothermal

According to the actual planning hydro and geothermal resources are to be exploited to the fullest extend. In particular geothermal activities are strongly pursued.

6. Other energy sources

Italy is expecting that in the long term solar energy could give a measurable contribution to its energy supply. Therefore, several major activities are underway both on the national and international level.

IX. Luxembourg

A. Overview

1. Overall situation

Luxembourg has no energy reserves and indigenous production is confined to a small amount of hydro-electricity with an installed capacity of 29 MW and an output of about 80 GWh per year. Main Consumer of energy is the iron and steel industry accounting for almost 70 % of all the energy utilized in the country.

2. Energy policy goals

Main goal of Luxembourg's energy policy is the security of supply. To achieve this goal the following measure should be performed:

- conservation of energy, in particular of oil products,
- rational utilization of energy,
- development of new alternative source energy technologies.

A specific energy programme has not yet been adopted by the government.

B. Energy demand: forecast and control

1. Forecast

The energy forecast for Luxembourg through 1990 is displayed on table IX.1. One of the most striking features is the decline in energy consumption from 1973 to 1978 and the relatively slow increase for the next 5 years. This recent decrease in energy demand is for one reason due to the strong coupling of energy requirements to the industrial sector mainly composed of iron and steel manufacturers. The recession in this sector of economy had, therefore, a direct influence on Luxembourg's energy demand in the years 1975-1978.

Table V.1

LUXEMBOURG
Key Energy Indicators and Data
(Mtoe)

	1960	1973	1978	1985	1990
General					
Energy Demand (A)	3.27	4.32	3.88	4.77	5.35
Energy Production (B)	0.01	0.03	0.03	0.03	0.03
B/A (%)	0.3	0.7	0.8	0.6	0.6
Net Oil Imports	0.23	1.69	1.40	2.00	2.00
Total Oil Consumption	0.23	1.69	1.40	2.00	2.00
TPE/GDP Ratio	4.30	3.51	3.13	3.22	3.13
TPE/GDP Elasticity	0.68		–	1.00	0.77	
Per Captita TPE	10.22	12.34	11.09	13.63	15.29
Total Final Consumption	2.81	4.09	3.61	4.49	5.08
Industry (including non energy uses)	2.42	3.22	2.53	3.16	3.58
Transport	0.13	0.29	0.42	0.51	0.57
Residential/Commercial	0.26	0.58	0.66	0.82	0.93
Production					
Nuclear	–	–	–	–	–
Hydro/Geothermal	0.01	0.03	0.03	0.03	0.03
Electricity (Twh)	1.46	1.38	1.14	0.72	0.72
Imports					
Coal Imports	3.03	2.41	1.76	1.85	2.16
Oil Imports	0.23	1.69	1.40	2.00	2.00
Gas Imports	–	0.22	0.46	0.55	0.70
Electricity Imports	–	0.18	0.22	0.34	0.46
Growth Rates					
TPE		2.2	-2.1	3.0	2.3	
GDP		3.2	–	3.0	2.9	
TFC		2.9	-2.5	3.2	2.5	
Net Oil Imports		16.6	-3.7	5.2	–	
Oil Consumption		16.6	3.7	5.2	–	
Reference:					
GDP (1970 $ US billion)	0.8	1.2	1.2	1.5	1.7
Population (millions)	0.32	0.35	0.35	(0.35)	(0.35)

There is no indication of a change in the consumption structure:
i.e. industry should prevail through the next decade as it was since
years. The other reason for the recent decline in energy requirements
was the improved efficiency of Luxembourg's iron steel industry due
to energy conservation measures and new technologies.

2. Control

Luxembourg is very interested in energy conservation and
government, therefore, is discussing to set up an administrative
structure to deal with the introduction of energy conservation and
rational energy use measures.

At present the following measures to control energy use are
implemented:

- mandatory thermal insulation standards for new public buildings,
  to be extended in future to all new constructions,
- limitation of heating temperature in public buildings,
- requirement for competent authoritie's agreement before utilisation of all new heating installations using oil,
- review of existing heating installations (including industrial)
  using oil every two years,
- provision as from 1st October 1979, of subsidies to private users
  for energy saving investments (25 % with a ceiling of 15.000 FL
  per house).

In addition vigorous efforts are being made by the iron and
steel industry to substitute coal for oil.

C. Supply situation

As mentioned before Luxembourg has practically no indigenous
energy resources. Security of supply should be achieved by switching
away from oil to coal and natural gas, both of which should be imported in an increasing percentage from neighbour countries. Nuclear
power has been open as an option. Plans to construct a first nuclear
power plant which existed a couple of years ago have been meanwhile
relinquished.

X. Netherlands

A. Overview

1. Overall situation

The Netherlands are blessed with great deposits of natural gas,
which is used both for domestic energy supply and large exports to

countries within Europe. However, a decline of the production rate
is anticipated for the medium term and the Dutch government is
therefore trying to preserve its gas reserves for a longer time.
This should be achieved by switching over to coal and to nuclear
power in the long term. However, the future of nuclear power in the
Netherlands is highly uncertain. Due to heavy public opposition the
Dutch nuclear energy programme was halted and a public debate is
just going to be launched which is anticipated to last about 2 years.
Decisions on the extension of nuclear power plants will not be taken
before this debate is finished.

2. Energy policy goals

     Main energy policy goal is to preserve the gas resources of
the Groningen gas fields. To accomplish this the Dutch government
is not hesitating to increase the oil imports in the near term and
even to embark on a gas import programme. However, in the long term
coal should replace gas.

     Securing the supply side seems to have first priority as com-
pared to conservation measures although comprehensive efforts to
safe energy are underway. A comprehensive and extensive documen-
tation of the Dutch energy policy will be the basis for the debate on nuclear power for 1980 and 1981.

B. Energy demand: forecast and control

1. Forecast

     The energy forecast of the Netherlands for the next ten years
is displayed on table X.1. Underlying to this forecast is the
assumption of an annual GDP growth of 3 %.

     For the time being the Netherlands is one of the rare countries
in Europe supplying most of its energy from indigenous sources. However, due to the decline of domestic natural gas production this
situation will change even within the next ten years. This decline
will also influence the energy policies of users from abroad as
for instance the Federal Republic of Germany which has to orient
its gas policy towards other supplies in the mid and late nineties.

     As the figures show the gap arising by decreasing gas production
should be filled by coal imports in the long term (i.e. increase of
imports by 215 % within 10 years) and in the medium term by enhanced
oil imports. In particular the latter measure is highly unusual in
view of all the efforts within the EEC and IEA to reduce oil consumption as fast as possible. It can be foreseen that the vigorous
coal policy will encounter some environmental (pollution) problems
and it might therefore be that the introduction of new coal-fired
plants will be impeded by the public. Therefore, in view of those

Table X.1

NETHERLANDS
Key Energy Indicators and Data
(Mtoe)

	1960	1973	1978	1985	1990
General					
Energy Demand (A)	21.9	61.7	64.8	83.5	91.8
Energy Production (B)	11.2	57.9	71.4	65.8	55.9
B/A (%)	51.1	93.8	110.2	78.8	60.9
Net Oil Imports (inc. bunkers)	11.5	40.2	36.8	51.6	56.3
Total Oil Consumption	10.4	29.5	26.8	41.6	45.3
TPE/GDP Ratio	1.16	1.70	1.57	1.64	1.58
TPE/GDP Elasticity		1.61	0.40	1.19	0.73
Per Capita TPE	1.91	4.61	4.66	5.80	6.29
Total Final Consumption	15.2	49.7	53.2	69.0	75.7
Industry (incl. non-energy uses)	5.7	21.8	22.2	32.5	38.0
Transport	2.8	7.5	8.3	10.4	11.5
Residential/Commercial	6.7	20.4	22.7	26.1	26.2
Production					
Solids	9.0	1.3	–	–	–
Oil	1.9	1.5	1.6	2.0	2.0
Gas	0.3	54.8	68.6	62.6	52.7
Nuclear	–	0.3	1.0	1.0	1.0
Hydro/Geothermal	–	–	–	–	–
Electricity	16.5	45.8	52.6	69.8	79.8
Other	–	–	0.2	0.2	0.2
Trade					
Coal Exports	3.4	1.5	0.5	0.5	0.5
Imports	5.3	2.8	3.8	7.6	12.0
Oil Exports	13.7	42.6	30.0	40.6	50.9
Bunkers	2.4	11.7	11.8	12.0	13.0
Imports	25.2	82.8	66.8	92.2	107.2
Gas Exports	–	25.7	36.9	35.4	26.0
Imports	–	–	1.8	6.4	7.1
Reference					
GDP (1970 $ US billion)	18.9	36.4	41.2	51.0	58.0
Population (million)	11.5	13.4	13.9	14.4	14.6

problems and perhaps of heavy delays or even abandoning of the
nuclear power programme, it will be of great interest to observe
whether the expected increase of energy requirements can be really
achieved.

Main energy consumers are the Dutch industry and the resi-
dential/commercial sector with both about 42 % of final energy
consumption. However, this structure is going to be changed towards an increase in the industrial sector to 50 % in 1990. This
increase is not only due to an expected increase in industrial
production but also to effects anticipated from current energy
conservation measures in the residential sector.

2. Control

Dutch government is using market mechanisms as well as conservation programmes to control future energy consumptions. Prices
for oil products and electricity reflect fairly well the international situation on the world energy market.

The conservation efforts comprise measures for all the three
consumption sectors. In industry government stimulates energy
saving measures along the following lines:

- grants for energy-saving investments,
- financial backing of R, D and D-programmes,
- grants for energy audits
- PR- and educational activities.

In the transportation area, incentives were given by the
government for enlargement of public transportation services and
systems. In the residential/commercial sector the following measures
have been set in force or are underway:

- raising of insulation standards for new buildings,
- assistance for retrofitting existing buildings in order to increase insulation measures,
- improvement of heating systems by large scale introduction of
  new highly efficient central heating boilers,
- improved utility load management,
  enhanced public education and information campaigns.

C. Supply situation

1. Oil

As of January 1979 total reserves of oil were about 46 Mio
tons. Production from these sources account for about 2 % of the
present energy demand of the Netherlands.

2. Natural Gas

As of January 1979 natural gas reserves were estimated for on shore deposits to be 1,844 billion cubic meters and for off-shore sources to be 367 billion cubic meters. Exploration activities are going on but are encountering increasing difficulties due to environmental problems. It is planned to supplement natural gas with LNG and, in the long term, with gas from coal gasification plants. As mentioned before aim of the Dutch energy policy is to switch to other energy sources in order to protect the domestic gas resources from an early exhaustion.

3. Coal

The Netherlands have no indigenous coal reserves.

4. Nuclear

The Netherlands are operating a 500 MW nuclear power plant since 1973, an enrichment facility which is currently enlarged and they are participating in the SERENA-cooperation on the development of fast breeder reactors. SERENA partners are Belgium, France, Federal Republic of Germany, the Netherlands and Italy. At present nuclear power is heavily debated in the Dutch public and this debate, now organised by the government, will last until the end of 1981. It is only after this debate that the Dutch parliament will decide on a future extension or abandoning of the nuclear programme.

5. Hydro

There are no hydro resources in the Netherlands.

6. Other energy sources

The Netherlands are strongly interested in the development and introduction of new energy sources, in particular solar, wind and geothermal power as well as waste heat. It is expected that in the long term these energy sources could contribute some percent to the total energy demand of this country.

XI. Norway

A. Overview

1. Overall situation

Norway is in a unique situation in Europe: It does not only entirely cover its energy demand by domestic sources but is also a major exporter of oil and natural gas. Oil and gas are found off

Table XI.1

NORWAY
Key Energy Indicators and Data
(Mtoe)

	1960	1973	1978	1985	1990

General

	1960	1973	1978	1985	1990
Energy Demand (A)	9.0	19.7	21.5	27.5	31.0*
Energy Production (B)	5.0	12.8	43.9	67.4	63.6
B/A (%)	55.6	65.0	204.2	248.7	211.6
Net Oil Imports	3.6	6.9	-8.3	-20.2	-16.0
Total Oil Consumption	3.5	7.8	7.6	11.8	14.0
TPE/GDP Ratio	1.31	1.54	1.33	1.37	1.33*
TPE/GDP Elasticity	1.26		0.38	1.11	0.99
Per Capita TPE	2.50	4.80	5.37	6.58	7.33*
Total Final Consumption	6.4	13.8	14.1	20.3	22.9*
Industry	3.2	6.5	5.7	8.0	9.2*
Transport	1.3	2.5	3.0	4.2	4.4*
Residential/Commercial	1.8	4.8	5.7	8.1	9.3*

Production

	1960	1973	1978	1985	1990
Coal	0.3	0.3	0.3	0.3	1.0*
Oil	-	1.6	17.3	31.0	30.0
Gas	-	-	14.2	23.0	18.0
Nuclear	-	-	-	-	
Hydro	4.7	10.9	12.1	14.1	15.6*

Trade

	1960	1973	1978	1985	1990
Coal Exports	0.1	0.1	0.1	0.1	-
Imports	0.7	0.6	0.6	0.9	0.4*
Oil Exports	0.4	4.3	18.7	18.8	16.0
Imports	4.0	11.2	10.4	-	-
Gas Exports	-	-	14.2	22.0	17.0
Imports	-	-	-	-	-

	1960	1973	1978	1985	1990*
GDP ($US billion)	6.9	12.8	16.1	20.1	23.3
Population (million)			4.05	4.18	4.23

*Estimate by IEA Secretariat

shore mainly in the Ekofisk, Frigg and Statfjord field. In addition, hydro power plays a major role for the domestic energy supply.

2. Energy policy goals

Main efforts of Norway's energy policy are directed towards oil and gas: New exploration activities north of 62° latitude and enhanced activities in the existing fields. However, this is not ment to increase production to an utmost extend but to get a better feeling for the resources and to hold production on a nearly constant level for the long term. In addition energy policy is dealing with the problems associated with an extension of hydro power utilisation. On the conservation side Norwegian government has launched several measures to improve energy use in particular in industry and transportation. A comprehensive programme for energy conservation, however, is still lacking.

B. Energy demand: forecast and control

1. Forecast

The energy forecast for Norway for the next decade is displayed on table XI.1. The figures for 1990 were estimated by the International Energy Agency since for that time no official data are available.

Main sources of supply are oil (about 35 %) and hydro (about 55 %) from indigenous resources. On the consumption side industry and residential/commercial sector are of equal importance with 40 %. As compared to the geographical location of Norway the percentage of residential/commercial consumption is rather low, indicating that insulation standards are already well developed.

2. Control

Energy prices in Norway are not subject to free market forces. In fact there was an overall wage and price freeze until January 1980 by which only small changes of energy prices could take place. The freeze is followed up by strict price controls in 1980. Nevertheless prices for oil products are rather high in Norway giving, therefore, goad incentives to conserve energy.

Conservation efforts by the government are for the time being more directed to the industry and transportation sector than to the field of residential and commercial consumption. Loans and loan guarantees are provided for energy conservation investments in industry. The use of waste heat and waste as fuel is promoted through tax exemptions.

In the transportation sector stringent speed limits are in-

force and substantial price subsidies for public transport are
provided which amount to 1/3 of total costs in public transportation.

In the residential/commercial sector new building codes for
reduced energy consumptions are under consideration. However, it
should be mentioned that mandatory thermal insulation requirements are already in place since the 1950's. Subsidies, grants,
special loans for improved insulation investments are not yet in-
force.

C. Supply situation

1. Oil

Norway's estimated production from producing fields and fields
under development (as of 1979) are displayed on table XI.2.

Table XI.2   Norway's oil and gas reserves from producing field-
sand and fields under development.

	1980	1985	1990
EKOFISK			
oil (incl. NGL) 10^6 tons	24	13	5
gas $10^9 m^3$	16	13	9
FRIGG			
gas $10^9 m^3$	10	8	4
STATFJORD			
oil (incl. NGL) 10^6 tons	2	14	23
gas $10^9 m^3$		2	4
VALHALLA			
oil 10^6 tons		4	2
gas $10^9 m^3$		1	2
MURCHISON			
oil 10^6 tons		1	
Total oil 10^6 tons	26	32	30
Total gas $10^9 m^3$	26	22	19

NGL: Natural Gas Liquid (by-product of gas production)

The total proven recoverable reserves (as of end 1978) are displayed on table XI.3.

Table XI.3 Norway's total proven recoverable reserves of oil,
NGL and gas (1978)

	Oil million tons	NGL million tons	GAS billion cubic	Total million tons o.e.
Ekofisk	190	20	220	430
Frigg			110	110
Statfjord	300	10	40	350
Valhalla	30		20	50
Murchison	10			10
Reserves which are being developed	530	30	390	950
Further proven recoverable reserves under evaluation	220		350	570
Total proven recoverable reserves	750	30	740	1520

The potential resource base, for both oil and gas as estimated by the Norwegian Petroleum Directorate is to 2500-3500 Mtoe
in addition to the proven reserves of 1500 Mtoe, thus giving a
total of 4000-5000 Mtoe.

Approximately 42 % of the Norwegian sector south of 62° N
latitude in the North-Sea has been licensed for exploration. A
portion of this area was relinquished to the State, and currently
about 14 % of the total area south of the 62° N is under licence
for exploration and production.

In May 1979, the Norwegian Parliament approved the plans for
exploration drilling north of 62° N. First drilling in this area
is expected for summer 1980.

Afterall, Norway is pursuing a relatively moderate pace of
oil activities adapted to the needs of the country with regard to
domestic consumption and foreign exchange revenues.

2. Gas

For the time being natural gas produced by the Norwegian fields is completely exported to other European countries on the continent. This will not be changed in the immediate future.

3. Coal

There are some coal deposits in the country. The resources, however, do not play a major role in Norway's energy supply.

4. Nuclear

Nuclear energy is considered as a very long term option. Whether this energy source will be tapped depents on public acceptance as well as on the future development of hydro sources and oil and gas fields.

5. Hydro

For decades, electricity in Norway is generated from hydro-power. Construction of new plants kept always pace with the growth of consumption and mostly there was a surplus of electricity which could be exported. However, future expansion of hydropower plants is publically controversal for environmental reasons.

6. Other energy sources

Other energy sources like wind and wave power are still in the R+D phase. In the long term they could potentially contribute to Norway's energy requirements in the order of some percent.

XII. Spain

A. Overview

1. Overall situation

Spain has one of the lowest TPE/capita in Europe and the government is trying hard to increase this ratio considerably during the next ten years. This should be associated with a GDP-growth rate of 5 % annually as compared with 2.5 % during recent years.

Import dependence of Spain is 70 % and should be decreased to less than 60 % within ten years mostly by a strong nuclear pro-gramme and enhanced use of coal from both domestic production and imports. Almost all of the oil, which accounts for 68 % of the energy demand has to be imported, since indigenous petroleum re-sources are rather limited.

Table XII.1

SPAIN
Key Energy Indicators and Data
(Mtoe)

	1960	1973	1978	1985	1990
General					
Energy Demand (A)	19.9	56.6	70.8	101.0	128.0
Energy Production (B)	14.8	15.9	21.5	42.4	56.1
B/A (%)	74.4	28.1	30.4	42.0	43.8
Net Oil Imports	6.5	40.5	48.6	52.9	58.1
Total Oil Consumption	4.8	37.8	47.0	55.7	61.9
TPE/GDP Ratio	1.27	1.48	1.37	1.39	1.38
TPE/GDP Elasticity	1.13		1.04	0.96	
Per Capita TPE	0.65	1.67	1.93	2.70	3.08
Total Final Consumption	13.3	43.4	51.0	71.7	90.0
Industry	5.8	24.6	26.5	38.1	47.8
Transport	4.0	11.3	14.5	21.3	27.8
Residential/Commercial	3.5	7.6	10.0	12.4	15.2
Production					
Coal	10.0	7.1	8.5	13.4	14.7
Oil	-	0.7	1.0	5.3	6.3
Gas	-	-		1.7	1.8
Nuclear	-	1.5	1.7	12.2	21.1
Hydro/Geothermal	4.8	6.7	9.8	9.3	10.4
Renewables	-	-	-	0.4	1.7
Electricity (TWh)	18.6	76.3	99.3	152.6	219.7
Trade					
Coal Exports	-	-	-	-	-
Imports	0.3	2.5	3.0	3.4	10.5
Oil Exports	0.8	4.3	3.3	-	-
Imports	7.3	44.8	51.9	52.9	58.1
Bunkers	1.6	2.1	2.5	2.5	2.5
Gas Exports	-	-	-	-	-
Imports	-	1.0	1.5	4.9	5.9
Reference					
GDP (US $ billion)	17.9	45.6	1.6	72.5	92.6
Population (million)	30.4	34.7	36.6	37.4	41.6

US$1 = 70.0281 Pesetas
(1970 exchange rate)

2. Energy policy goals

Spain has a National Energy Plan (NEP) which was approved by Congress in mid 1979. It should cover Spain's energy policy for the next ten years. The two general goals of the NEP are:

- Reduce energy consumption and adopt its use progresively to the resources existing in the country.
- Foresee an energy supply able to allow increases of the Gross Domestic Product, consistent with the international and foreign balances of national economy.

From these two general goals a variety of subgoals have been established covering the entire energy buisiness including R+D-programmes. They will be mentioned later in more detail.

B. Energy demand: forecast and control

1. Forecast

The energy forecast for the next ten years is displayed on table XII.1. They represent government forecasts as of 1979. As can be seen on the supply side Spain is undertaking all efforts to enhance energy production from domestic sources: coal should be increased by 73 % through 1990, oil by a factor of 6 and nuclear by a factor of 12. It is not at all sure that this ambitious supply programme from own resources can really be realized because production from hydrocarbon sources might encounter some economic, technical and environmental problems and the nuclear programme was several times delayed in the past.

On the consumption side the most striking observation is the low figure of 19.6 % of the residential/commercial sector of the total final energy consumption which to a large extend is due to the geographical location of this country. Conservation measures should even decrease the share to 17 % in 1990. The industrial sector with more than 50 % consumption indicates the importance of industry in Spain's economy. This share will remain about constant, however it might be influenced by Spain's future attendance to the EEC.

2. Control

The NEP stipulates that the prices and tariffs of energy products and services be the real costs of supply, including all costs and including the financial charges of the investments made. Prices, therefore, do more or less reflect world market levels and give good incentives for voluntary conservation measures.

The Spanish conservation programme has been carefully designed and is pursued with great efforts. Main target is apparently the industrial sector. Therefore, all medium and large consumers have been examined in order to gain the basic information for appropriate conservation measures. As a result a law on energy conservation in industry was formulated which is currently due to the legislative process. The draft text of this law contains the following provisions:

- preferred access to credits with lower interest than normal banking rates,
- special tax reductions,
- grants for investments in energy saving technologies or renewable energy sources.

Spanish government estimates that these provisions could lead to savings of about 7 % of current industrial consumption. Expectations for energy savings in transportation are much lower. This is mainly due to the fact that car ownership is low compared to other European countries (164 cars per 1000 people in 1977) and that consumption with 71/100 km is also relatively low. Main measures of the conservation programme are speed limits, support of the national railway company and promotion of public transportation systems in the large cities.

Action to reduce consumption in the residential/commercial sector include:

- reduction of heat oil allotments to 90 % of the 1977 consumption,
- issuing new standards for thermal insulations,
- strengthening of measures regulating public lighting.

C. Supply situation

1. Oil

Domestic oil production is not very high but should be enhanced for the next years. Proven reserves are 153 million barrels. There are, however, some prospects for new discoveries in the Mediterranean offshore area but unfortunately in deep water and in geologically difficult locations.

2. Gas

Domestic gas production is insignificant. Proven reserves are only 35 Mtoe, which were discovered only recently in 1978. However, the potential for future discoveries is assessed to be promising.

3. Coal

Coal production is scheduled to rise from 8.6 Mtoe in 1978
to 14.7 Mtoe in 1990. Out of this increase of 6.1 Mtoe about
5 Mtoe is to come from open air lignite and sub-bituminous mines.
This production is to be burnt in power plants directly located
to the mines. There is a possibility that high sulphur content
(4-5 %) and the presence of clay in the coal seams may become
major constraints on the future development of indigenous coal
resources. At present, total estimated reserves are about 3750
million tons.

4. Nuclear

Spain has a very comprehensive nuclear energy programme. Its
target is to have an installed capacity of 11 GW by 1987. At present 1 GW is on line and 6.6 GW are under construction some of
them heavily delayed. The construction of three new plants was
approved by the Spanish government in 1979. If the plans of the
government could be pursued, 37 % of Spain's indigenous energy
could be supplied by nuclear in 1990.

Spain has own uranium deposits. Proven reserves are 20.000 t.
Current production level is 230 t per year. This level should be
raised to 800 t/a in 1987 so that by that time 45 % of Spain's
uranium requirements could be covered by own resources. Spain is
shareholder of the Eurodif enrichment company in France.

5. Hydro

Spain has some hydro resources which cover about 13 % of the
current demand. The potential of these resources, however, is almost completely exhausted so that its share will decrease to 8 %
in 1990.

6. Other energy sources

Spain is heavily interested in use of solar power and, therefore, has an own comprehensive programme on this energy source
and is, in addition, strongly involved in international cooperation,
in particular in the frame of the IEA.

XIII. Sweden

A. Overview

1. Overall situation

Sweden's only indigenous energy sources of importance are
hydro power and to a much less extend some coal. Sweden, therefore,

Table XIII.1

SWEDEN
Key Energy Indicators and Data
(Mtoe)

	1960	1973	1978	1985[*]	1990[*]
General					
Energy Demand (A)	27.2	47.1	49.0	57.9	61.6
Energy Production (B)	12.1	17.2	21.5	29.0	32.7
B/A (%)	44.5	36.5	43.9	50.1	53.1
Net Oil Imports	13.5	28.8	26.2	28.3	26.6
Total Oil Consumption	12.6	28.3	26.0	26.8	25.0
TPE/GDP Ratio	1.30	1.34	1.31	1.21	1.12
TPE/GDP Elasticity		1.08	0.62	0.69	0.43	
Per Capita TPE	3.6	5.8	5.92	6.98	7.33
Total Final Consumption	18.8	36.0	34.0	39.2	40.5
Industry (incl. non-energy uses)	9.5	16.7	13.8	19.2	20.6
Transport	2.9	5.2	6.1	6.7	7.0
Residential/Commercial	6.4	14.1	14.1	13.3	12.9
Production					
Solid Fuels	2.7	3.2	3.2	3.4	4.9
Oil	0.1	-	-	-	-
Gas	-	-	-	-	-
Nuclear	-	0.5	5.4	10.8	12.4
Hydro/Geothermal	9.3	13.5	13.0	14.4	14.6
Others	-	-	-	0.4	0.8
Electricity (Twh)	35.0	78.1	89.1	112.0	120.0
Trade					
Coal Exports	-	-	0.1	-	-
Imports	2.6	1.7	1.4	2.1	3.9
Oil Exports	0.2	1.4	3.0	-	-
Imports	13.7	30.2	29.2	28.3	26.6
Bunkers	0.5	1.1	1.1	1.5	1.6
Gas Exports	-	-	-	-	-
Imports	-	-	-	-	-
Reference:					
GDP (1970 $ US billion)	21.0	35.2	37.5	47.7	54.8
Population (million)	7.48	8.14	8.28	8.30	8.40

has very clearly embarged on a strong nuclear programme in order
to reduce its heavy dependence on energy imports, in particular
oil. The combination of a strict oil substitution and an active
conservation policy should reduce oil dependence from 51 % in
1978 to 41-46 % in 1985 and to 31-41 % in 1990. It is obvious that
in such a situation nuclear power plays a crucial role both for
the promotors and the opponents since it is one of the keystones
for the future economic development of that country. The public
debate on nuclear energy has meanwhile been ended by a referendum
in March 1980 with the result that the nuclear programme with 12
reactors in operation or under construction should be pursued.

2. Energy policy goals

      The energy policy goals as outlined by the Swedish Minister
for industry in 1979 could be summarized as follows: To increase
the security of supplies by reducing dependence on oil, to minimize the pollution and security problems connected with energy
conversion, and to follow an energy policy which will leave a free
hand in the future. These aims demand strict energy conservation,
an active oil policy and efforts to develop durable, preferably
renewable and indigenous energy sources with minimum of environmental impact.

      As can be seen now, these general goals were not altered by
the outcome of the referendum and will be underlying to the energy
policy of Sweden for the next years.

      The Swedish energy programme derived from the general goals
is very comprehensive and transparent and is vigourously pursued.

B. Energy demand: forecast and control

1. Forecast

      There is no official forecast available for the next 10 years.
However, some projections are provided by the Energy Bill of 1979.
Table XIII.1 presents the figures of the scenario I, which assumes
a GDP-growth of 3.5 % and 2.8 % in 1985 and 1990, and TPE-growth
of 2.4 % and 1.2 % for the same years.

      As can be seen from the figures nuclear energy will be play
a considerable role in Sweden's future energy supply. Even at
present nuclear accounts for 24 % of the total electricity supply
which is the highest figure in the western world.

      The consumption structure shows an equivalence between industry
and residential/commercial sector in the order of 40 %. According
to the scenario underlying to the shown projection this situation
will change considerably during the next ten years. Since for 1990

the residential/commercial sector will only use 31.9 % of the total
final consumption leaving the industrial sector with 50.8 %. This
change in balance should be first of all achieved by a strict
conservation programme and second by an increased production output of
Swedish industry, which is expected to grow on an average of 4.3 %
from 1977 to 1990.

2. Control

High priority is attached by Swedish Government to energy conservation in order to control future demand growth. Due to some
Government interventions during recent years the prices for
energy do not fully reflect the situation on the world market.
Swedish Government obviously does more rely on its conservation
programme than on free market forces influencing the price structure for energy supplies.

The conservation programme for Sweden's industry entails a
broad grant and subsidy system for energy saving investments. This
system is in force since 1974/75 and has already contributed to a
remarkable saving of oil. This is of particular importance since
60 % of the energy consumption in industry is due to energy-intensive branches like iron, steel, pulp and paper. In addition to
these grants and subsidies an advisory service was established for
small and medium sized enterprises. Energy consultants have been
placed in few of the regional development officies.

Special attention is paid to the residential/commercial sector.
Strong building codes and generous retrofitting incentives are in
place. Due to these measures energy growth in this sector is
expected to increase only between 0.7 and 1.6 % per year from 1978
to 1990. For 1980 a new energy labeling programme for refrigerators
and freezers is to be launched.

In the transportation sector efforts are mainly based on
voluntary measures like price incentives for use of public transportation systems or millage efficiency commitments. Speed limits
are introduced all over the country.

C. Supply situation

1. Oil

With exception of minor deposits on Gotland there are no oil
fields in Sweden. Therefore, Sweden is very actively trying to
participate in exploration and exploitation of oil fields in the
North-Sea or other countries, in particular in North Africa.

2. Gas

The situation for gas is the same as for domestic oil resources. Gas imports were almost completely excluded in Sweden's energy programme. However, due to recent announcements this could be altered during the next years.

3. Coal

Technically recoverable reserves of hard coal amount to around 30 million tons, however, the calorific value of 3 600 - 5 000 kcal/kg is rather low. For the comming years some coal should be imported, but compared to hydro and nuclear coal will play a minor role in the supply structure at least during the next decade. However, Swedish Government is trying hard to convince municipalities, utilities and major companies to switch to a greater extend over to coal instead of oil. But, for the time being, the reactions to this policy have not been very promissing.

4. Nuclear

Sweden has its own nuclear industry. Capacity of about 3 600 GW at present will be gradually extended by 9.4 GW by the next years. Uranium reserves of Sweden are estimated to be 300 000 tons which would be more than sufficient for the nuclear programme.

5. Hydro

Electricity production from hydro power is about 61 TWh per year. Capacities should be expanded by 1990 to generate about 66 TWh annually. Sweden is aware of the problem, that extended exploitation of hydro power is associated with some environmental influences so that it might occur that the actual plans might be somewhat delayed.

6. Other energy sources

Sweden has some peat and wood, which is used for energy production. However, resources are not high enough to play a major role. R+D work is conducted on wind and solar power, however, even in the long term contribution by these energy sources will be rather limited.

XIV. Switzerland

A. Overview

1. Overall situation

In Switzerland the only indigenous energy source of abundance

is hydropower. Swiss's energy dependence, therefore, is rather
high. For the future, a strong nuclear power programme is
scheduled to help decrease this dependence. Switzerland unlike
other countries is not planning to embark on a major coal programme for the next ten years.

2. Energy policy goals

Due to the special federalistic structure of Switzerland
Swiss Government is not in a position to formulate energy policy
goals which are compulsory for the entire country. This makes it
very difficult for Switzerland to act coherently. In particular
in the field of energy conservation this fact makes it almost
impossible for the time being to implement all the measures
necessary for a successful conservation policy.

Origins of the Swiss energy programme go back to 1974 when
the Federal Commission for an Overall Energy Strategy (GEK) was
founded. It submitted its final report at the end of 1978. According to this report the basic goals of the Swiss energy programme
(Federal Government) are:

- to use energy resources more rationally and to develop energy
  sources in order to reduce the country's heavy reliance on imported oil;

- to secure energy resources at economically acceptable costs;

- to reconcile economic and environmental considerations both to
  protect environment and preserve resources for future generations.

Main issue of the whole energy policy, however, is a proposal
for a constitutional amendment which was submitted to Parliament.
This amendment was drafted in order to obtain a basis for the
implementation of a comprehensive long-term energy policy.

B. Energy demand: forecast and control

1. Forecast

The energy forecast of Switzerland for the next ten years is
displayed on table XIV.1. The figures correspond to the scenario II
prepared by the GEK, which assumes that a stronger energy policy
could be conducted in the future, mainly on cantonal level (the
26 cantons are the constituting parts of Switzerland) but based
on the existing legal framework.

The consumption structure of Switzerland shows a very large
share of the residential/commercial sector. This is first of all
due to insulation codes from those times when energy costs were

Table XIV.1

SWITZERLAND
Key Energy Indicators and Data
(Mtoe)

	1960	1973	1978	1985	1990
General					
Energy Demand (A)	11.8	23.4	23.4	26.7	30.1
Energy Production (B)	6.3	8.0	9.4	11.7	13.0
B/A (%)	53.4	34.2	40.1	43.8	44.2
Net Oil Imports	4.2	15.1	14.0	13.2	14.4
Total Oil Consumption	3.9	15.2	13.5	13.2	14.4
TPE/GDP Ratio	0.87	1.00	1.05	0.97	0.96
TPE/GDP Elasticity	1.28		0.62	0.88	
Per Capita TPE	2.18	3.66	3.71	4.17	4.63
Total Final Consumption	6.8	17.4	16.9	18.4	20.5
Industry (including non energy uses)	2.6	4.8	4.1	5.0	5.7
Transport	1.5	4.3	4.2	4.9	5.2
Residential/Commercial	2.7	8.3	8.6	8.5	9.6
Production					
Solid fuels	-	-	0.4	0.6	0.6
Oil	-	-	-	-	-
Gas	-	-	-	-	-
Nuclear	-	1.4	2.1	4.1	5.2
Hydro/Geothermal	6.3	6.6	7.0	7.0	7.2
Electricity (Twh)	20.9	38.0	46.6	49.2	55.7
Trade					
Coal Exports	-	-	-	-	-
Imports	1.8	0.2	0.2	0.4	0.5
Oil Exports	-	0.2	-	-	-
Imports	4.2	15.3	14.0	13.2	14.4
Bunkers	-	-	-	-	-
Gas Exports	-	-	-	-	-
Imports	-	0.2	0.7	1.7	1.9
Reference:					
GDP (1970 $US billion)	13.6	23.3	22.2	27.4	31.4
Population (millions)s	5.4	6.4	6.3	6.4	6.5

marginal. Conservation activities,therefore, are predominantely
directed towards this sector.

The nuclear component on the supply side is very strong and
will almost be equal to hydropower within the next 10 years, so
that by 1990 Switzerland will be mainly supplied by only three
energy sources: oil, hydro and nuclear energy. There will be also a small component of natural gas which will be increasingly
used for room heating pruposes.

2. Control

Switzerland is primarily relying on free market forces in
the energy field. For the conservation policy mandatory measures
are, therefore, of second priority.

As mentioned before the field of main concern is the residen-
tial/commercial sector. However, implementation of appropriate
conservation measures is due to the cantons which have different
legislative tools and different motivations. Federal Government
is limited to information campaigns, elaboration of base studies,
recommandations, coordination and promotion of cantonal policies.
Canton activities are mainly focussed on:

- courses for heating controllers,
- thermal insulation codes (six out of twentysix have released
  satisfying minimal standards for new buildings)
- minimum standards for burners and boilers of new installations,
- compulsory regular heating controls.

Measures for the industry and the transportation sector are
very limited. ·

C. Supply situation

1.-3. Oil, Gas, Coal

There are practically no indigenous oil, gas or coal reserves
in Switzerland.

4. Nuclear

Switzerland is strongly in favour of nuclear energy which
is planned to become the long term energy sources number two.
However, the nuclear programme is not without opponents and it
is uncertain whether nuclear power capacities can be expanded on
schedule.

5. Other energy sources

Switzerland has great interest in the development of new
energy sources, in particular solar and waste heat. It is anticipated that both these energy sources could cover almost 2 % of the
primary energy requirements in 1990.

XV. United Kingdom

A. Overview

1. Overall situation

The United Kingdom is blessed with great amounts of oil and
gas from reserves in the North-Sea. Together with indigenous coal
resources and nuclear power the UK is well on the way to achieve
energy self-sufficiency on the very near term, probably even in
1980. This four-fuel economy is the center piece of the British
energy policy for the years to come. However, the contribution
of each of these four energy sources will vary over time since
it is expected that production from oil and gas fields in the
North-Sea will decline towards the year 2000 and the arising gap
will have to be filled by nuclear energy and coal.

2. Energy policy goals

The UK's national energy policy is aiming at reducing dependence on imported energy through economic home production. This
should be achieved by exploiting indigenous oil, gas and coal
reserves, by an increasing share of nuclear energy, and by enhanced efforts in energy conservation. Considerable importance,
however, is attached to flexibility in order to maintain options
for the years to come so as to meet future uncertainties and
changing circumstances at minimum cost.

This energy policy has already been formulated by the former
Labour Government. The new Conservative Government has adopted it
in principle but has put special phasis on the following issues:

- energy conservation should be the center of energy policy,

- nuclear power plays a vital role in meeting long term energy needs,

- offshore oil licensing and the arrangement of licences should be
  reviewed to increase the speed of exploration activities,

- activities of the British National Oil Corporation (BNOC) should
  be more limited in the future and some priviliges of BNOC should
  be ended.

B. Energy Demand: forecast and control

1. Forecast

A projection of UK's development of energy requirements is displayed on table XV.1. The figures shown on the table do represent neither national plans nor targets. They were derived from a model designed by British experts based on the assumption that GDP grows at an average rate of about 3 % annually to the end of this century. The projections take account of all energy policies and programmes in place.

As can be seen from the figures the model underlying to this projection shows energy self-sufficiency for the eighties with a clear decrease towards the end of this decade, which is due to an anticipated decline of oil production during the next ten years. This decline will not be surpassed by other energy sources as for instance nuclear power which is expanded by more than 100 % from 1978 to 1990.

The consumption side shows a pattern which is typical for a major European industrialized country: Industry consumption dominates with more than 40 % followed by the residential/commercial sector consuming about 35 % and the transportation area with some 20 %. Consumption in the residential/commercial sector is expected to decrease due to enhanced efforts in energy conservation. On the other hand, industry will have an increasing share of the consumption as a consequence of a growing economy.

In electricity generation coal and nuclear power will play the dominant role on the long term. However, there is and there will be a considerable capacity of oil-fired power plants and four new units are under construction.

2. Control

Energy prices are one of the tools to control energy growth. Prices do in general reflect world market levels. There was a problem in the past in relation to natural gas pricing in the domestic sector. However, due to recent developments it is anticipated that gas prices will go up by 10 % per year over the next three years, in order to reflect more accurately the situation on the world market.

The energy conservation policy of the UK has been extensively developed and strengthened. The programme is designed to yield savings of 11 Mtoe a year after 10 years.

Table XV.1

UNITED KINGDOM
Key Energy Indicators and Data
(In Mtoe)

	1960	1973	1978	1985*	1990*
General					
Energy Demand (A)	169.7	223.0	211.5	232.0	250.0
Energy Production (B)	125.5	112.3	170.2	248.5	227.0
B/A (%)	74.0	50.4	80.5	107.1	90.8
Net Oil Imports	50.7	116.1	42.2	19.5	19.0
Total Oil Consumption	44.0	110.9	93.9	102.5	105.0
TPE/GDP Ratio	1.065	0.984	0.863	0.800	0.746
TPE/GDP Elasticity		0.69	(-)	0.55	0.51
Per Capita TPE	3.26	4.0	3.79	4.12	4.38
Total Final Consumption	122.7	154.3	150.8	167.0	178.0
Industry	52.7	69.7	62.8	74.0	80.0
Transport	21.1	30.8	32.9	37.0	40.0
Residential/Commercial	48.9	53.8	55.1	56.0	58.0
Production					
Coal	123.7	78.5	71.7	71.0	77.0
Oil	0.2	0.4	55.3	125.0	90.0
Gas	0.1	25.0	33.2	37.3	40.0
Nuclear	0.7	7.2	9.0	14.0	19.0
Hydro/Geothermal	0.9	1.2	1.0	1.0	1.0
Electricity (Twh)	111.3	282.1	271.0	308.0	348.0
Trade					
Coal Exports	4.5	2.0	2.1	3.0	3.0
Imports	0	1.2	1.7	1.0	1.0
Oil Exports	14.7	25.8	42.1	N.A.	N.A.
Bunkers	5.5	5.4			
Imports	59.9	136.5	81.5	N.A.	N.A.
Gas Exports	-	-	-	-	-
Imports	-	0.7	4.4	10.0	10.0

References:
| GDP (1975 $US billion) | 159.4 | 235.3 | 245.1 | 290.0 | 35.0 |
| Population (million) | 52.0 | 55.8 | 55.8 | 56.3 | 57.1 |

* Average values of actual figures which are as follows (in Mtoe)

	1985	1990
Energy Production	221-276	202-252
Oil	100-150	70-110
Gas	35-40	35-45
Oil Imports	8-(-47)	44-(-6)

For the industrial sector grants are given to help industry replace or improve inefficient boiler plants, insulate premises and improve combined heat and power systems. Industrial combined heat and power production is well established in the UK and it might be that in the future district heating will also be introduced on a large scale since there is strong support for this technology by a report of a group of specialists submitted to the Department of Energy in July 1979.

Conservation efforts in the transportation sector is mainly focussed on gasoline savings in the automobile traffic. However, these efforts are most of all of a voluntary nature as for instance the announcement of car manufacturers to achieve a 10 % improvement in the average gasoline consumption of new cars by 1985.

Efforts in the residential/commercial sector are concentrated on improved thermal insulations and heating controls for public buildings. In addition subsidies are available for improved insulations in private housing. Since more than half of the energy used in this sector is for room heating this area is very promising for futive energy savings.

C. Supply situation

1. Oil

Oil reserves of the UK are displayed on table XV.2. Production projections indicate a decline towards 1990:

1980 : 85-105 million tons,

1985 : 100-150 million tons,

1990 : 70-110 million tons,

However, forecasting production from a region where much production will come from new fields and where exploitation demands the development of novel technology is very uncertain. In addition, the contribution of future discoveries is of course also especially uncertain.

Table XV.2

Estimated oil reserves on UK continental shelf as of December 1978

(a) Remaining recoverable reserves in present discoveries.

(million tonnes)

	Proven*	Probable*	Possible*	Possible Total
(1) Fields in production or under development	1,121	124	173	1.418
(2) Other significant discoveries not yet fully appraised	276	385	432	1,093
(3) Total for present discoveries	1,397	509	605	2,511

(b) Recoverable reserves in future discoveries

(million tonnes)

(1) Reserves in future discoveries under present licences 350-800 including the Sixth Round

(2) Reserves on the remainder of the UK Continental Shelf 550-1000

(3) Total for future discoveries

---

*   The terms 'probable' and 'possible' are given the inter-
      nationally accepted meanings in this context –
  (i) Proven – those reserves which on the available evidence are
      virtually certain to be technically and economically producible.
 (ii) Probable – those reserves which are estimated to have better
      than a 50 % chance of being technically and economically
      producible.
(iii) Possible – those reserves which at present are estimated to
      have a significant but less than 50 % chance of being techni-
      cally and economically producible.

(c) Total of recoverable reserves originally in place on the UK Continental
    Shelf

(million tonnes)

(1) Cumulative production to the end of 1978	106
(2) Remaining reserves in present discoveries (from table 2a)	1397-2511
(3) Total of reserves originally in place in present discoveries	1503-2617
(4) Reserves in future discoveries (from Table 2b)	900-1800
(5) Total of reserves originally in place in the UK Continental Shelf (rounded)	2400-4400

Overall remaining reserve estimated are in range 2,300-4,300 million tonnes.

2. Natural gas

UK's proven gas reserves at the end of 1978 were estimated
at 706 billion cubic meters. Total known reserves were estimated
at 1980 cubic meters at the end of 1977. Gas production will be
increased for the next ten years by about 16 %. This expansion
is not very high and is due to the current policy of UK's Government. However, growing interest in the use of gas as a substitute
for oil could perhaps cause UK authorities to reasses future policy
towards gas development and usage.

3. Coal

Coal continues to be a major source of primary fuel within
the UK with total inland consumption amounting to about 34 % of
total consumption of primary energy in 1978. Current production
levels of coal are about 120 million tonnes per year. It is estimated that at this production rate technically recoverable coal
reserves could probably last for over 300 years.

Scheduled expansion of coal productions might be constrained
by environmental problems relating to coal mining and pollution.

4. Nuclear

UK's nuclear programme is presently based on advanced gas
cooled reactors (AGR). However, for the future deploiment of
additional light water reactors is envisaged. Nuclear technology
is highly developed since the UK is operating for the civil programme not only reactors but also a reprocessing plant for thermal reactors, an enrichment facility on the basis of the gas cen-
trifuge process and a fast breeder. Current expansion plans call
for 12.3 GW installed in 1990 and 27.6 to 40.2 GW installed in
2000.

5. Hydro

Hydro power is of minor importance in the United Kingdom.

6. Other energy sources

UK is conducting R+D programmes on renewable energy sources
like solar, wind and wave power. In addition, importance is
attached to nuclear fusion. Presently, the Joint European Torus
(JET) experiment is set up in the UK.

XVI. Summary

A. Overall situation

        Energy demand in Europe is steadily increasing over the next
ten to twenty years as indicated by table XVI.1 which is a compilation of the figures given within the country chapters. The
continously rising energy consumption is due to Europe's structure
as a composition of heavily industrialized nations with an essential need to export. This demand cannot be supplied entirely by
own energy resources. Therefore, dependence on imports to Europe
will remain in the order of 50 % for many years.

        A view on the historical development shows for Europe as a
whole the same trend as it was observed for many of the individual nations: Through the fifties and the first half of the
sixties coal was the prevailing energy source and dependence on
foreign imports to Europe was less than 40 %. Then oil came and
the dependence increased to more than 60 %. During recent years,
however, oil and gas production from fields in the UK, Norway,
Denmark and the Netherlands had contributed more and more to
Europe's energy demand. The dependence could be reduced to about
50 %. For the future it is indicated that the steepest increase
in terms of capacities can be expected for nuclear energy which
towards the end of the century will be the fourth energy source
of importance in Europe. However, such a development depends on
howfar public acceptance problems can be solved for the years to
come. Extrapolating the present trends one might arrive in the
long term at a supply structure where coal, oil and nuclear play
the dominant role, closely followed by natural gas. Hydro/Geo-
thermal sources and renewables might then have a share of 5 %
to 10 % of the total supply system. Thus, the objective away from
oil is not easy to achieve and a realistic view into the future
will see oil as a major energy source for at least a further
20 or 30 years.

        In parallel to the supply side, the demand structure has
changed and will continue to be changed. Whereas in the past the
demand structure was mainly governed by the availability of
domestic sources, geographical location and cheap oil, for the
future this structure in many states will be deliberately changed
due to conservation programmes. Efforts to conserve energy are
different in Europe. However, all the Governments have attached
high importance to this issue. After all, two tools are available
and are exercised: price mechanisms and conservation programmes
with grants, subsidies, etc. for the introduction of energy
saving measures. Price mechanisms as for instance to allow free
market forces to rise the level of energy products to that of
the world market are not used in all the countries since such
a policy is directly connected with the basic political philosophy
of a nation.

Table XVI.1

EUROPE[*)]
Key Energy Indicators
(Mtoe)

	1960	1973	1978	1985	1990
Demand (A)	597.5	1118.7	1154.8	1428.5	1620.5
Production (B)	393.2	422.1	549.4	738.5	814.5
B/A (%)	65.8	37.8	47.6	51.7	50.3
Total final Consumption	428.4	843.7	860.4	1063.8	1172.6
Total Oil Consumption	176.7	666.3	628.8	788.1	749.4

Production

	1960	1973	1978	1985	1990
Solids	315.5	213.3	199.5	210.8	226.0
Oil	12.9	14.9	84.2	176.9	140.3
Gas	7.8	110.1	145.6	147.3	224.6
Nuclear	0.7	17.3	37.6	128.1	208.8
Hydro/Geothermal	55.9	67.4	80.4	85.0	92.0
Electricity (Twh)	506.9	1268.5	1480.1	1855.5	2472.5
Population (millions)	308.6	335.8	338.4	340.1	346.8

[*)]Without Finland, Iceland, Portugal

Table XVI.2

EUROPE*)
Key Indicators (1978/1990)

	$\dfrac{\text{TPE}}{\text{GDP}}$	$\dfrac{\text{TPE}}{\text{GDP}}$**)	$\dfrac{\text{TPE}}{\text{Capita}}$	Industry***)	Transp.***)	Residential/***) Commercial
Austria	1.26	1.04	3.3	38.8	23.6	37.6
	1.24	0.20	5.1	37.0	20.5	42.9
Belgium	1.36	0.78	4.7	48.6	14.9	36.2
	1.17	0.60	6.4	51.2	17.2	31.6
Denmark	1.05	0.16	4.0	22.0	21.3	56.7
	0.77	0.64	4.7	25.8	24.7	49.5
France			3.0	36.3	23.3	40.4
			4.5	37.0	23.0	40.0
Fed.Rep.Germ.	1.19	0.25	4.4	41.4	18.6	40.0
	1.03	0.54	6.4	14.1	16.2	39.7
Greece	1.16	1.38	1.6	41.5	33.0	25.5
	1.38	1.23	3.2	44.0	26.3	29.7
Ireland	1.46	1.31	2.3	30.5	28.8	40.7
	1.58	0.91	3.8	41.7	27.8	30.4
Italy	1.11	1.13	2.5	44.9	22.1	33.0
	1.12	0.88	3.8	47.2	20.6	32.2
Luxembourg	3.13	1.00	11.1	70.1	11.6	18.3
	3.13	0.77	15.3	70.5	11.2	18.3
Netherlands	1.57	1.19	4.7	41.7	15.6	42.7
	1.58	0.73	6.3	50.2	15.2	34.6
Norway	1.33	1.11	5.4	39.6	20.8	39.6
	1.33	0.99	7.3	40.2	19.2	40.6
Spain	1.37	1.04	1.9	52.0	28.4	19.6
	1.38	0.96	3.1	53.2	30.9	16.9
Sweden	1.31	0.69	5.9	40.6	17.9	41.5
	1.12	0.43	7.3	50.8	17.3	31.9
Switzerland	1.05	0.62	3.7	24.2	24.9	50.9
	0.96	0.88	4.6	27.8	25.4	46.8
UK	0.86	0.55	3.8	41.6	21.8	36.6
	0.75	0.51	4.4	44.9	22.5	32.6

*)Without Finland, Iceland, Portugal

**)Elasticity

***)Percentage of total final consumption

Conservation programmes have been formulated or are in execution in almost all the countries. Volume and effectivity are different. The latter one in some cases depends on the internal administrative strucure of some European countries. A view into the long term future might show that the now prevailing direct coupling between GDP and energy growth will become much weaker. Therefore, the increase in energy demand will slow down. Estimates of the overall effectivity of energy conserving measures vary considerably from country to country, however, in total it will result in savings of about 20 % in 2000 compared to our present situation and to 30 % - 40 % in 2025 or 2030.

To achieve such a goal would require large efforts not only by Governments but also by many million individuals. Thus, there are many people doubting that we in Europe will realy arrive at a considerable change in our energy requirements.

However, even the figures available through 1990 show, that a "do-nothing" philosophy would be the worst thing European people could do.

The general problem faced by Europe is that it is formed by many individual and sovereign nations with partially completely different domestic situations. For the time being they mostly try to solve their energy problem individually although the EEC and in a wider frame IEA attempt for coherent actions. For the future, it is of vital importance for all the nations in Europe that this coherence is realy achieved.

B. Energy demand and forecast

Energy demand and forecasts available from the countries are mostly confined to 1990. However, during spring of this year the Nuclear Energy Agency (NEA) of OECD undertook a new assessment of energy demand in its member states which was extended to the year 2000. The respective aggregated figures which take also into account the information gathered during INFCE 1977-1980 are attached in table XVI.3. They contain also energy requirements of Finland and Portugal.

In recent years, comprehensive forecast efforts for Europe were made by the World Energy Conference (1977), the CEC (1980), IIASA (1979) and IEA (1980). The work by CEC, IIASA and IEA was conducted in a very close cooperation during the years 1977 to 1980. The results of the CEC-study were published in spring 1980 in a small booklet called "Crucial Choices for the Energy Transition".

Table XVI.3

Total Primary Energy Requirements (Mtoe) based on NEA-questionnaire
1980 and INFCE 1977-1980

	1979	1980	1985	1990	1995	2000
Europe	1238.7	1278.7	1480	1720	1980	2200
		1271.0	1420	1580	1730	1890

Table XVI.4

Comparison of Total Primary Energy Requirements (Mtoe) of Different
Studies

	2000	2020	2025
IEA-report + France	1940	2480	
World Energy Conference (OECD-Europe)	2430 2090	4110 2950	
NEA 1980 + INFCE (OECD-Europe)	2200 1890		
This lecture (OECD-Europe)	2200 1750	2700 2300	
EEC (CEC 1980)	1230 875		1688 1072

Main objective of the work by CEC and IEA was to find out, by what kind of strategies including R+D future energy demands could be met. They came to the general conclusion that only nuclear energy and coal can take over that great fraction of Europe's energy supply that is necessary to replace oil through the next 20 to 30 years. In addition, great efforts for energy conservation are absolutely necessary since otherwise a supply demand balance under reasonable conditions is impossible to achieve.

The figures given in the report by the CEC mentioned above cover a wide range: for the EEC-countries they vary in 2000 between 875 and 1230 Mtoe and in 2025 between 1072 and 1688 Mtoe. For all these figures a high degree of energy conservation was assumed.

The IEA-report will be published in fall this year. Unfortunately the figures will not be split into regions but given only for the IEA as a whole. Nevertheless, the share of IEA-Europe can be estimated from the preliminary figures. Using this information and the figures given by France to the NEA-questionnaire of 1980 one arrives at an energy demand for the countries discussed in this report of about 1940 Mtoe in 2000 and of about 2480 Mtoe in 2020.

The IEA-report is working with more than a dozen of different scenarios. Therefore, a wide range of demand figures is covered. The figures given here correspond to the upper half of this range.

The report of the Conservation Commission of the World Energy Conference of 1977 gives also demand figures for OECD-West Europe. They amount to a range of 2090 to 2430 Mtoe in 2000 and to 2950 to 4110 Mtoe in 2020. The higher values correspond to a scenario with only modest energy conservation whereas the lower figures assume enhanced energy conservation measures set into force within the next 20 years.

Taking all the information given in this report and in addition available in the literature, one can assume that the energy demand for Europe in 2000 might be in the order of 1750 Mtoe to 2200 Mtoe and that it might increase to 2300 Mtoe to 2700 Mtoe in 2020.

Bibliography

1. Report of the Workshop on Alternative Energy Strategies,
    McGraw-Hill 1977

2. Report of the Conservation Commission of the World Energy
    Conference, 1977

3. International Nuclear Fuel Cycle Evaluation IAEA-publication
    1980

4. Overall report of the Standing Group on Long-Term Co-operation
    of the International Energy Agency (IEA), 1980

5. Energy Policies and Programmes of IEA countries, Review 1977
    and Review 1978

6. An Initial Multi-National Study of Future Energy Systems and
    Impacts of some Evolving Technologies, 1977
    Brookhaven National Lab. - 50641
    Kernforschungsanlage Jülich - 1406

7. Crucial Choices for the Energy Transition, Commission of the
    European Communities, EUR 6610, 1980

8. The Energy Programme of the European Communities, COM (79) 527,
    October 1979

9. Scenarios for the Community for the year 2000, Puplication
    XVII/448/78-EN, October 1978

10. Energy Research, Development and Demonstration for the IEA:
    A Strategy View, to be published in 1980

11. Energy Policy Goals of the European Communities for 1990,
    COM (79) 316, June 1979

12. Nuclear Engineering International, January 1980

13. Nuclear Engineering International, March 1980

14. NUKEM market reports 1977-1979

15. Energy Policy Programme for the Federal Republic of Germany

16. "Third International Institute for Applied Systems Analysis
    Energy Programme Status Report", January 1978 (IIASA)

17. Fuel Cycle Requirements, OECD (NEA), 1978

18. Statement of the Swedish Energy Minister to the Folketing, February 13, 1980

19. Energy, The Swedish Energy Commission, 1978

20. A fuels policy for the UK into the 21st century, A.F. Pexton, Nuclear Energy 19, 19-35, 1980

21. Nature 284, p. 203, 204, March 1980

22. Nuclear Energy in Great Britain, Lecture by Sir Francis Tombs, March 5, 1980

23. Figures of Oil-Industry, BP, 1979

DATA COLLECTION METHODOLOGIES - INTRODUCTION

Richard Bending

Energy Research Group
Cavendish Laboratory
University of Cambridge, U.K.

ENERGY CONSUMPTION DATA - USERS AND SUPPLIERS

It may seem self-evident that to collect information on the way energy is used is a valuable activity. Before we embark on massive statistical compilations, however, it is wise to reflect on the purposes for which the energy data are required. We shall find that such data serve many different purposes, and different uses may require different data and different approaches to data collection.

The collection of data is not without cost. Both manpower and physical resources are required, and we need to ensure that these resources are used wisely. Perhaps more important, the collection of data usually depends on mutual confidence between those who provide the data and those who collect and, perhaps, publish it. This goodwill should not be taken for granted.

We may explore the different uses to which energy data may be applied by considering the different groups who might be expected to use such data. Such groups are likely also to be important as sources of data, and the information which they are most willing and able to provide will reflect their own use for similar information.

Fuel Users

Fuel users will be interested in data for the insights which it may provide into opportunities for reduced losses, enhanced efficiency or changes in technology which will be to their benefit. We would not expect energy users to have more than a general concern for efficiency or conservation as such; their motivation is more likely to be economic pressure or the security of fuel supply. Highly aggregated sta-

tistics on energy use will be of little direct value to most fuel
users, except where they contribute to an informed view of energy
prospects in general, which then influences decisions on fuel use.
Of more direct interest to the individual user will be case studies
of the costs and benefits of changes in fuel use in similar establishments, whose conclusions can readily be applied to his own situation.

As suppliers of data, individual users present two major problems. Firstly, the number of users is very large, and in some
sectors (particularly industry) there are wide variations in the
pattern of fuel use. Secondly, many small users have little idea
of how they use energy, so that the task of data collection is a
very substantial one. These problems will be considered later.

Equipment Suppliers

The second group of potential users of energy consumption data
are those who supply fuel-using equipment. They will be interested
in information which enables them to discover new markets (in the
present or the future), to modify their products or to choose between
new technological options in the development of their company strategy. They will be able to make use of highly aggregated data covering
a wide range of users, but they will also have an interest in special
situations - geographical, technical or social - which represent an
opportunity for their products, even where these are of little national significance.

Because energy consumption data is of considerable value to
equipment suppliers, we would expect them to be active in collecting
such data. However, they may well see their best interests in keeping
that information to themselves. For a manufacturer in a competitive
field, the best situation is to have results of market research which
are not possessed by competitors. The second best situation is for
all competing suppliers to enjoy a similar degree of knowledge, and
this is preferable to the situation in which all suppliers are equally
ignorant. Well-informed manufacturers may become willing to release
information if they feel that other manufacturers are equally informed, while those whose market research is weak may (if they realise
the value of such information) be willing to provide data in the hope
that they will in so doing gain access to a wider body of knowledge.

Fuel Suppliers

The third group which make use of fuel consumption data are the
fuel suppliers themselves. For most fuels in most countries, there
is an element of monopoly in fuel supply (though there is competition
between fuels, and in large countries such as the U.S. monopoly may
be regional rather than national). There is often also a degree of
government influence or control. Oil suppliers are an exception to
this rule, retaining a substantial degree of competition between

companies and in most cases transcending national boundaries.

The interests of fuel suppliers in energy use data are similar to those of equipment suppliers; they need to understand the present in order to assess the future demand for their products and to make investment decisions. Their monopoly position makes such knowledge particularly important. A small supplier competing in a large market may be assured that he will sell all he can produce if the price is right; the size of the total market need not concern him. A monopoly supplier, by contrast, needs to know the size of the market, particularly if (as is usually the case where fuels are concerned) he will be held publicly to account if he is unable to supply the demand, or if he invests to meet a demand which fails to materialise.

Where a fuel supplier covers a whole country or region, his main concern will be total fuel use, and he will be less concerned with limited special markets unless they represent the beginnings of more extensive demand.

Because of their monopoly positions, the release of information on fuel use will not damage the interests of fuel suppliers by informing competitors. Even where there is competition between fuels, the disclosure of information carries little risk since the relations between different fuels are usually considered an area where government intervention in the national interest is justified. Indeed, secretiveness may well reflect a dislike of government intervention rather than fear of competing fuels.

In general, therefore, we would expect fuel suppliers to be a useful source of data on the way in which fuel is used.

Policy Makers

The fourth group of users of fuel consumption data, are, of course, the policy-makers, including those in government and those, for example in the academic world, who help to create the knowledge on which energy policy decisions may be based.

The requirements of policy makers for energy use data depend on the purpose for which the data are needed. We distinguish five categories:

a) Assessment of long term trends. This is usually the first action of a government wishing to frame an energy policy. Energy use data are collected to answer the questions 'Where are we now?' and 'Where, in the absence of specific policy actions, are we going?'. The main requirement is for aggregate data, and it is useful to obtain such data for as long a period into the past as possible in order to gain insights into the causes of change in the past, and the speed at which change occurs.

b) Policy appraisal. Energy policy comprises targets (on substitution or conservation, for example) together with the measures by which the targets will be achieved. Knowledge of energy use is necessary in order to identify targets which are consistent and realistic, and to determine the choice and scale of policy measures. A government has many options in the choice of policy instruments, including public education, taxes and subsidies, the funding of research, development and demonstration, and direct regulation. In order to choose the most effective combination of policy instruments for a particular target, the government must have a quantified assessment of the effects of different measures. Such an assessment can only be based on knowledge of how fuel users have responded to similar constraints and incentives in the past.

c) Once policy measures have been put into effect, the resulting changes in fuel use need to be monitored so that the measures can be adjusted as necessary. This requires detailed knowledge of fuel use in the areas which the policy measures affect. In practice, policy monitoring is often very difficult owing to changes in fuel use due to other factors. For example, a substantial fall in overall energy use may have resulted from economic recession and the market response to increased fuel prices, rather than from any government energy conservation campaign.

d) Some policy measures involving formal regulation may require the collection of data for their enforcement. For example, a government anxious to reduce long term oil consumption might specify that all new cars should have gasoline consumption rates below a set value; enforcement of such a policy would require data collection, and the data might have further uses in forecasting oil consumption, gaining an insight into the preferences of new car buyers, and possibly assessing the reduction of vehicle efficiency with age.

e) Part of the energy policy of most governments is to maintain public awareness of the need to conserve fuels, and to inform the public of how best to do so. If such a policy is to be more than fine-sounding rhetoric, it must be based on detailed knowledge of fuel use, on realistic assessments of the costs and benefits of fuel conservation measures, and on a sympathetic understanding of the way in which fuel users make decisions.

OVERCOMING OBSTACLES

The Pursuit of Perfection

Several obstacles to data collection have already been mentioned. Such obstacles are a constant source of frustration to analysts, but they need not be a cause for despair. Effective policies or wise investment decisions do not require perfect data, and there is little to be gained from the pursuit of complete knowledge. We should be guided by three principles:

a) Keep in mind the relevance of the data which is sought to the purpose for which it will be used.

b) Remember that the resources available for data collection are not unlimited.

c) Be perceptive of, and sympathetic towards, the interests of those from whom data is sought.

The first and second of these principles can be expressed in a quantitative way which illustrates the futility of seeking total knowledge. Suppose, for example, that we wish to assess total oil use by measuring use in n sectors of different sizes. Denote by c_j an initial estimate of the oil consumption in sector j. Suppose (somewhat artificially) that for any sector j, expenditure of data collection resources r will yield an estimate of c_j with fractional standard deviation σ, and that the total resources available are R. How should we divide the resources between the sectors to obtain the best estimate of total oil use?

By analogy with the simple rules relating standard deviation to sample size, we shall assume that an increase of resources by a factor x will reduce the standard deviation by a factor $\sqrt{x}$. Then if we were to devote resources r_j to sector j, the absolute standard deviation of the estimated oil use in that sector would be

$$c_j \; \sigma \; \sqrt{r/r_j}$$

The variance of the estimated total oil use would be the sum of the variances for each sector, i.e.

$$V = \sum_j c_j^2 \; \sigma^2 \; \frac{r}{r_j}$$

The best allocation of resources will be that which minimises V subject to the constraint that $\Sigma r_j = R$, and a little simple calculus yields the result that for minimum variance,

$$r_j = R c_j$$

Thus we should allocate resources in proportion to the estimated share of each sector in the final result. (This result is, of course, an illustration of a general principle and real situations are far more complicated; one complexity which should be mentioned is that poor initial estimates of the sector sizes c_i may damage the results, particularly if some of the sectors are small).

Let us consider some of the practical hindrances to data collection.

Number of Users

In some sectors, the number of users is far too large to obtain information directly from more than a small proportion. For fairly homogeneous sectors, such as households and transport, a well-designed survey may produce a clear picture of the whole sector with adequate statistical accuracy. Techniques for the design of such surveys are well-established (e.g. Raj, 1972).

For more heterogeneous sectors, particularly industry, general surveys may be less useful. If the aim of data collection is to draw conclusions about total fuel use, then it is quite acceptable to concentrate attention on major energy users as long as we do not claim that the resulting data give an unbiassed view of the sector as a whole. It is not uncommon to find that a large proportion of energy use is concentrated in relatively few establishments, making the task of data collection much easier. In the U.K. in 1967, for example, almost 50% of solid and liquid fuels (taken together) was used in the 363 largest establishments (0.2% of the recorded fuel-using establishments), while the largest 66 establishments accounted for 27% of all fuel use (Ministry of Power, 1968)

On the other hand, there are arguments for considering small users. Such users are likely to be those for whom energy costs are a small proportion of total production costs, and knowledge of energy use and of energy conservation opportunities are limited. This has three implications:

a) Energy use by small establishments is unlikely to be typical of the whole sector.

b) Opportunities for short term conservation are likely to be greater than for large users (who will already have taken up such opportunities).

c) The contact established for the purpose of data collection may usefully be used to educate small users in energy matters.

If the large number of fuel users remains a major obstacle, the

best course is probably to concentrate effort on fuel or equipment
suppliers rather than users. If good information is available from
both these sources, a comprehensive picture of fuel use may be
constructed. Neither source is likely to be adequate by itself; fuel
supply data will leave many questions on the detail of how fuel is
used unanswered, while fuel use cannot be inferred from equipment
data without additional information on equipment load factors.

Fuel or equipment suppliers are points where fuel use data may
be concentrated and therefore readily accessible. Other points of
concentration may be sought, which will provide incomplete, but
nevertheless useful, data. These include insurance companies (where
insurance of equipment, such as boilers, is compulsory or general
practice; a major U.K. study - Chesshire and Robson, 1979, uses this
data source), national or local government bodies dealing with
environmental control, planning or fire regulations, and so on.

Lack of Users' Knowledge and Management Time

It has already been pointed out that many users, especially
those for whom fuels represent a minor expenditure, are often
unaware of how much energy is used or where the fuels go to. This
problem is best overcome by combining the task of data collection
with the identification of opportunities for improved fuel use, and
general education on energy matters. Indeed, direct help may be the
main aim of the contact, with the collection of data a by-product.
This approach should make the contact worthwhile from the fuel users'
point of view, helping to overcome the other common obstacle, lack
of management time. It also has the benefit, from a policy viewpoint,
of short-circuiting the cycle of data collection, model development,
policy appraisal and policy implementation, achieving worthwhile
aims of improved fuel use directly.

It would be expected that the main energy conservation benefits
of this type of direct contact with users would be immediate reduction
of unnecessary losses, improvements in efficiency through better
control, and similar changes requiring little or no investment.

Confidentiality

Information on fuel use may be of commercial value to competing
users, either by drawing the attention of the latter to possible
improvements in their own fuel use, or by revealing details of production processes. In these circumstances, an understanding must be
reached between the supplier and collector of data to respect the
confidentiality of the information. In practice, this means that any
published data must be sufficiently aggregated to prevent even a
well-informed reader from gaining useful information about specific
establishments.

One way of avoiding problems of confidentiality is to use a mutually respected third party, such as a trade or research association or consumer group, to collect the data. This approach is likely to ease the initial contact, but in some circumstances it may be counter-productive; a trade association's desire to retain the trust of its members could cause it to withhold information which individual companies would be willing to release.

Energy use is by no means the first field in which the desire for private gain by an inventor has had to be set against the value of public disclosure. The usual mechanism used in such circumstances is the patent, which combines public disclosure with effective, but limited rights to profit from an invention. The role of patents in the promotion of improved fuel use techniques deserves to be considered more carefully.

Ultimately, the key to the problem of confidentiality must be the creation of a situation in which all gain from the existence of a comprehensive body of knowledge, and this should be one of the aims of energy policy.

Bias

Where information on past fuel use is sought, problems of deliberate or accidental bias are likely to be minor. Data on expectations for the future, however (which may be very relevant for the formulation of energy policy) are subject to distortion. We need to distinguish three types of bias:

a) Deliberate falsification of data. This may occur with reference to data on past energy use if the company is afraid of public criticism of its record of fuel use. It may also apply to expectations if the company depends on the government for investment funds and wishes to offset expected cuts in the funds allowed, or if it wishes to prevent unease among shareholders.

b) Deliberate targetry. It is perfectly proper for an organisation to set targets which are beyond its honest expectations in order to encourage effort or reduce the probability of a damaging shortfall (this is particularly true of the fuel supply industries, where the consequences of under-investment are usually perceived as more serious than those of over-investment). Such a tactic may be recognised by supplier and collector of data without embarrassment.

c) Unconscious bias. It is common for the members of an organisation to be optimistic about its market performance or energy efficiency, simply through a habit of loyalty to the firm. Less commonly, a general mood of pessimism may

prevail. Such moods change slowly, as is shown by the
tendency of most fuel industries in recent years to produce
high demand forecasts. Data gatherers may recognise, and
wish to offset, such moods. Such a procedure is not unrea-
sonable, thoush it must be remembered that governments and
other data-collecting bodies are prone to similar bias.

DATA STRUCTURE

It is often necessary to compromise between the detail in which
fuel use data are recorded, and the quality of the data obtained.
The argument for detail is straightforward. If all data exist and
are accurate, then the greater the detail in which they are available,
the more complete can be our understanding of the factors influencing
fuel use, and the more confident we can be of formulating a realistic
energy policy and choosing practical and effective measures to bring
it about.

Unfortunately, we are more likely to find ourselves working
with incomplete and approximate data, and as we disaggregate further,
the data become more fragmentary and less accurate. In addition, the
effort required for data collection expands enormously as we attempt
more detailed structures, and we are wise to ask, in view of the
intrinsic uncertainties of energy issues, whether the practical
knowledge gained is worth the effort.

It is doubtful whether there is any 'best' level of disaggregation. Highly aggregated data are convenient for the construction of
quantified models, while more detailed data, because they are often
incomplete, are more useful as a source of qualitative insights. The
wisest course would seem to be to collect data at different levels
of detail according to the ease or difficulty of collection, and to
formulate energy policies in the light of insights gained from
different levels of aggregation.

A further complication arises from the fact that energy use may
be classified in many different ways. This is particularly true of
industrial energy use, and we may ask whether it is more important
to classify such use according to the type of fuel, the process in
which it is used, the product, the geographical area, the size of
establishment, or any other factor.

Regrettably, the choice of a particular classification scheme
may limit our ability to analyse specific policy issues. For
example, if we omit geographical area from our classification, we
shall be unable to discern the effect of proximity to coalfields on
fuel choice, and unable to draw conclusions on that question for
the future. This type of problem will be discussed further in the
content of industrial energy use. For the present, the most that
can be said is that the way in which data are stored or retrieved

should not prevent analysis in terms of categories about which
details were known when the data were collected.

REFERENCES

Chesshire, J., Robson, M., 1979, Boilers - ancient and modern,
    Energy Manager 2:1 pp 46-47
Ministry of Power, 1968, "Statistical Digest 1967", HMSO, London
Raj, D., 1972, "The Design of Sample Surveys", McGraw-Hill,
    New York

DATA COLLECTION METHODOLOGIES - INDUSTRY

Richard Bending

Energy Research Group
Cavendish Laboratory
University of Cambridge, U.K.

DATA AVAILABLE IN THE UK

Serious compilation of energy use data in the UK started (with some exceptions such as the iron and steel sector, where earlier data are available) after the Second World War. A review of the main sources illustrates the compromise necessary between the scope of the data and the degree of detail provided.

Digest of United Kingdom Energy Statistics (Department of Energy, 1979)

This is an annual publication issued (with various titles and by various government departments) since 1948. It provides detailed statistics of fuel supply, consumption, prices and financial flows. As far as industrial fuel use is concerned, the data are provided for ten sectors:

- iron and steel
- engineering and other metal trades
- food, drink and tobacco
- chemicals and allied trades
- textiles, leather and clothing
- paper, printing and stationery
- bricks, tiles, fireclay and other building materials
- china, earthenware and glass
- cement
- other trades

Though this breakdown may appear fairly detailed, several sectors show substantial changes of product mix during the period covered by the data, which creates problems for analysis. Notable examples of such sectors are engineering, chemicals, paper, and bricks.

The fuels for which data are given for each of these sectors are:
- coal
- coke and breeze
- other solid fuel
- coke oven gas
- town gas
- natural gas
- electricity
- petroleum
- creosote/pitch mixtures.

Analysis of data given under petroleum supply allows the petroleum use data to be further disaggregated into fuel oil, gas/diesel oil and lighter oils.

Few data are given on the way in which fuels are used, except in the 1953 issue which contains a survey of steam and power plant (Ministry of Fuel and Power, 1954).

Census of Production Reports

Information on the use of fuels in smaller sectors than those given above may be obtained from recent Census of Production reports (e.g. Department of Industry, 1974 onwards; censuses were also carried out in 1951, 1954, 1963, 1968 and 1979). Detailed reports are issued on specific industries. Data are only given for the census years themselves, and in the earlier issues it may be difficult to infer physical (as opposed to financial) quantities. Nevertheless, this information is a useful aid to understanding changes in the patterns of fuel use in larger sectors.

End-Use Data

Information on the purposes for which fuels are used in industry is far more fragmentary than data on fuel deliveries. The Energy Technology Support Unit (ETSU) at Harwell, Oxfordshire, has compiled tables of energy use in a single year, 1976, which are extensively quoted in Leach, 1979. These cover the same sectors as the Energy Digests (see above) except that the iron and steel sector is excluded, and bricks, china and cement are treated as a single sector.

For each sector, energy use is classified by fuel - coal, other solid fuel, oil, natural gas, other gases, heat from combined heat and power schemes, and electricity - and by the following end-use categories:

- electricity generation (including combined heat and power schemes)
- process energy via central steam/water system, classified by temperature at end use
- process energy via direct-fired plant, classified by temperature at end use
- electrochemical process energy
- space and water heating, via central steam/water systems, local boiler plant or direct-acting plant
- motive power, in stationary plant or off-road vehicles
- other uses

A separate survey by the Confederation of British Industry (1975) though limited in coverage (around 11% of industry covered in terms of employment, but only 1% of establishments) gives additional data not included in the sources mentioned so far. The sector breakdown is more detailed than in the ETSU compilation (19 sectors, with most additional detail in the engineering sector) and there is information on establishment sizes, fuel stocks and stocking capacity, age and capacity of electricity generating plant, recent and planned changes in fuel use, recent investigation into energy use, and energy expenditure in relation to total expenditure. The data relate mainly to the single year 1973.

More detailed information on specific industries is available in some cases in the reports of the Energy Thrift Scheme and the Energy Audit Scheme (National Physical Laboratory, 1977a and 1977b). The Thrift Scheme is based on one-day visits to a large number of establishments, and is aimed primarily at short term energy conservation measures. The Audit Scheme uses more detailed studies, and is intended to quantify energy use and identify research and development requirements for improving energy efficiency.

CLASSIFICATION OF INDUSTRIAL FUEL USE

It will be clear from what has been said about U.K. sources of industrial fuel use data that many classifications are possible, and those who collect and present the data must choose a classification scheme on the basis of the use which will be made of the data. Regrettably, it is common to find that information which was implicitly available at the time of data collection (for example on geographical location) is lost in the published form of the data.

For much of the following analysis, I am indebted to Dr. R. K. Cattell, of the Cambridge Energy Research Group.

The aim is to collect and present data in a way which helps to show how users might make energy-related decisions in the future, so that ideally the data would cover all the factors which have influenced such decisions in the past.

Industry and Time

Before considering energy-related factors, we need to identify
the data as applying to a particular industry (or sector) and to a
particular year.

Fuel

Fuel choice is primarily an outcome of the decision-making
process, and no compilation of industrial energy data which excludes
it can give an insight into past or future choice decisions. One
classification of fuels - that used in the U.K. Energy Digests -
has already been given. A more complete breakdown is:

Solid fuels - Coal (may be classified by volatile matter,
                caking tendency, ash content and heat content)

             Coke (gas coke or metallurgical coke)

             Other coal-based solid fuels

             Wood

             Charcoal

             Solid wastes from agriculture or industry

Liquid fuels- Petroleum-based (liquefied petroleum gas, naphtha,
                aviation spirit, wide-cut gasoline, aviation
                turbine fuel, motor spirit, burning oil,
                vaporising oil, gas/diesel oil, fuel oil,
                creosote/pitch mixtures)

             Coal-based liquids

Gases -   Natural gas

             Coal-based gas (low or medium heat content gas,
                including coke oven gas, and substitute
                natural gas)

             Blast furnace gas

Electricity - (may be divided into that sold on unrestricted
                and off-peak tariffs)

Fuel End-Use

A particular industry is likely to use fuels for a variety of
purposes, and decisions will usually relate to a specific use

(though on occasions a systems approach - looking at the establishment as a whole - may be appropriate). A possible qualitative form of this classification is:

Process use   -   chemical reaction induction

              heat treatment

              cleaning

              working (e.g. casting, machining)

              heat separation (drying, distillation etc.)

Personnel use -   space conditioning (heating, cooling and
                  ventilation)

              lighting

              communication

              transport

              hot water

              cooking

In some instances the end-use may be further categorised. For example, uses involving heat may be classified by the temperature concerned, or uses involving motive power by the size of load. Such a classification may well be important to an understanding of fuel choice. In addition, some end-uses may be associated with particular fuels, such as non-substitutable electricity use, electrochemical processes and the use of fuels as feedstocks.

Characteristics of Fuel

For some purposes, properties of the fuels which are incidental to their role as energy sources may be important factors in fuel choice. Such properties include physical strength (important, for example, in the use of coke for iron smelting), chemical purity and flame emissivity in furnaces, controllability, impact on the working or external environment, and any practical requirements for storage, handling, operation, waste disposal and equipment maintenance.

Fuel and Other Prices

Clearly linked to the previous classification are fuel prices, with which should be associated other costs incurred in fuel use, such as those involved in fuel storage, handling, plant purchase, maintenance and operation, and waste disposal.

Equipment

The type of equipment in which the fuel is used is in part an outcome of the decision-making process, though where equipment (a boiler, for example) is already in place, this will clearly affect fuel decisions. Equipment type is usually closely related to the nature of the end-use and (for heat applications) to the temperature required, but often there remains a choice of equipment type (for example between internally or externally fired furnaces) which is relevant to fuel choice. Equipment capacity is a particularly important element in historical data, as the rate at which patterns of fuel use change in the future will be related to the stock of equipment and to factors influencing modification or retirement.

The Manufacturing Establishment

Thus far we have identified factors affecting fuel choice by looking at the point of final consumption only. A broader view reveals other, sometimes less tangible factors. Consideration of the establishment as a whole may show that a particular end-use could be provided by waste heat from another process (perhaps with no additional fuel use, or with electricity for pumping, or with greater electricity use through a heat pump to upgrade low temperature waste heat). Several end-uses commonly share a single steam supply system, and a requirement for both steam and electricity may point to the use of a combined heat and power scheme. For a large establishment, circumstances may arise in the future when it is economic to buy coal and convert it on-site to low heat content gas for process heating.

The Local Situation

A view from outside the fuel-using establishment may reveal other relevant factors, which we may summarise by the following questions:

- Is the establishment close to a coalfield, port or oil refinery, so that supply and transport considerations, or local traditions, favour a particular fuel?

- Does the establishment's position relative to roads or railways favour (or inhibit) the use of a particular fuel?

- Is there sufficient room for storage on the site, and if not could additional land for fuel storage be acquired?

- Is the position of the establishment, relative to residential areas or other factories, such that constraints on environmental pollution will limit fuel choice?

- Is the position of the establishment, in relation to other
  energy-using sites, such that it might be possible to make use
  of a local centralised steam supply system, combined heat and
  power scheme, or coal-based low or medium heat content gas
  scheme?

STRUCTURE OF INDUSTRIAL FUEL USE

To summarise the previous section, we need to specify two
parameters - industry and time - to indicate the scope of the data.
We may then classify fuel use in terms of three variables: type of
fuel, end-use (in connection with which we need to bear in mind any
special characteristics of different fuels in different uses), and
type of equipment. We also need to be aware of the place of specific
end-uses within the overall pattern of energy requirements in the
establishment, and any special implications of the local situation.
In addition to this classification of fuel use, we need data on fuel
prices and other costs incurred in energy use.

Matrix Formulation

A 'complete' set of historical data on fuel use might comprise
a matrix of six or seven dimensions (industrial sector, time, three
parameters for fuel use, and one or two parameters characterising
the type of establishment and its local situation), together with a
matrix of fuel prices and other costs.

Such a formulation would be ideally suited to the analysis of
most energy policy issues. It would also be extremely bulky, it
would require a quite impracticable amount of data collection effort,
and the statistical accuracy of individual data items would be very
low.

Developing a Practical Structure

There are three ways to change the conceptually ideal matrix
formulation into something more practical. Most collections of
historical data use all three techniques.

The first method is to condense particular dimensions of the
matrix. For example, we might use four fuels - solids, liquids, gases
and electricity - instead of the many categories listed earlier.
We might use only three end-use categories - steam-raising, process
heat and other uses - in place of the eleven or more suggested. Such
condensation reduces enormously the bulk of the data, reduces collection problems and increases accuracy; it is both necessary and useful.
However, we must not ignore the loss of information which this action
involves. Grouping natural gas, purchased manufactured gas and site-
manufactured gas together will disguise the fact that these fuels have
different price structures, different patterns of local availability

and in many cases different, and s cific, uses; we should not be
surprised if economic models based on this aggregation fail to behave
as we expect. Similarly, condensation of end-uses draws our attention
away from a complex web of relationships between fuels, equipment and
applications, leaving us to assume broad patterns of economic determination which may be quite unrealistic.

The second technique, rather more extreme than condensation, is
to eliminate some matrix dimensions altogether. Equipment type is
often omitted from compilations, usually because of difficulties of
data collection. Similarly, the characteristics of the establishment
and the locality are often ignored, despite their practical importance
to fuel decisions. Here there may also be problems in collecting
data, though some information, such as geographical location, must be
implicitly available at the data collection stage.

The third technique does not reduce the bulk of the data, but
imposes on it a hierarchical structure which is of great value in
presentation and discussion. The most common approach is to subdivide first into industries, then into fuels, and then (though this
classification may be eliminated) into end-uses. For each element in
the resulting scheme, a time-series of values is given. Since this
method does not destroy data, it is not subject to the criticisms
which may be directed towards other techniques. However, care must
be taken, in using hierarchically structured data, to avoid the
assumption that the chosen hierarchy has some special significance.
This will be illustrated by the examples in the following section.

POLICY ISSUES AND ENERGY DATA

In this section we give four examples to illustrate how different policy questions make use of energy consumption data in different
ways.

Changing Industrial Structure

How will changes in the structure of a nation's industrial
production affect its use of fuels? If it is planned to move away
from heavy industry towards high value-added goods, for example, this
will affect both total energy requirements and fuel mix. The conventional data structure, with industrial sector as the first subdivision, is suited to the examination of this question. Fairly
simple assumptions could be made about trends in energy use and fuel
mix for each sector, geographical variations and end-use details could
be neglected, and the results would probably be quite informative.

New Markets for Coal

An important policy issue in many countries is the extent to
which coal could replace oil and gas in industrial markets, and how

fast such a change could occur. In studying this question it would
be important to distinguish between fuels used for steam-raising,
where coal use is feasible with appropriate prices, and fuels used
for process heat, where the prospects for using coal directly are
much more limited. Thus the classification of energy consumption
by end use might be more important than classification by industry.

Opportunities for Low Heat Content Gas

An alternative route for the use of coal in industry is the
production of low heat content gas, similar to the manufactured gas
used before the introduction of natural gas. Because of its low
heat content and resulting high distribution costs, this is most
likely to be produced close to the point of use. An examination of
the prospects for this option would need to consider end-uses in
detail (the gas would find markets in process heating and in boilers
where direct use of coal was not practicable). It would in addition
be necessary to consider the size of the fuel-using establishment,
since only a large factory could consider on-site gasification, and
also the local situation, in order to identify the scope for single
gasification plants serving groups of factories.

Internal Trade in Fuel

Another important policy issue is the changing pattern of fuel
movements within a country, and its implication for the development
of ports, roads and railways. Here it would be necessary to use data
which was classified on a regional basis, and also to take into
account local factors such as the proximity of manufacturing areas
to railways.

DATA COLLECTION STRATEGY

In some respects, decisions on data collection strategy are
easier today than in the past. Today the use of computers allows a
degree of flexibility in data storage and retrieval which was not
previously possible, and database design is a specialised field
(e.g. Date, 1977). In addition, it is to be hoped that energy policy
issues are more clearly perceived than they were before the 1974 oil
crisis.

Some guidelines for the collection of industrial energy use data
(drawing also upon material from the previous lecture in this series)
are:

   a. Concentrate effort on major users, but do not neglect small
        users.
   b. For small users especially, combine data collection with
        education.

c. Do not neglect parameters which may be relevant, especially local factors and equipment stocks.
d. Retain as much data as possible (subject to agreements on confidentiality) in computer storage.
e. Provide a flexible retrieval system without commitment to a specific hierarchy.

REFERENCES

Confederation of British Industry, 1975, "A Statistical Survey of Industrial Fuel and Energy Use", CBI, London
Date, C. J., 1977, "An Introduction to Database Systems", Addison-Wesley, Reading, Massachusetts and London
Department of Energy, 1979, "Digest of United Kingdom Energy Statistics 1979", HMSO, London
Department of Industry, "Report on the Census of Production 1974", HMSO, London (separate reports for each industry published in the years following 1974)
Leach, G., 1979, "A Low Energy Strategy for the United Kingdom", Science Reviews Limited, London
Ministry of Fuel and Power, 1954, "Statistical Digest 1953", HMSO, London
National Physical Laboratory, 1997a, "Industrial Energy Thrift Scheme First Progress Report", NPL Report CHEM 68 (EU 2), NPL, Teddington, Middlesex
National Physical Laboratory, 1977b, "Energy Audit Scheme First Progress Report, NPL Report CHEM 72 (EU 3), NPL, Teddington, Middlesex

DATA COLLECTION METHODOLOGIES - TRANSPORT AND COMMERCIAL SECTORS

Richard Bending

Energy Research Group
Cavendish Laboratory
University of Cambridge, UK

ENERGY POLICY AND THE TRANSPORT SECTOR

The transport sector is of particular importance for energy policy. Perhaps the most important policy issue for the next quarter century, in both the developed and the developing nations, is how to manage an orderly and timely transition away from short-lived resources - oil and gas - towards longer-lived energy sources, in a way which is consistent with aspirations for economic growth and development.

Transport is heavily dependent on oil. In the UK in 1977, 99% of transport energy was oil, and this accounted for 48% of all oil use. In most developing countries, the establishment of a transport infrastructure, implying growing future dependence on oil, is an essential part of the development process. The technology to replace oil by coal-based liquids exists, but (except in countries such as South Africa where coal costs are very low) it is very expensive. Other forms of vehicle (notably electric vehicles) have not thus far become established except in specialised markets.

For these reasons it is very valuable to have a clear picture of future trends in the transport sector, and in many countries this can best be obtained from knowledge of the past.

DATA CLASSIFICATION IN THE TRANSPORT SECTOR

Though the transport sector is less complex and diverse than industry, we find a similar problem of classification. Six parameters which may be used for classification are outlined below.

Mode of transport

A possible subdivision would be:

- Air (inland and overseas)

- Water (inland, coastal and overseas)

- Rail

- Road

- Pipelines

Different modes of transport have very different energy requirements. Air travel is the most energy-intensive, followed by roads, with rail and water transport the most economical in energy terms. Furthermore, air and road travel are almost wholly dependent on oil, whereas other modes present an element of fuel choice. Clearly a switch from high-energy to low-energy modes would be advantageous, but in countries such as the UK such a switch seems unlikely; in 1977 93% of passenger transport and 81% of freight transport (excluding air, water and pipelines) was by road, and the historical data show a growing commitment to road transport.

Freight/Passenger Transport

A second important division is that between freight and passenger transport. The former is usually measured in tonnes or tonnes-km, the latter in number of passengers, number of journeys or passenger-km. The purpose and lengths of journeys may be important in vehicle or modal choice; this will be considered later.

Locations

For passenger transport in particular, it is important to distinguish between urban and rural transport. In urban areas in the UK, for example, the availability of public transport and the inconvenience of the private car make public transport a natural choice, whereas in rural areas the opposite is the case. Furthermore, the efficiency of a specific vehicle is likely to be lower under urban driving conditions, and urban/rural differences are also found in vehicle load factors (passengers per vehicle).

Vehicle

The mix of vehicle types is important for an understanding of energy use. For rail transport, a suitable division might be:

- Long haul electric traction

- Long haul diesel traction

- Short haul electric traction (surface)

- Short haul electric traction (underground)

- Short haul diesel traction

For road transport, the structure of the vehicle stock is more complex, and ideally we might use a classification such as:

- Private cars

- Motorcycles

- Bicycles (relevant for energy use because of substitution to and from fuel-using vehicles)

- Public transport vehicles (classified by size)

- Freight vehicles (classified by size)

- Others

In addition to transport activity, it is valuable to collect historical data on the stock of vehicles in order to foresee future changes.

Fuel

We should expect a classification of transport activity by type of vehicle to be closely related to a classification by fuel (though some vehicle categories may be a mixture of, for example, gasoline and diesel-powered vehicles). The link between these two classifications is, of course, vehicle efficiency, and any inferences which can be drawn about this variable from historical data are particularly valuable.

In practice, however, it is likely that the classification of transport energy use by type of fuel will be straightforward (i.e. the data will be readily available) while data on vehicle use will be more fragmentary.

One area of fuel use data which often causes confusion is bunkers, that is, fuel used in international sea and air transport. Different conventions for dealing with bunkers may be found in different countries, or between modes in the same country. For those compiling data, it is important that the conventions used should be clearly stated, and that any data on the volume of international passenger and freight transport should be consistent with the fuel assigned to the same end-use.

Fuel prices are another important element if we are to understand past changes in transport energy use. Data are likely to be
readily available, classified by transport mode and fuel, from fuel
suppliers.

Characteristics of Journey

The choice of mode or vehicle may well depend on the purpose of
the journey (whether it is regular travel to work, a leisure trip,
a holiday or a visit to a shopping centre, for example), as well as
on the distance travelled. For a short journey with little requirement for luggage, foot or bicycle may be the preferred mode; the
need to carry shopping will lead to the use of the family car. The
vehicle load factor is an important parameter both in determining
energy use per passenger-km and in modal choice. For an individual
contemplating a journey of, say, 100 km, rail transport may well be
more economical than car transport; for a family, the reverse will
often apply.

For freight transport, relevant parameters include the length
of the journey, the need (if any) for intermediate stops, and the
location of the end-points in relation to such transport facilities
as railways or ports. The effect of these variables may be seen in
the contrast between an urban delivery vehicle (where short distances
and frequent stops may favour electric traction) and long distance
inland transport of raw materials, which often favours rail freight.

When we come to look at the future, we find many questions,
relating to journey characteristics, which may be important for
future transport energy use. These include:

- To what extent will telecommunications reduce the need for
    daily travel to work?

- Will increased leisure create new transport requirements?

- How will the changing distribution of population between
    rural, suburban and urban areas affect transport needs?

- How will the changing structure of industrial production
    change the pattern of freight transport?

- Where will industrial production be situated and how will
    this affect freight transport?

Hierarchy

Figure 1 shows a possible hierarchy for a dababase of fuel use in transport. It should be noted that this is neither the only nor the best structure, and there is no particular significance in the order in which the subdivisions are introduced. In practice, most data collection and storage systems will use a simpler structure, probably omitting the distinction between long and short-distance (or urban) travel, and simplifying (or omitting altogether) the breakdown into types of road vehicle. As always, the choice of data structure is a compromise between the data collection effort required and the practical usefulness of the information.

DATA COLLECTION IN THE TRANSPORT SECTOR

Fuel Use

Information on fuel use by mode of transport may be obtained from records of fuel deliveries. The split between gasoline and diesel oil in the road sector gives an initial estimate of the division between passenger and freight fuel use. A better estimate of this division, or of that between classes of vehicle, must be obtained by means of surveys.

Length and Purpose of Journeys

This aspect of transport data must also depend on statistical surveys. Such surveys are also of value to vehicle manufacturers and those responsible for planning the road system, so that there is scope for co-operation in data collection.

For modes other than road, collection of this type of information is usually easier than for road transport, owing to the smaller number of operators involved.

Vehicle Stocks and Efficiency

Historical data on vehicle stocks and new vehicles, classified by type of vehicle, are readily available through whatever vehicle licensing system is in use.

Some information on vehicle efficiency may be obtained from tests on new vehicles or from manufacturer's data, though the usefulness of such data in relation to older vehicles must be questioned. Trends in effienciey may also be inferred by relating vehicle use (based on surveys) to fuel deliveries.

Fuel Prices

Fuel prices, as noted earlier, should be obtainable from the fuel supply industries.

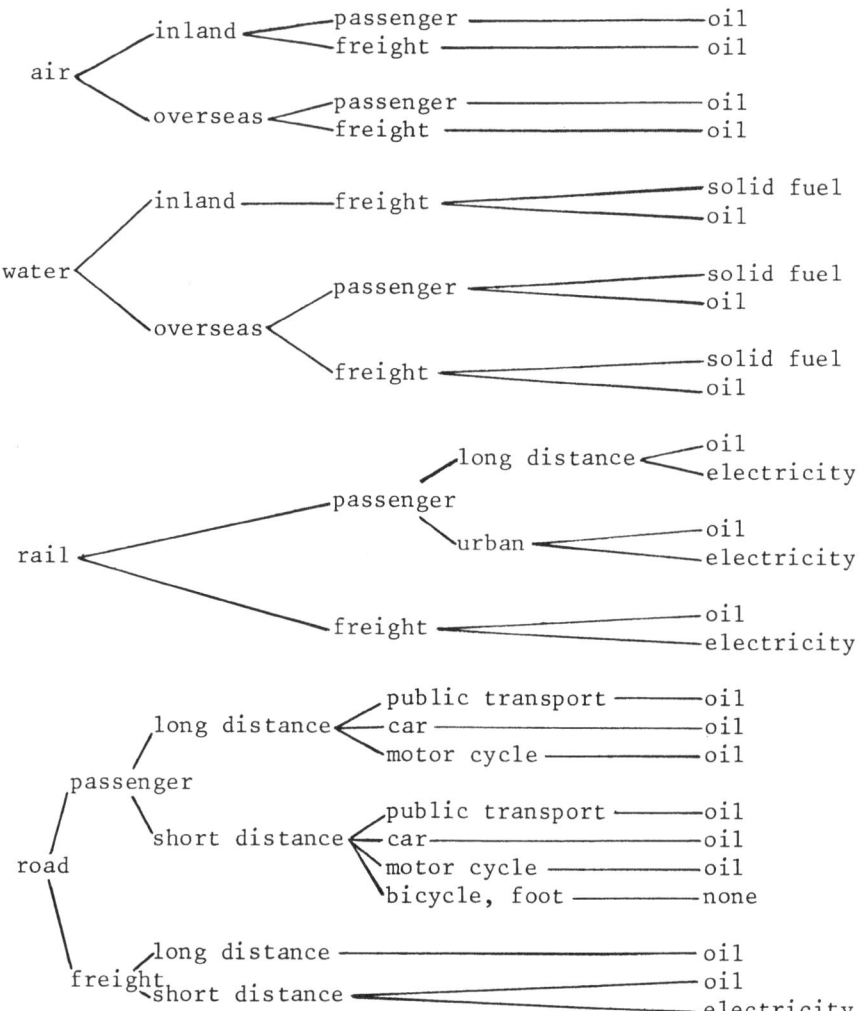

<u>Figure 1</u> A Hierarchy of Transport Energy Use

    Note: For each entry in the right-hand column, time-series
           would be provided of passenger-km or tonne-km, and of
           fuel use.

ENERGY USE IN THE COMMERCIAL SECTOR

Data Structure

The structure of fuel use in the commercial sector (offices, shops, public buildings, schools and so on) is simpler than in either the industrial or the transport sector. Energy use may be classified by:

- time

- type of establishment; for example a hospital is in use continuously and is likely to use a substantial system for the supply of steam or hot water, in contrast to a small shop which is characterised by intermittent use and probably has a simple space heating system based on gas or electricity

- end-use; heating, cooling, ventilation, lighting, communications, movement (e.g. escalators and lifts), hot water and cooking

- type of fuel; usually closely related to the end-use, though there is an element of fuel choice for heating

- size and type of equipment

- local factors, such as proximity to fuel supplies and constraints on environmental pollution.

In practice, data on some of these parameters is likely to be difficult to obtain, and a simpler structure, arranged in the usual hierarchical form, is shown in figure 2.

In addition to fuel use date, information on fuel prices is necessary to gain an understanding of the pattern of fuel use in the past.

Energy Efficiency

The commercial sector introduces some problems for the assessment of energy efficiency. In industry, fuel use may be related to the volume of production to give an indication of changes in efficiency. Similarly in transport, total passenger-km or freight te-km provide a measure with which energy use may be compared. No such measure exists in the commercial sector, but possible measures which may be of some value include the following:

- number of employees; this is based on the fact that most energy use in the commercial sector is to meet personnel

needs. It is probably appropriate for offices, but less so
for hospitals, schools, shops and public buildings where the
number of employees is not a realistic indication of the
number of people using the building.

-   man-hours worked; this may be an improvement on number of
    employees in that it takes into account the different require-
    ments of continuously and intermittently occupied buildings.

-   floor area; this is based on the observation that most
    commercial energy use is for space heating, and floor area
    gives some measure (though a very crude one) of building heat
    loss.

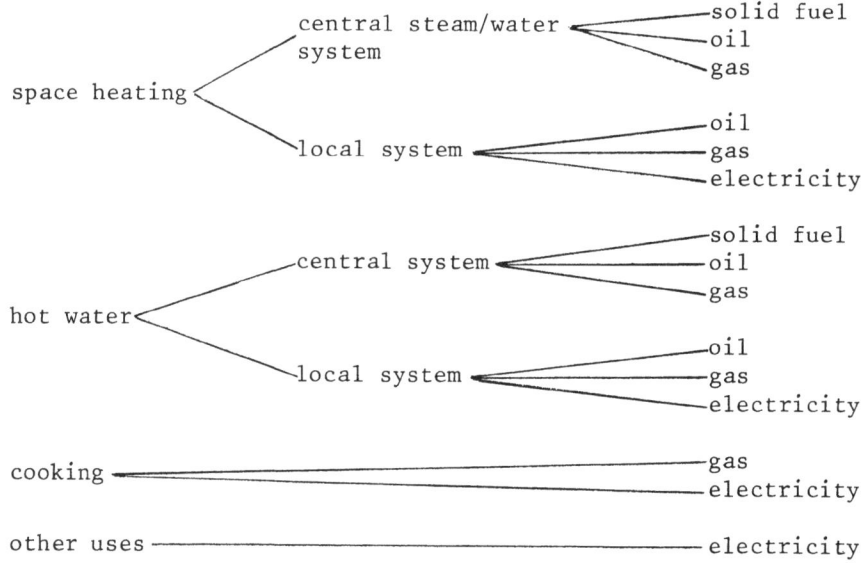

Figure 2   A Hierarchy of Commercial Energy Use

    Note: For each entry in the right-hand column, time series
           of fuel use would be given.

The Building Stock

Because space heating is the main energy use, knowledge of the heat loss characteristics of commercial buildings is a major factor in understanding past and future patterns of fuel use. These characteristics vary greatly, from the very poor performance of the steel and glass office blocks popular some years ago to the much better performance of some modern buildings which have been designed with energy conservation in mind. In such buildings it is not uncommon to find that, because of the low heat loss through the building fabric, incidental heat gains from lighting, sunshine and the occupants make a mjor contribution to space-heating needs.

Changes in the building stock, resulting from both new construction and demolition, are one cause of the changing pattern of fuel use, both in the past and in the future, and need to be taken into account in any description of commercial fuel use.

DATA COLLECTION IN THE COMMERCIAL SECTOR

Fuel Supply Industries

The fuel supply industries are likely to be able to provide data on fuel deliveries (possibly classified by type of establishment) and fuel prices.

User Survey

For more detailed information, surveys of commercial users are necessary. Such a survey could cover the type of establishment, fuel use classified by fuel type, end-use and type of equipment, any local factors relevant to fuel choice (including fuel prices where these show local variations), the pattern of occupancy of the building, and the heat loss characteristics of the building structure. In most instances, data collection would involve monitoring and calculation, and could usefully be combined with education of the users and identification of potential savings.

Government Buildings

Government and local government offices, as well as schools and hospitals where these form part of a state system, form a significant part of the commercial sector. The collection of data from such users should present fewer problems than collection from other users, not least because a government which includes energy conservation among its aims will want to be able to demonstrate that its policies are practicable and effective in the areas over which it has some measure of control. An initial concentration of survey effort on such users may be worthwhile.

Employment Data

It is likely that data on patterns of employment will be collected already, and these may indicate the number of employees, or man-hours worked, in the commercial sector.

Construction Industry

Information may be available from the construction industry on the scale and characteristics of new commercial buildings. If regulations are in force governing energy conservation standards in new buildings, these will give a direct indication of the heat loss characteristics of new buildings.

THE DOMESTIC SECTOR - DATA, HISTORY AND PROSPECTS

Richard Bending

Energy Research Group
Cavendish Laboratory
University of Cambridge, UK

ENERGY USE DATA

In this lecture we shall consider briefly the data necessary for an understanding of the domestic sector, and then review the history of energy use in this sector, and the prospects for changes in the pattern of fuel use.

Classification of Fuel Use

Fuel use in the household sector may conveniently be classified in terms of four variables:

- Time
- Type of dwelling; in the UK a suitable subdivision is into detached, semi-detached and terraced houses, and flats. Other subdivisions, such as between cavity-walled and solid-walled houses (which corresponds closely to a classification by date of construction) may be relevant.
- Fuel; the main subdivisions are solid fuel, oil, gas and electricity, and the last may in the UK be divided into electricity sold on restricted-time tariffs (off-peak electricity) and unrestricted electricity.
- End-use; the main end-uses are space heating, cooling and ventilation (the last two are of little importance in the UK), hot water supply, cooking, and specific uses of electricity such as lighting, communications and motive power.

In addition to fuel use data, an understanding of the domestic energy market depends on fuel price data, knowledge of the housing stock, and knowledge of the stock of fuel-using equipment.

It should be noted that fuel prices, fuel availability and the
distribution of types and age of houses may show substantial local or
regional variations. Taking fuel availability as an example, natural
gas (currently the dominant fuel in this sector) is unavailable in
parts of the UK, wood fuel may be cheap and readily available in
specific areas, and the use of solid fuel may be constrained in some
areas by smoke control regulations.

Data Structure

It is natural to present household fuel use data in a hierarchical structure, though a structure covering the above classification
(which is not exhuastive) in full might well be too complicated for
practical use. A possible hierarchy might be:

First level: Type of dwelling (detached, semi-detached, etc.,
               and perhaps with an indication of type of
               construction)
Second level: End-use
Third level: Fuel
Fourth level: Time

Data Collection

Information on fuel prices, and on fuel use in the sector as a
whole, is likely to be available from the fuel supply industries.
Some data on end-uses, on equipment stocks and on the variation of
fuel use with type of dwelling may be available from the same sources,
though such data is of great value in marketing and may be regarded
as confidential.

Statistics on the housing stock may be available from government
sources, or from the construction industry. Any regulations on
building standards may allow some inferences to be drawn on the
heat loss characteristics of recently-completed housing.

More detailed information on end-uses, equipment stocks and
building characteristics, as well as local variations, may require
the use of special surveys.

HISTORY OF THE DOMESTIC SECTOR

Total Energy Use

The total energy delivered to the UK household sector
(measured in terms of its thermal content) is shown in figure 1
(Department of Energy, 1978). It may be seen that over the last 20
years there has been little change in the total, and this initially
surprising result is altered only slightly if we express the same
data in terms of energy per household.

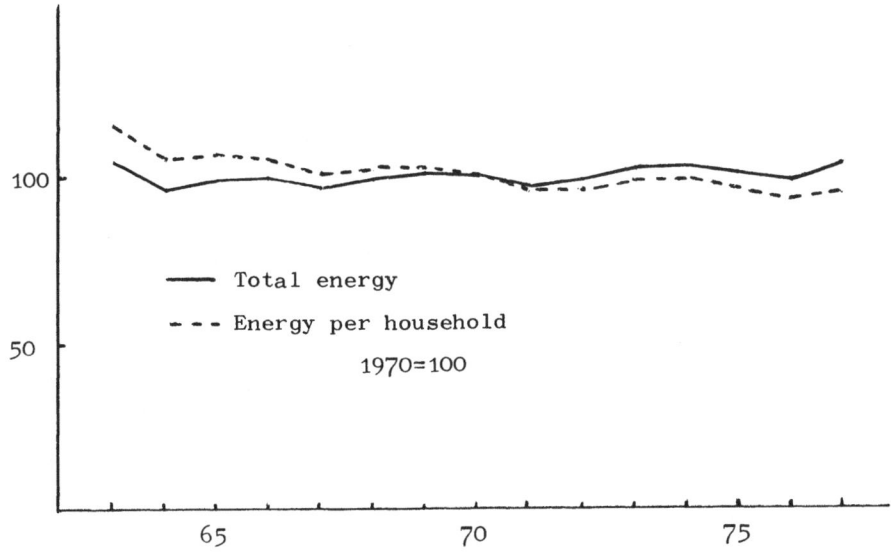

Fig. 1   Energy delivered to the UK household sector, 1963-1977

More detailed examination reveals that this apparent stability was the result of two opposing changes; an improvement in fuel use efficiency, offset by an increase in consumer comfort. Both of these changes result principally from the use of central heating (in which most or all of the house is heated from a central boiler or furnace). Before about 1960, the main sources of heat in most houses were open fires in one or two principal rooms, which were very inefficient and left most of the house unheated. By 1969, 25% of households had full or partial central heating, and by 1977 this has risen to over 50% (Central Statistical Office, 1978).

A second factor which has helped to keep fuel use higher than might have been expected has been the growth in electricity use, mainly brought about through the increasing ownership of electrical appliances. Figure 2 shows domestic electricity use together with ownership levels of some appliances (Anscomb et al, 1975). It will be seen that ownership has stabilised for some appliances, whereas for others there is room for further growth.

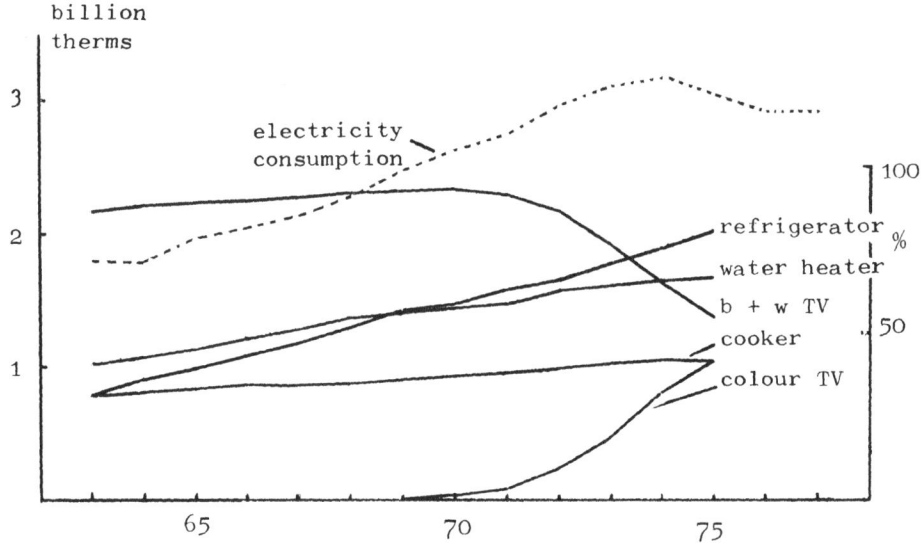

Fig. 2 Electricity use and appliance ownership in UK households, since 1963

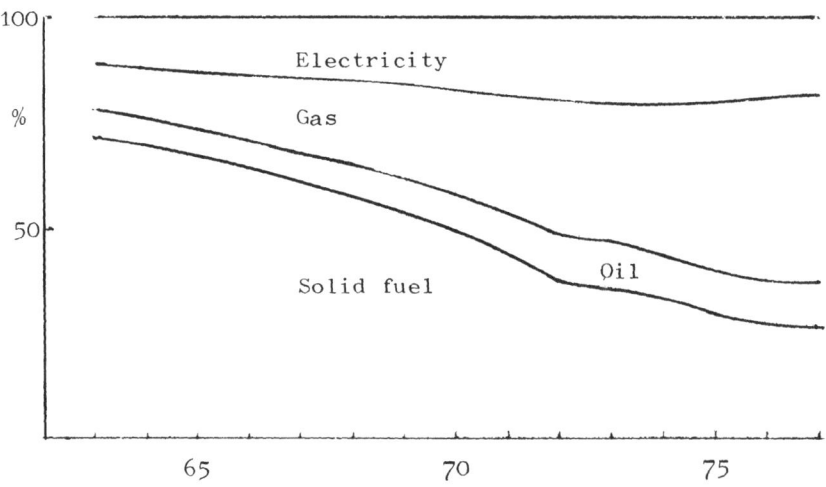

Fig. 3 Shares of UK household energy use by fuel, 1963–1977

Market Shares

Figure 3 shows the fuel shares in the domestic sector. Central heating may use solid fuel, oil, gas or (by means of storage heaters) off-peak electricity supplies. The convenience and low capital cost of gas (initially manufactured gas, but from about 1970 natural gas) made this the preferred fuel for most consumers, so that the pattern is one of falling solid fuel use and growing gas use. Oil was chosen mainly in areas not served by the gas network.

Fuel Prices

The changing shares of gas and solid fuel may readily be explained by the change in relative prices. Between 1962 and 1977, the real price of solid fuel rose by 29%, whereas the real price of gas fell by 29%, so that the gas/solid fuel price ratio almost halved during the period. Electricity prices rose by 25% over the same interval; the growing market share for electricity is the result of the introduction of off-peak electricity (introduced in 1963 and peaking at 22% of all domestic sales in England and Wales in 1973/4 - Electricity Council, 1979) and the growing stock of electrical appliances. Domestic energy prices as a whole rose by 15% in real terms between 1962 and 1977.

The fraction of total household expenditure going to energy purchases has remained almost constant at just over 6% since the early 1960s, apart from a temporary fall to just over 5% during the early 1970s (Central Statistical Office, 1978).

These figures make it clear that there has been little pressure to save energy in the domestic sector, and help to explain why the benefits of improved energy efficiency (of central heating boilers compared with open fires) were taken in the form of increased comfort rather than reduced expenditure.

FUTURE DEVELOPMENTS

The Housing Stock

At present, 300,000 new houses per year are being built in the UK, in a total stock of about 20 million (Central Statistical Office, 1978). It is clear from these figures that energy conservation in the sector as a whole will depend mainly on changes in fuel use in existing houses. Conservation prospects depend on the type of structure; houses with solid walls (probably some nine million in the UK - Mould, 1977) are more difficult and costly to insulate than those with cavity walls. There has been an active programme to replace substandard dwellings in the UK, but more recently the emphasis has moved to renovation and improvement; such changes of

policy make it difficult to forecast the future structure of the
housing stock.

Insulation

There is no doubt that the scope for energy savings in households is very high. The main fuel use is for space heating, and
though there is some scope for improvements in boiler efficiency (by
better control or improved design), the main opportunity is in the
heat lost from the building itself. The rate of heat loss from a
typical semi-detached house maintained at 20°C above ambient temperature, and some of the potential savings, are shown in figure 4 (from
Watt Committee on Energy, 1979). The figure shows that the rate of
heat loss from a typical house can be halved by the use of well-
established insulation techniques (draught-proofing, loft insulation
and the filling of wall cavities with insulating foam). The poor
thermal performance of solid-walled houses is also clear; internal
or external insulation of such walls is possible, but more
difficult than for cavity walls.

Energy Balance

A halving of the building heat loss will lead to a reduction of
space heating fuel bills by more than half, as is demonstrated in
figure 5. The building heat loss is balanced by the space heating,
together with incidental heat gains from other energy uses (cooking,
hot water, lighting, etc.), from the occupants and from sunshine; the
latter items causes the seasonal variations in incidental heat gain.

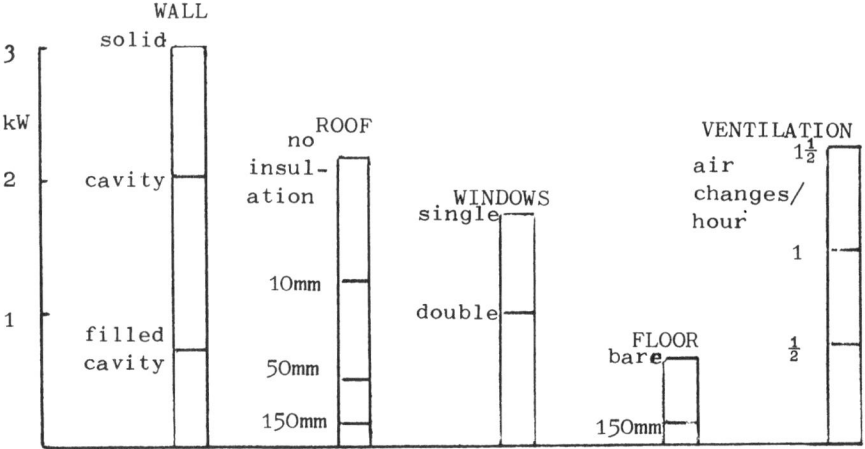

Fig. 4   Rate of energy loss from a typical semi-detached house
         (for 20°C temperature difference)

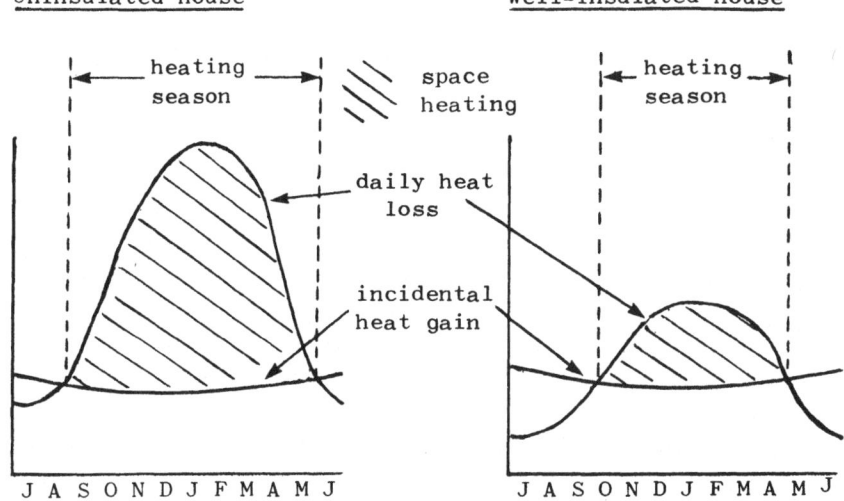

Fig. 5 House energy balance and space heating requirements

The total energy used in space heating is represented by the areas between the building heat loss curve and the incidental heat gain curve, and it is clear that the proportional effect of incidental gains is greater when the building loss is reduced. Further, the length of the heating season is reduced as a result of reduced heat losses.

The Economics of Conservation

House insulation is the substitution of capital for energy. It requires a single payment, made in anticipation of reduced expenditure in future years. The ratio of capital cost to annual saving, or payout time, (Beijdorff, 1979) is typically 3-5 years for insulation measures in an existing house (nearer 15 years for double glazing, though this gives other benefits such as noise reduction and reduced draughts). This is a short period in comparison with the life of the insulation, and even at current high interest rates it would pay the householder to add the capital cost to an existing mortgage.

We would therefore expect householders to be eager to insulate their homes. In fact, progress seems to be very slow, and we may suggest several reasons for this:

a. Lack of knowledge - householders do not know the options available to them.

b. Prices - the market has not provided clear price signals; as a result of general inflation, fuel prices rose in real terms by only 15% between 1973 and 1977, while the real price of gas (the marginal fuel) fell by 11% in the same period (Department of Energy, 1978).

c. Capital availability - capital is in short supply, and building societies - the traditional lenders in this sector - may be unwilling to lend money for home improvements.

d. Ownership time - unless a householder is confident that he will remain in his house for many years, or that when he moves his investment in energy conservation will be reflected in the market value of the house, he may be unwilling to invest.

e. Rented property - in rented property (46% of households in 1977 - Central Statistical Office, 1978) investment in energy conservation by the landlord will benefit the tenant, but the market (or regulations) may not allow the landlord to claim a return on his investment through increased rents.

Heat Sources for Low Energy Housing

If the problems noted above can be overcome, it seems likely that most houses in the future will have much lower space heating requirements than at present. This has important implications for the choice of heat source. It is difficult to design small solid-fuel boilers, and the accurate control needed in low-energy houses is also difficult for solid fuel. We would therefore expect a continuation of the move away from solid fuel, even though at the boiler sizes common today solid fuel has a similar efficiency to gas or oil. As total energy requirements fall, the capital cost of the boiler or furnace becomes more important relative to fuel costs, which will favour gas or electricity even if (as is likely) real gas prices rise more quickly than in the past. Controllability also favours these fuels. If the contribution of nuclear energy to electricity generation becomes substantial, the cost advantage of off-peak electricity will increase. Heat pumps (driven by electricity or gas) provide a means of substantially reducing fuel use, though at the expense of higher capital cost.

Other Fuel Uses

Substantial savings may also be made by insulation of hot water tanks and pipes, design of hot water systems to reduce long pipe runs, and by the re-design of electrical appliances. Apart from the first, these changes depend on the replacement of capital goods and are likely to take place fairly slowly. Such changes do, of course, reduce incidental heat gains and so increase space heating requirements slightly.

CONCLUSIONS

The prospects for fuel saving in the domestic sector appear more promising than in most other sectors. The potential saving is large, and much of it is attainable with well-established technology. At least for simpler options such as draught-proofing and loft insulation, there is much that can be done by the householder himself. The economic case for domestic energy conservation is strong. Changes which might help the potential savings to be realised include:

a.   Public education so that adequate insulation becomes an expected feature of any home; this will help insulation investment to be reflected in property values and (since most householders are more influenced by fashion than by cost-benefit analysis) will encourage investment directly.

b.   Easing of the constraints on the supply of capital; this may result from economic recovery, or could be brought about directly by low-interest loans for conservation investment, or by building society policy in favour of energy conservation.

c.   Steps to offset the conflict of interest between landlord and tenant in rental property; since most rented property in the UK is owned by local government authorities, recommendations on standard of insulation could be effective.

REFERENCES

Anscomb, A., Hatch, D., Junankar, S., 1975, "Domestic Sector Analysis 1954/55 to 1974/75 (England and Wales)" Electricity Council Report EF 61, Electricity Council, London

Beijdorff, A. F., 1979, "Energy Efficiency" Shell International Petroleum Company, London

Central Statistical Office, 1978, "Annual Abstract of Statistics 1979", HMSO, London

Department of Energy, 1978, "Digest of United Kingdom Energy Statistics", HMSO, London

Electricity Council, 1979, "Handbook of Electricity Supply Statistics", Electricity Council, London

Mould, A. E., 1977, "External Insulation - A Practical Way to
    Improve the Thermal Performance of Existing Solid external
    Walls", Report ECRC/MI033, Electricity Council Research
    Centre, Capenhurst, Chester

Watt Committee on Energy, 1979, "A Warmer House at Lower Cost",
    Watt Committee on Energy, London

PROSPECTS OF ENERGY CONSERVATION IN TRANSPORTATION

Ugo Lucio Businaro - Aldo Fedrighini

FIAT Research Center S.p.A.

Strada Torino, 50 Orbassano (Italy)

INTRODUCTION

The oil crisis has put on the foreground, among others, the problem of energy conservation in transportation. Several agenda for actions for decision makers have been prepared in the last years, touching more or less all the possible interventions from increasing the efficiency of each transportation means, especially the automobile, to increasing the use of public transportation.

To predict to which extent and with which dynamics the complex transportation system will respond to the several intervention is a difficult job. As is well known, complex systems do not respond to action as quickly and as linearly as the decision makers would like. "What-if" studies and analysis using system models are performed and hopefully could serve as an indication for actions. However the more complex and the more similar to reality the model is, the less clear is the answer.

Having accepted the task to present to you here the perspective of energy conservation in transportation, we found ourselves trapped in the tentative to get some general answers from the analysis of data and from comparing the many studies performed by experts of transportation systems.

After careful perusal of the literature one finds himself still unconfortable when asking simple, or apparently simple questions, such as: what could be the impact on energy conservation by shifting urban transportation from car to public modes?, or, could we really convince the market to shift to cars of lower fuel consumption but also of lower performance in terms of speed and acceleration?

The crux of the matter is that we are dealing with complex systems and that we try to understand the behaviour and dynamic response of the system which, because of its complex (homeostatic) mode of reaction and of the interconnection and non linearity of the effects, usually follows complex patterns difficult to predict. The temptation is large in this case to ask for some external compulsory feed-back to the system to assure desired changes but at the same time requiring to pass from a homoestatic system to a simplistic, external ("dictatorial"?) type of control.

It is therefore important to look for a more modest approach and try to understand the intrinsic dynamics of the system by observing its behaviour. We refer to the same approach of the astronomer in understanding the dynamics of the universe. Such an approach could be used as a paradigm for studying any complex system and in particular the transportation system.

The astronomers observe the universe by means of "snap shots" taken at the same terrestial time, and (apart from some very special cases of historical record of some past event like the Chinese observation of Super Nova) there is no hope to observe the time behaviour of the universe. Nevertheless the astronomer is able to gain insight in the dynamics of the universe because these "snap shots" refer to stars or stellar systems which are at different stages of their development. It is therefore possible to recognize different patterns of evolution, and with them as a guide to predict the future development of single stars or stellar systems.

By analogy, when we observe, at a certain time, the transportation system in different countries and different towns, we are considering different stages of development of the systems. The analysis of other complex systems, such as the economic development of a country, have shown that they follow typical patterns of development.

For instance a country GDP (Gross Domestic Product) developes along with a shift of society from agriculture to industry and to service. As another example, it is indeed fortunate for the automobile marketing people that the number of automobile per family in industrialized countries be strictly related to the GDP per capita.

By analyzing the data of transportation systems one could therefore hope that realistic answer on "what-if" analysis can be obtained, more than by analytical model, by observing the countries that have already realized the "if" condition.

Making decision that goes against the natural pattern of development of complex system requires, as already said, the capability of forcing strong global interactions changing the homoestatic development of the system, otherways they are bound to fail.

Actions that do not require a total system change but are limited to single components of the system (such as increasing the energy efficiency of the automobile) have higher probability to produce results as expected than when one tries to modify the entire system behaviour (such as shifting from one mode of transportation to another) against its "natural" pattern of development.

When observing "terrestrial" complex systems one might have the illusion that it is possible to keep track of the dynamics of the system by observing it at different times. Unfortunately the data available, collected at different times, are often not consistent and cannot be repeated. Standardized observation in the future with more reliable techniques and more detailed data will help the student of the system behaviour in the same manner as new observations of the universe with new techniques help the astronomer to refine his models of the universe's dynamics.

According to the described approach, in these lectures we will:

1 - present the data on transportation systems world wide in different countries

2 - try to reconstruct from such data, patterns of dynamic behaviour of the system in terms of some macro economic variables

3 - by reporting the case of a complex analytical model point out how interrelate is the transportation system with other societal systems

4 - examine two specific cases of actions at "component level" on energy conservation in transportation and specifically: impact of optimal traffic control and car efficiency improvement.

1 - THE TRANSPORTATION SYSTEM

According to Forrester "a system means a grouping of parts that operate together for a common purpose". In other words one can define a system as a "set" finalized to obtain a given purpose, subdivided into parts (subsystems), each one performing a different function harmonized to reach the system purpose.

The purpose of the transportation system is to satisfy the need of mobility as expressed by the mobility demand.

Mobility is one of the fundamental need of human being referring both to the need to move people and goods. It could be influenced by subjective motivations (like the desire to escape from a given environment or life style) and/or by objective motivations (such as the need to reach the work place or school) ranked according to a scale of priority. The mobility need is subdivided according to the subjective/objective motivations and it is structured and articulated by scale of values and priority.

The offer of transportation to match the mobility demand has to take into consideration the why, where and how which varies according to the individual, the society, the environment, etc. The difficult matching between demand and offer, because of the complexity of the transportation system, has to be taken into due consideration when analyzing the possible actions for energy conservation by modifying the transportation offer.

1.1 Transportation system morphology

We should first distinguish between the transport system and the transported people or goods. Three principal subsystems characterize the system (see Fig. 1.1):

. infrastructure (roads, railroads, ports, airports, waterchannel, etc.)

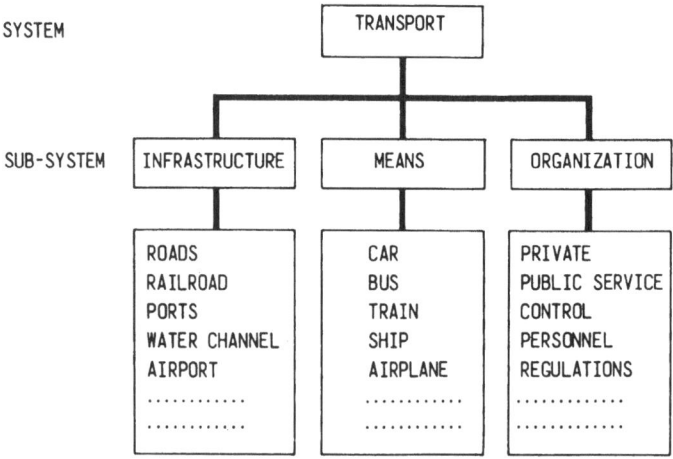

Fig. 1.1 Transportation System Morphology

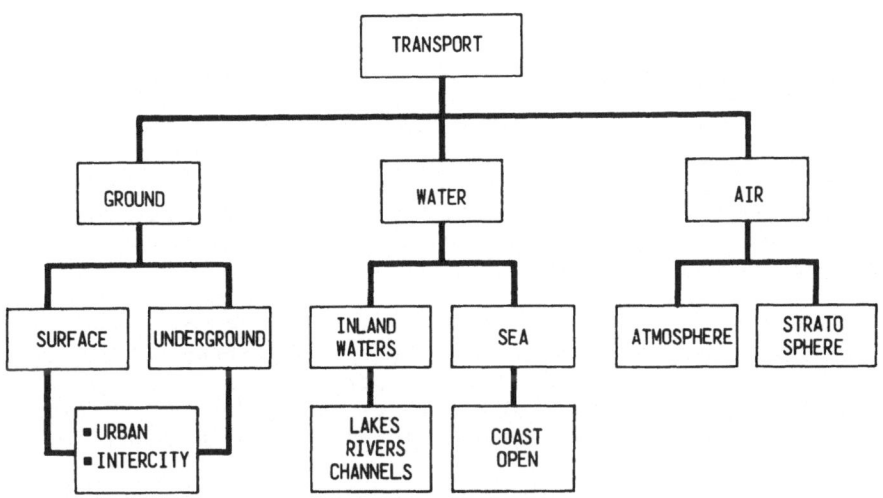

Fig. 1.2 Operational Environment of the Transport System

transport means or vehicles (car, bus, train, ship,
aircraft, etc.)
   . services and organization (management, personnel, control,
etc.)

The morfological analysis of the transport system should
continue at lower level from the subsystems, including components,
sub-components, materials. For the analysis reported below we will
limit to the first level of subdivision.

From the "transported" point of view we can distinguish
between:

   . passengers
   . goods
   . mixed.

Not always there is a well clear-cut subdivision between
passengers and goods but we will keep to it for the purpose of
this analysis. The mixed passengers and goods case is not
considered in all the literature analyzed either because it is of
low importance or it is reconducible to one of the other two.
The transport system operates in the three specific
physical environments: ground, water, air, which can, on their
term, be subdivided in a more or less unique way (see Fig. 1.2)

Combining the different types of infrastructures, means, and
organizations, one obtains different modes of transportation, each
one able to perform one or more transport mission (see Fig. 1.3).
The transportation mode is often, in practice, designated through
the name of the transport means (car, bus, train, airplane, etc).

Among all the possible transportation modes we will limit
ourselves to consider only those for which there are consistent
data. The concept of transportation mode is useful to emphasize
the possibility to perform a given mission with different modes.

MISSION	MODE
PASSENGER-URBAN	ROAD × CAR × PRIVATE
PASSENGER-URBAN	RAILROAD × STREETCAR × PUBLIC SERVICE
GOODS-INTERCITY	ROAD × TRUCK × PRIVATE
GOODS-INTERCITY	RAILROAD × TRAIN × PUBLIC SERVICE

Fig. 1.3 Transportation Modes and Missions

Each mode is characterized by a set of parameters which indicates the degree by which a given mobility need could be satisfied. Among such characteristics parameters, there are:

. the cost of tariff
. the door-to-door time
. the frequency of the services
. the comfort
. the investment and the operation and maintenance costs
. the fuel consumption
. the type of energy used
. etc.

The choice of a transportation mode is not always based on quantifyable parameters, but also on such "imponderable" as habit, status symbols, etc.

1.2 Transport system data

The data used for the analysis here presented refer to a given year, namely 1975[3]. We did not take into consideration data at different times, to get more direct information on the system dynamics, for the reason given in the introduction. We will refer to data with time variation only for specific cases.

The type of data analyzed are very aggregated, at country level, and disaggregated for different transportation modes. To get more acquainted to the type of data used we present the case of Italy[8,11,13,46]:

. Fig. 1.4 gives the total passenger travelled distance in terms of passengers x km in the year 1975 and the tons x km of goods transported limited to routes internal in the country. Pipeducts are excluded. We considered not only the data on ton x km, but also that on the total tonnage transported, because they might reveal different aspects of the productive structure of a country. As a matter of fact the role of the different transport mode appears quite different if one looks at the tonnage or ton x km figures. In the case of Italy, as an example, the raylroad represents 3% of the total tonnage transported but 15% of the ton x km.
. Figs. 1.5 and 1.6 give the repartition of the primary energy sources (respectively in 10^6 TOE* and percentage wise) on the different final uses, including transport and its three main

---

* TON OF OIL EQUIVALENT

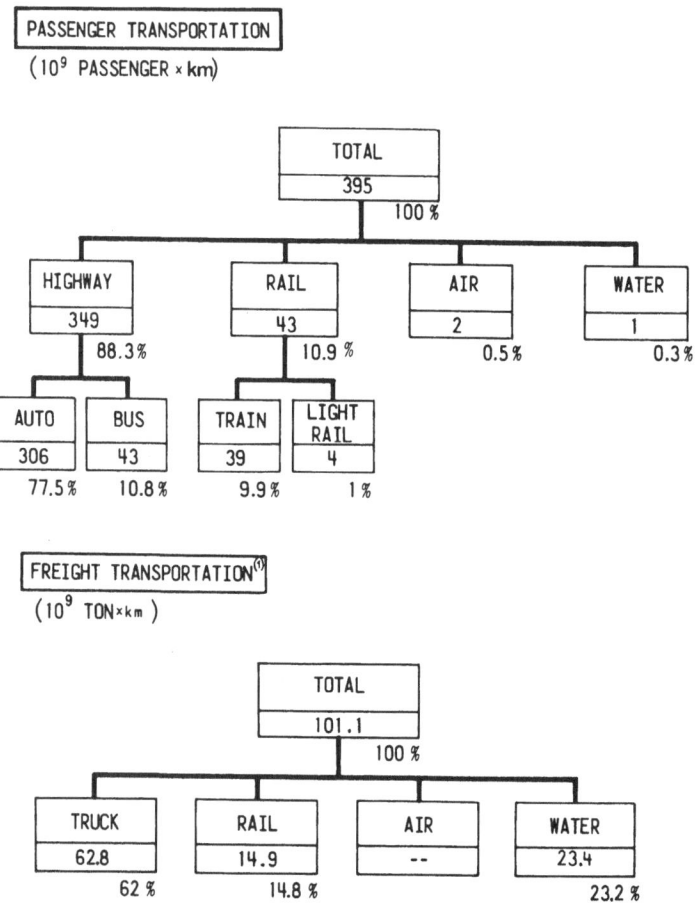

Fig. 1.4 - Italy 1975 - Passenger and Freight Transportation

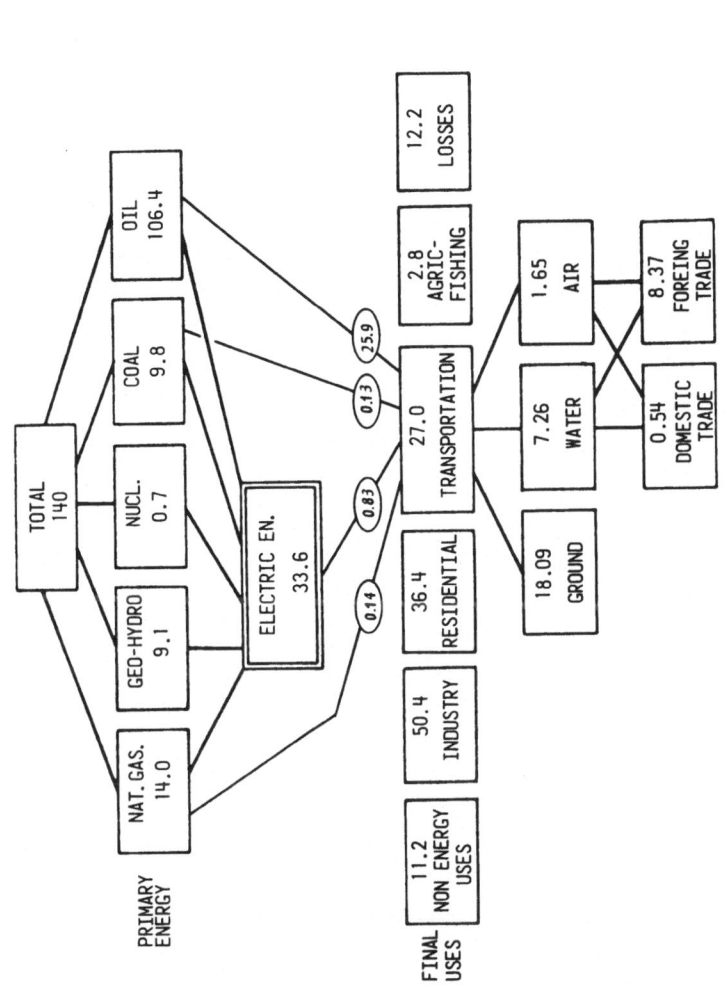

Fig. 1.5 - Italy 1973 - Repartition of the Primary Energy Sources (10^6 T.O.E.)

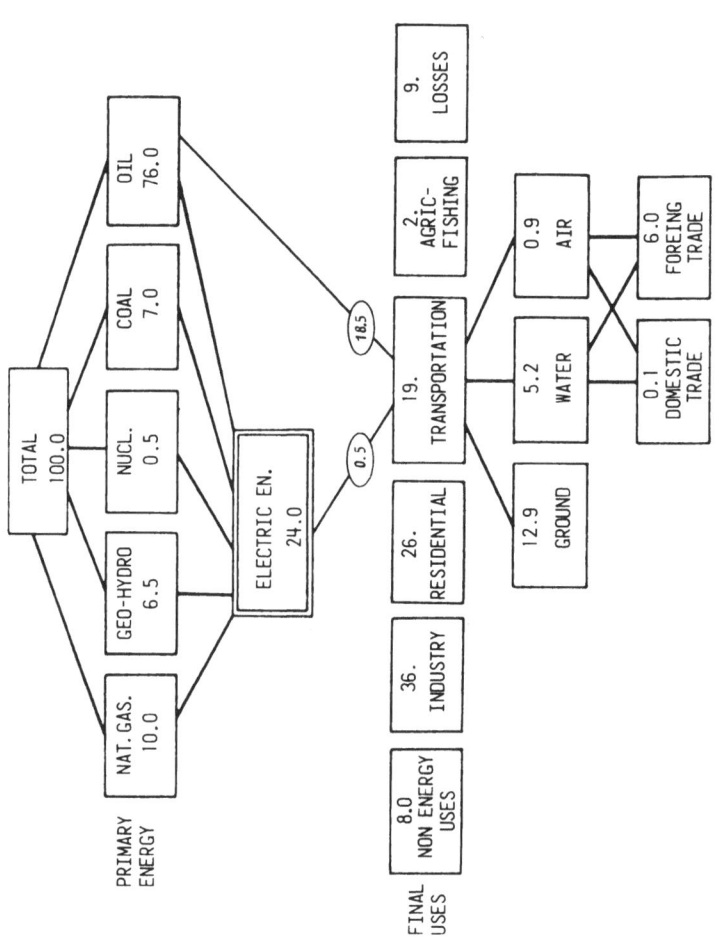

Fig. 1.6 - Italy 1973 - Repartition of the Primary Energy Sources (%)

domain: ground, air, sea, (international sea transport is not included)

. Figs. 1.7 and 1.8 (respectively in 10^6 TOE and percentage wise) give the energy used for transportation, subdivided into the main modes, (including international sea transport)

The data on energy consumption on Italy[11] (see Fig. 1.8) refer to 1973. Transport uses 19% of total energy, and it depend on oil for 97%. Among the oil derivates, gasoline has the largest share as transportation energy, followed by heavy oil and diesel oil at a level which is approximately half that of gasoline; 98.8% of gasoline and 85.7% of diesel oil are used in road transport, the rest being natural gas and LPG. Heavy oil and jet fuel are used only in sea and air transport.
Railroads depend for 75% on electricity and the rest is shared equally by coal and oil. Among the energy user, car is at the first place, (44%, including also motocycles), followed by international sea transport (26%) and trucks (17.4%).

From the total passenger travelled distance or total tonnage x distance and the energy consumed in each mode, one can determine the "energy intensity" as kcal/passenger x km or kcal/ton x km, or its reciprocal: the energy efficiency.

Regarding the data on energy consumption in transportation, one should remember that one could alternatively refer to:

. the direct energy use for transport, which is measurable in term of the amount of energy "burned in the engine" to move the vehicle
. the direct plus the indirect energy use, the latter being defined as the quantity of energy needed to manufacture the vehicle, to build the infrastructure and to manage the transportation service. This indirect energy use could represent a substantial part of the total energy for transportation. In the case of USA[10] (see Fig. 1.9) the ratio of indirect to direct energy use averaged on all modes in 1977 gives a value of 42%, with a peak of 112% for the railroad mode.

The data used in the present analysis refer only to direct energy data, with exception when analyzing special cases. Fig 1.10 and Fig. 1.11 report the data on energy used in transportation in USA in 1974, while Fig. 1.12 and Fig. 1.13 refer to the case of the Fed. Rep. of Germany for the same year.

Comparing data from different bibliographical sources one find variations of approx 13% for the base year 1975[10]. This is due to different definitions of the "transport system". For instance, some authors do not include the energy used for pumping in oleoducts and gasducts or energy used in military transport.

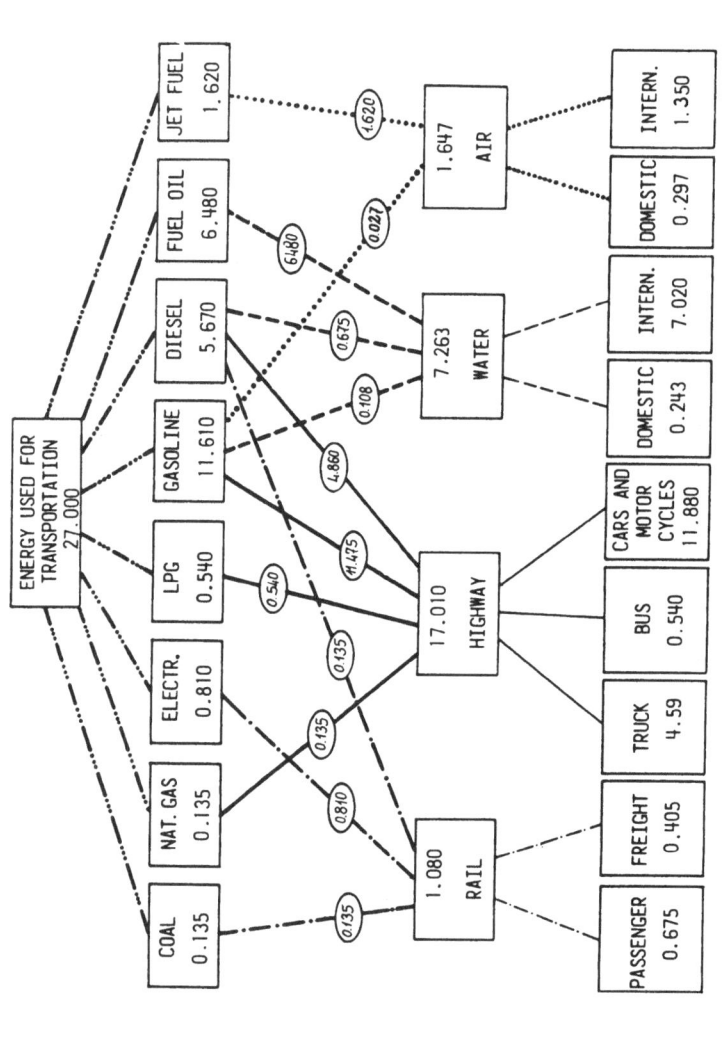

Fig. 1.7 - Italy 1973 - Repartition of the Energy used for Transportation (10^6 T.O.E.)

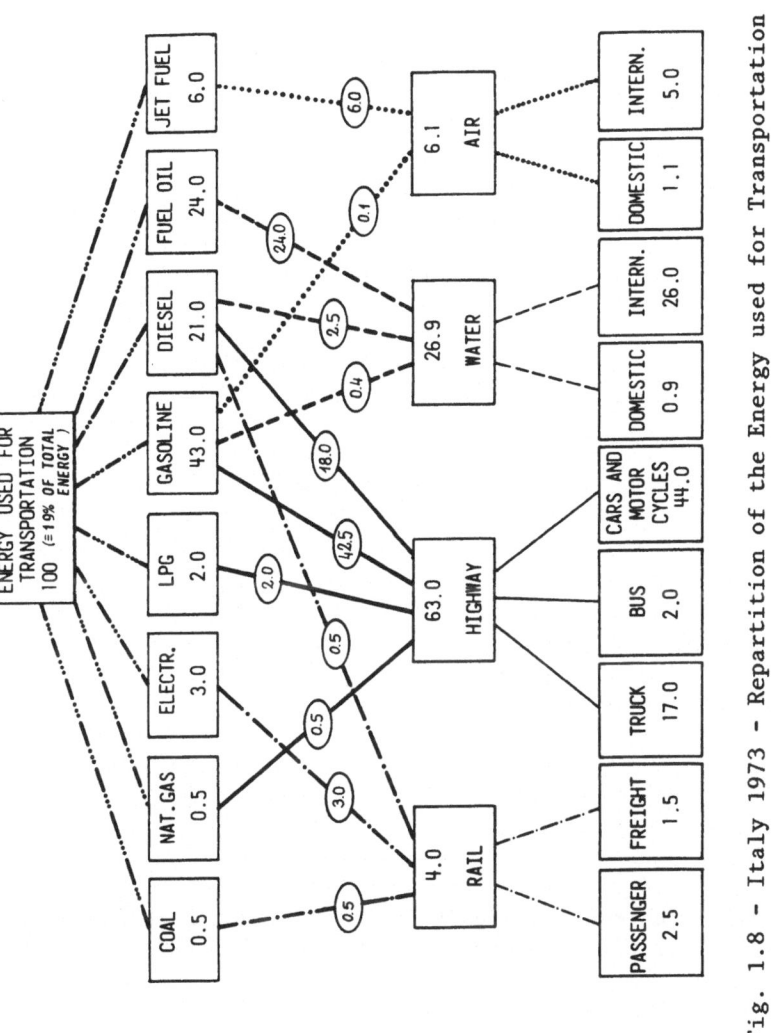

Fig. 1.8 - Italy 1973 - Repartition of the Energy used for Transportation (%)

INDIRECT ENERGY (U S A)

(10^{15} Btu)

	1977
TOTAL DIRECT AND INDIRECT TRANS.	28.4
DIRECT TRANSPORTATION	20.0
INDIRECT TRANSPORTATION	8.4
EQUIPMENT MANUFACTURE	(4.1)
SERVICES	(3.2)
INFRASTRUCTURE	(1.1)
INDIRECT AS PERCENT OF DIRECT	42 %

INDIRECT ENERGY USE AS PERCENT OF
DIRECT UNERGY USE BY MODE, 1977 (U S A)

MODE	INDIRECT ENERGY USE
AIR	63.2
AUTOMOBILE	37.9
BUS	100.0
MARINE	85 7
PIPELINE	7.1
RAIL	116.7
TRUCK	42.9
TOTAL	42.0

Fig. 1.9 - Indirect Energy (USA) and Indirect Energy Use as Percent
of Direct Unergy Use by Mode, 1977 (USA)[10]

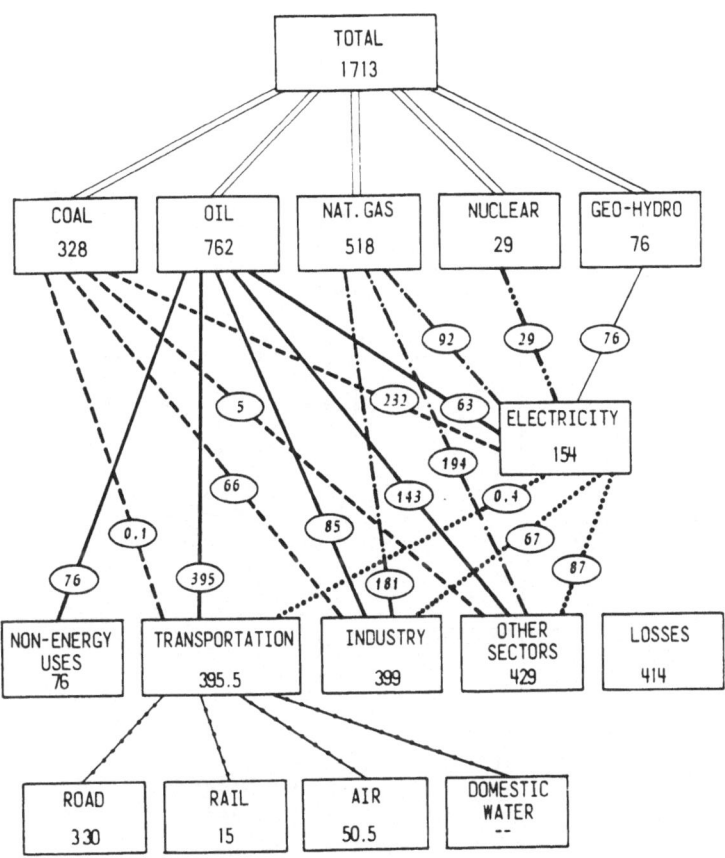

Fig. 1.10 - U.S.A. 1974 - Repartition of the Primary
Energy Sources (10^6 T.O.E.)

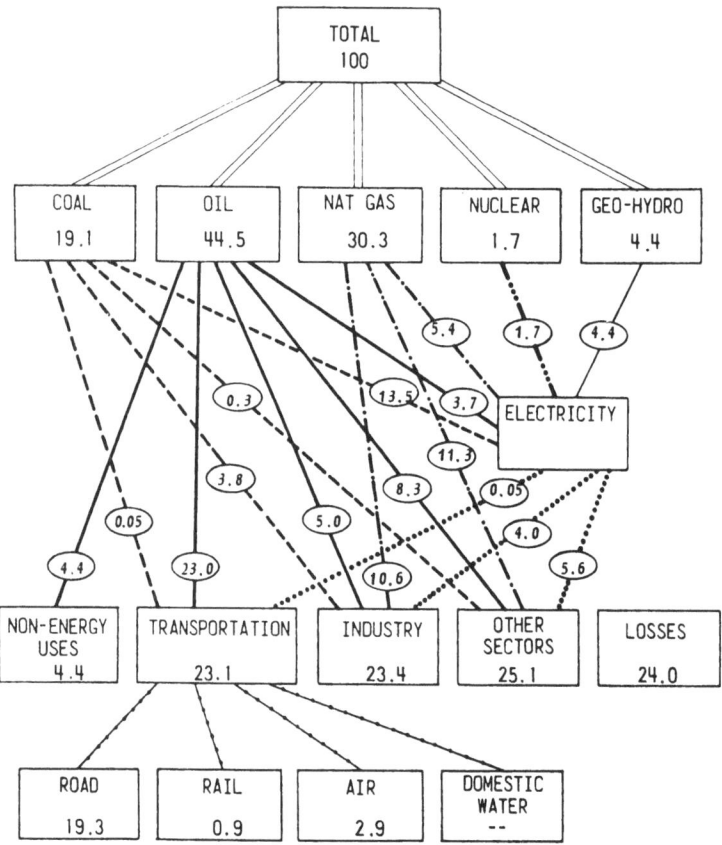

Fig. 1.11 - U.S.A. 1974 - Repartition of the Primary
Energy Sources (%)

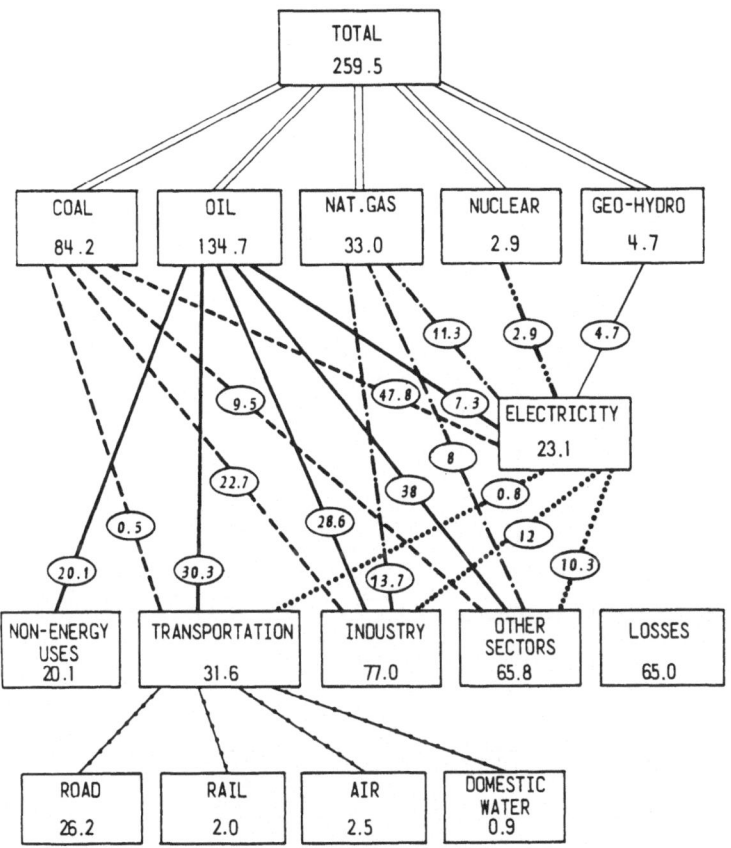

Fig. 1.12 - West Germany 1974 - Repartition of the Primary
Energy Sources (10^6 T.O.E.)

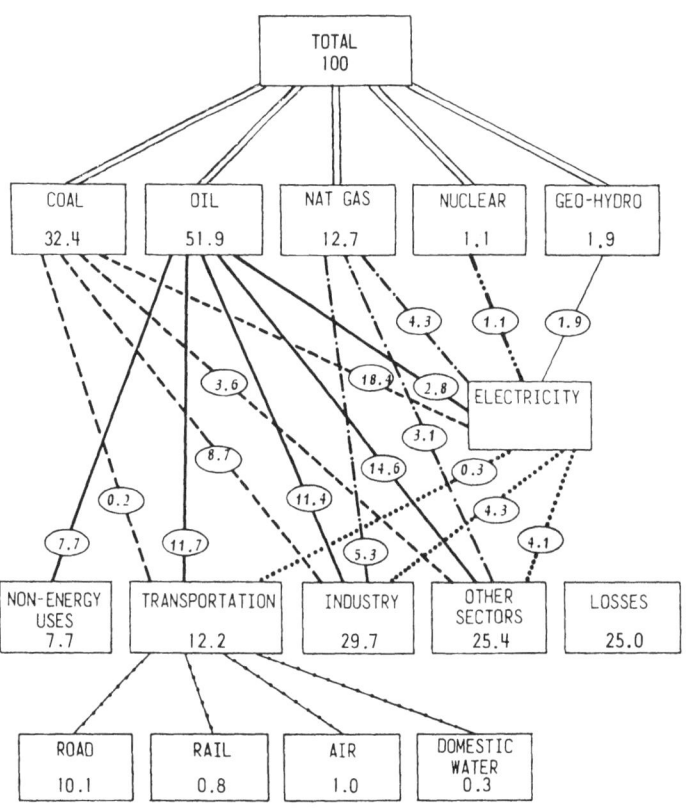

Fig. 1.13 - West Germany 1974 - Repartition of the Primary
Energy Sources (%)

2. PATTERNS OF DEVELOPMENT OF THE TRANSPORTATION SYSTEM

As discussed in the introduction we will try to have some insight in the transportation system dynamics by comparing the repartition of mobility among different modes of transportation and the related energy consumption in different countries at different levels of social and economical development.

Firstly we will perform a very qualitative analysis by comparing industrialized Western and Eastern countries, and the Less Developed Countries. From this qualitative analysis it will appear evident that the transport system is structured according the geographic and socio-political characteristics of each country, with influence from its history and tradition. Secondly, the more ambitious task to correlate the transport data with some macroeconomic data will be very preliminary attempted simply by plotting the data versus GDP, GDP/capita etc. We could remember here that the referred data represent, if not the whole world, at least 90% of it measured in term of GDP, population and energy consumption.[3]

2.1 Comparison of transportation systems

a) Passenger transport

As a first subdivision we separate individual transportation (mainly car) from collective (bus, train, air) transportation. These two different modes usually correspond to two different types of management: private and public, respectively.

By looking at Fig. 2.1 one sees a dichotomy which, as a first approximation, reflects the subdivision of the world into two ideologycal blocks, Eastern and Western.

The countries have been divided into three large regions:

- the West, which includes the industrialised countries, whith a prevailing free market economy

- the East, which includes the countries with centralised economy
- the Less Developed Countries (LDC)

The parameter used in Fig. 2.1 to characterise the structure of the passenger transport system is the percentage distribution, among the different modes, of passenger x km.

The structural difference between West and East regions appears very neatly. The car mode predominates in the West countries, while it gives a very small contribution in the East countries. Intermediate is the case for the LDC.

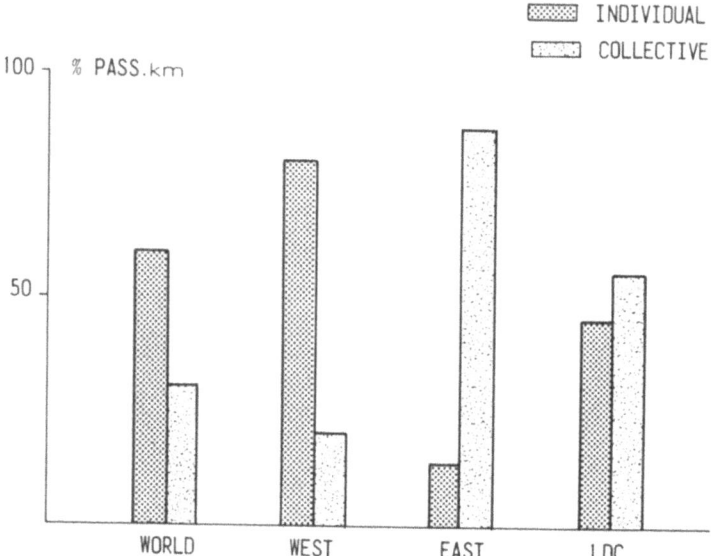

Fig. 2.1 - Structural Patterns of the Passenger Trasport System:
Individual and Collective Modes (1975) - Passenger*km

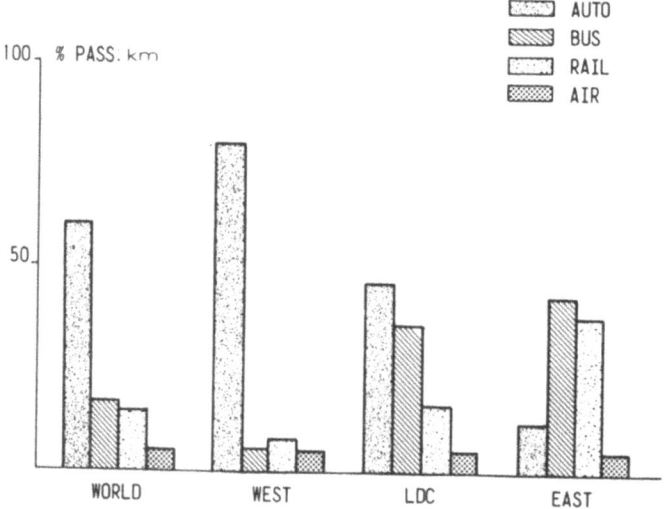

Fig. 2.2 - Structural Patterns of the Passenger Transport System:
Modal Splitting (1975) - Passenger*km

In Fig. 2.2 the "collective" mode is splitted into its components; bus, train, air. One can perceive a certain regular trends of decreasing role of car and increasing role of bus and train, if one puts the data, as in Fig. 2.2, into the order: West, LDC, East countries.

By furtherly disaggregating the three regions one can add further remarks. The trends are better grasped by plotting the data pictorially by means of bar-chart as in Fig. 2.3:

- the West area is subdivided in four subareas: West Europe, USA, Australia-Canada, Japan. For comparison purpose the case of Italy and U.K. is also plotted
- the East area is subdivided in three subareas: East-Europe, URSS, China
- the LDC are subdivided into: Latin America, Mead East-South Africa, other Africa, South Asia.

Within each of the three large areas of Fig. 2.1, one can, from this more disaggregate data analysis, find analogies and discrepancies in the structure of the passenger transportation system, such as:

a) car mode.
- While Western Europe, Australia-Canada and USA show a strong analogy (even though the share of the car mode is increasing in that order) in Japan the case is largely different;
- in the East area car mode is almost irrelevant, whereas it increases in importance from URSS to Eastern Europe;
- in the LDC area, Latin America has a passenger transportation structure more similar to that of the Western countries.

b) train.
- In Japan it is the dominant mode, whereas it has a negligible role in USA and Australia-Canada;
- in China, as in Japan, it is the dominant mode;
- among the LDC, it has the least importance in Latin America.

c) air.
- While, as a general role it ranks as the last mode in the sharing of the passenger x km, the case is different for URSS where it has the same share of the car mode, while in USA and Australia-Canada it is responsible for more passenger x km than the bus and train modes added together.

The case of Italy shows a structure very similar to that of the other Western Europe countries.

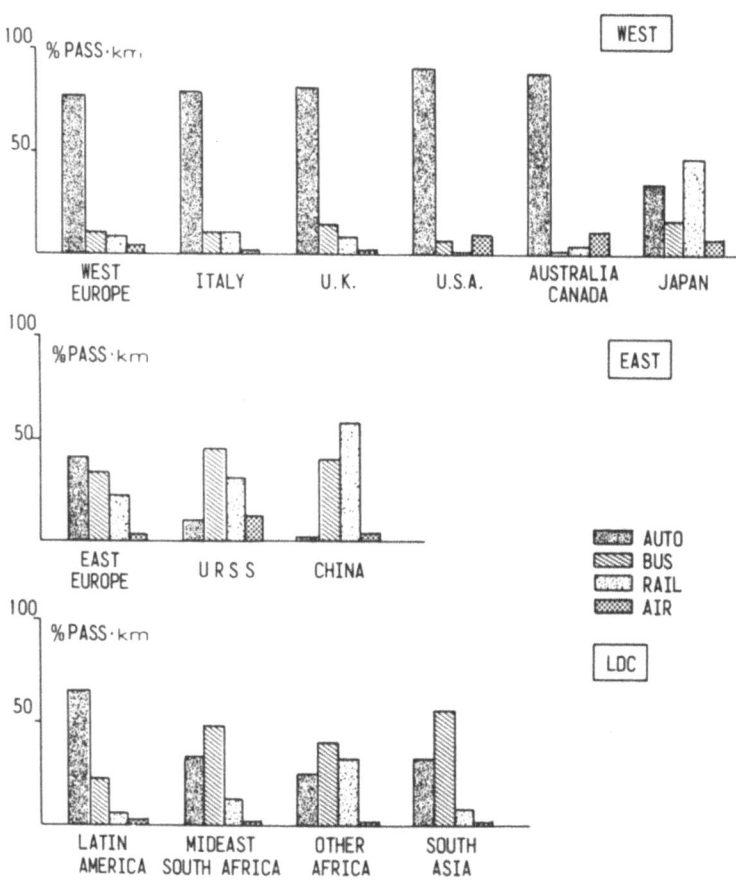

Fig. 2.3 - Structural Patterns of Passenger Transport System:
Modal and Regional Splitting (1975)-Passenger*km

In the analysis one has omitted the data related to passenger transport on water ways (rivers, lakes, channels, sea), which after the Second World War have drastically changed both qualitatively and quantitatively. The transoceanic passenger transport has pratically disappeared (especially between Europe and America) substituted by the much faster air transport mode. Meanwhile vacation sea cruising traffic has developed, and innovative transportation means (hovercraft, airfoils, car-ferry) have increasingly taken a share of internal-water transport. In any case the relative importance of all the water-ways modes is small so that having not included them will not change the very rough type of analysis here attempted.

The structural analogies and differences illustrated above, for the passenger transportation system depend obviously by a multiplicity of reasons and causes very difficult to distinguish one from the others. We are very conscious of how "un-scientific" a procedure is to try to esplain them on a cause-effect basis. Among other caveats, one should, for instance, beware of the bias included in the countries aggregation which has been used.

Nevertheless as a contribution to provoking thoughts we will try at this stage some very general and qualitative remarks.

First of all, it seems obvious to relate the large difference between Western and Eastern countries to their different social-political-economic systems. The very strong development of the car transportation mode could be related to the aspect of privatistic economy, free initiative, free markets, individualistic values which characterise the West area. On the other hand the stronger role of collective transportation (bus and train) in the East area could be related to the aspect of planned economy, centralised burocracies, and less individualistic autonomies that characterise those countries.

As attractive as it might, one should refrain to jump to the first conclusion, that there are two different patterns of development of the transportation system, one, based mainly on car, for the "liberal democracy" countries and the other, based on collective transport, typical for the "social-democracy" countries.

As a matter of fact one should notice that:

- the car transportation mode in the western european countries has developed at an accelerated pace after the Second World War with a time lag with respect to the USA which seems more related to the increase of the per-family income than to socio-economic differences.
- in the East area the highly industrialised european

countries show a larger use of the car mode with respect to URSS, as indicated also by the data on number of car owned per inhabitant[14]:

. East Germany   1/10
. Czechoslovakia   1/15
. URSS   1/91

- Japan, a country that from the economic-industrial--political point of view strictly partakes to the Western world, has an anomalous passenger transportation structure more related to East and LDC areas. To which extent this is attributable to the different society values with respect to the Western countries, notwithstanding the industrial-economic similarity, is matter of subjective evaluation.

One could think that the lesser development of the car mode more than an intrinsic characteristic of the Eastern countries - based on a supposedly negative attitude against the private mode of transport - is based to a priority choice which tends to use the industrial resources for other objectives than car manufacture and related infrastructure realization.

Whenever the socio-economic conditions are favorable (as f.i. in East Germany and Czechoslovakia) there seems to be non "ideological" obstacles to the development of private "individualistic" transportation mode.

The LDC area is in an intermediate position with respect to East and West. It seems to have a more equilibrated sharing between private and collective transport modes. It might appear to be strange the important weight of car in countries of low-per-capita income, if one compares them, for instance, with the case of China. Life-style imitation of rich western countries had certainly a large impact in the now emergent countries without the long historical tradition of China.

Moreover the LDC area is a very aggregated one, putting together very disomogeneous countries. From the 100/1000 car per inhabitants of Argentina, one goes to the 3/1000 in Nigeria and 2/1000 in India[4]. The disaggregation of the LDC area into four subareas, illustrated in Fig. 2.3, shows already large differences in the passenger transportation system's structure.

The remarks above might be considered improper or of no relevance in an analysis which is motivated by looking at the potentiality of energy conservation in the transportation system. Neverthless we think that any proposed actions, to be effective, should be based on the understanding of how complex are the interrelations of the transportation system with the society characteristics. To insist on the need for a deeper insight, we add here other remarks:

- historical, political, economic, psicological motivations, intermingled with geografical, urbanistic, demographic characteristics, have contributed to form the structure of each of the transport system taken into consideration, but no one of these motivations can be singled out to explain the remarked differences. Certainly the wealth of a country has an overwhelming influence, but not everywhere. For instance, Japan is wealthier than Italy but the Japanese, notwithstanding their large car park (21 millions in 1978, almost 1 every five inhabitants), do use the train modes to a greater extent than italians;

- geographical characteristics seems to have favoured the development of the air mode in the large spaces of North America, Australia and Russia;

- the existence of an extended and well performing railroad network before the Second World War, explains the importance of the train mode in Japan, South-East Asia, East Europe, but not its decline in USA;

- walking, bikes, animal carts are very important in transportation in poor countries. They ..."represent the most important means for mobility in the Third World. In India 12 million of animal carts are ten time larger than the car stock. But even these carts goes beyond the financial resources of many people of the Third World. Therefore walking has an importance since long forgotten in the industrialized work"[7]. It seems strange to note, in contrast, the importance of bicycle for urban transportation in very affluent countries, like Holland. Why bicycle is less used in Italy notwithstanding the lower income and better climatic conditions?

b) Goods transport

The analysis is extended to the same countries (with the exception of U.K) and geographical areas considered for the passenger transport.

Three modes of goods transportation are considered: road (trucks), railroad (train) and internal waters (channels, rivers, lakes, coastal navigation).

We have not considered the air mode, because its weight is very low in the transportation of goods (f.i. in USA in 1975 approx 10.5 x 10^9 ton x km was transported, which is 0.3% of all the modes), and international sea transport, oleoducts, gasducts because of lack of complete data.

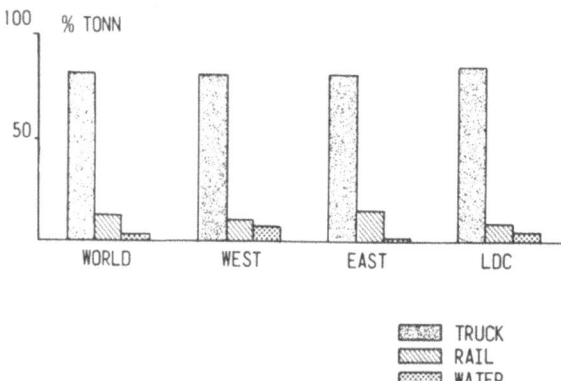

Fig. 2.4 - Structural Patterns of the Freight Transport System:
Modal Splitting (1975) - Freight - Tons

While in passenger transport the private mode (car) is
opposed to collective mode, in the case of goods transport the
dichotomy is represented by the road and the railroad mode.

As already said in paragraph 1.2 we have analysed both the
data referred to the total tonnage transported and the ton x km.
When considering the total weight (metric tons) of transported
goods one sees a large analogy in the three areas examined (see
Fig. 2.4). The road mode has the lion's share, with percentage
ranging from 80 to 90%, in the three areas. This characteristics
is maintained even at a more disaggregated level (Fig. 2.5).

It is difficult to understand these data especially if one
considers the different relevance of transporting raw materials in
contrast to finished goods in LDC or in industrialised countries,
and that rails and water should provide cheaper transportation
modes for raw materials. More insights on how the accounting is
done is needed to try to understand the reasons behind.

Since energy consumption is proportional to the "transport
work" (weight x distance) we will now divert attention to the ton
x km data. The structure of the goods transport system changes
completely when looked from that angle (see Fig. 2.6). This is due
because the average distances on which goods are transported by the
three modes are very different (see Fig. 2.7).

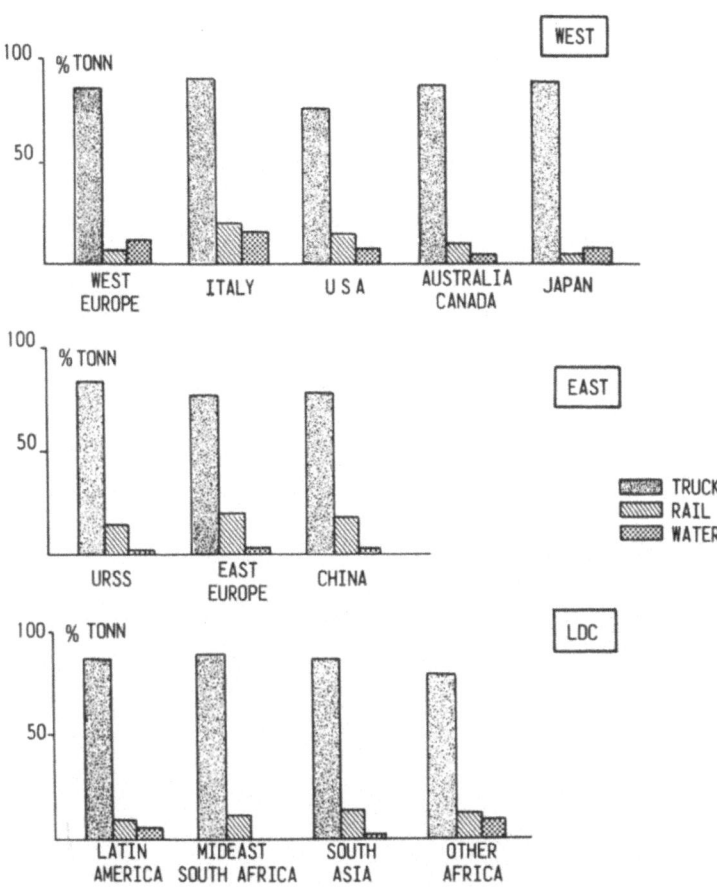

Fig. 2.5 - Structural Patterns of the Freight Transport System:
Modal and Regional Splitting (1975) - Freight-Tons

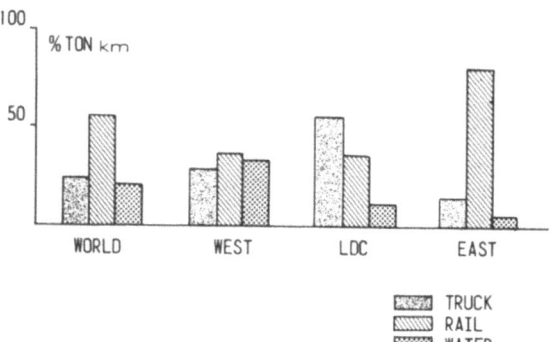

Fig. 2.6 - Structural Patterns of the Freight Transport System:
Modal Splitting (1975) - Freight-Tons *km

	TRUCK	RAIL (km)	WATER
WEST EUROPE	53.1	199	156
ITALY[1]	60.0	347	508
U.S.A.	84.4	1011	1328
AUSTRALIA-CANADA	32.3	527	900
JAPAN	29.4	243	419
URSS	16.2	894	466
EAST EUROPE	22.4	252	224
CHINA	62.5	746	1250
LATIN AMERICA	107.5	394	452
MIDEAST-SOUTH AFRICA	48.9	517	--
SOUTH ASIA	66.8	633	994
OTHER AFRICA	100.	208	250

DATA ELABORATED FROM TAB. 10 - REF. 3

(1) REF. 13

Fig. 2.7 - Freight Shipment Distance by Mode - 1975 (km)

In every area trucks are used for much shorter distance than rails and internal waters. LDC departs somewhat from the other two areas, and road distances are longer and almost the same as the rails distances. Goods travel on very long distance on rails and internal waters in large countries such as USA, URSS, China.

In Fig. 2.8 the data of Fig. 2.6 are disaggregated in subareas and single countries. In the Western countries the sharing among the three modes changes from West Europe to USA where rail is prevailing. In Japan, while the train has highest share in pass. x km, it has half the weight of trucks in goods transportation. In the East area the rail mode has the greatest share, while in the LDC area the road is responsible for more than 50% of the ton x km, with large variations in the mode's relative weights from one region to the other.

2.2 Energy use in transportation

Total energy uses and the share for transportation in different geographical areas are compared in Fig. 2.9.

Energy use for transport averages, world wide, 20% of total energy with a peak of 25% in USA and a minimum of 7% in East Europe. Italy, with its 19%, is aligned with the world average. Dependence on oil is very high all over the world ranging from 92% in USA to 72% in URSS.

The data are shown pictorially as bar-chart in Fig. 2.10. The specific countries cases shown are not the same of those included in the charts for discussing the passenger x km data because of difficulties on data collecting. LDC are divided into two: Advanced LDC and Less developed LDC. At the large area level we maintained the same geographic and socio-political aggregates and we have integrated the data with estimates, when needed. The approximations due to this integration of data with estimated values are not such to modify the analysis substantially.

Before analysing separately passenger and goods transportation we will consider the energy use in the entire system, passenger and goods. The data are referred in table form in Fig. 2.11.
On the left hand side the energy use is given as absolute value in 10^6 TOE and on the right hand side as percentage distribution. The latter data are plotted as bar-chart in Fig. 2.12, for the three major aggregation areas on the left of the figure (West, East, LDC) and for specific countries on the right side of the figure.

At the world wide level, twice as much energy is used for passenger than for goods transportation. Responsible for this are the countries of the West area, where energy in passenger

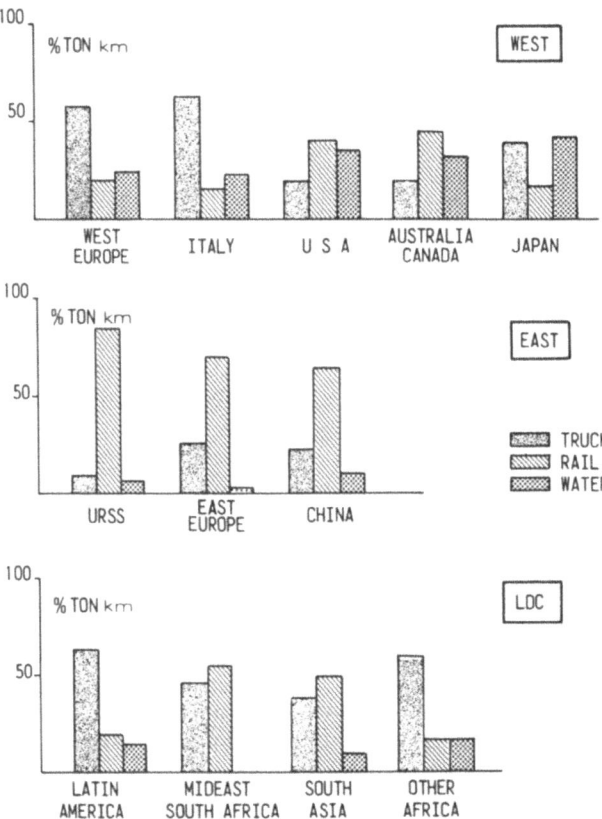

Fig. 2.8 - Structural Patterns of the Freight Transport System:
Modal and Regional Splitting (1975)-Freight-Tons*km

	TOTAL ENERGY (10⁶ T.O.E)	TRANSPORT ENERGY (10⁶ T.O.E.)	TRANSP. TOTAL (%)	TRANSPORT DEPENDENCE FROM OIL (%)	TRANSPORT ENERGY PER CAPITA (T.O.E.)	TRANSPORT ENERGY PER G.N.P. (10⁻³ T.O.E.)
WEST EUROPE	1128	221	20	95.5	0.71	0.24
ITALY [1]	140	27	19	96.		
U.S.A.	1668	409	25	[3] 96.2	1.90	0.33
JAPAN	319	60	19	96.8	0.55	0.21
URSS	961	109	11	71.6	0.43	0.24
EAST EUROPE	376	26	7	87.7	0.31	0.14
ADVANCED DEVELOPING COUNTRIES [2]	371	90	24	97.1	0.33	0.38
LESS DEVELOPED COUNTRIES	721	57	8	85.5	0.02	0.21
WORLD [4]	5855	1159	20	94.4	0.33	0.28

(1) (1973)
(2) INCLUDING CHINA
(3) (1976)
(4) SEE NOTE 6 - FIG. 2.3

Fig. 2.9 - Direct Transport Energy as % of Total Energy - 1975

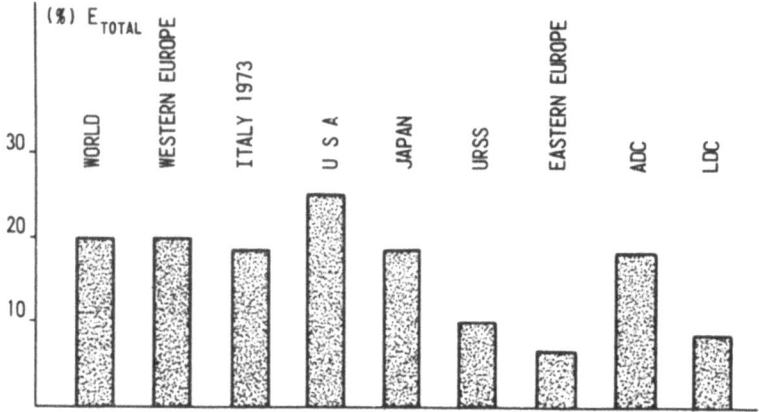

Fig. 2.10 - Direct Transport Energy as % of Total Energy - 1975

transport is 3.5 times larger than in goods transport. In the East and LDC area more energy is used in goods transport than in passenger.

Looking at the data on specific countries, an exception with respect to the West area comes from Japan where more energy is used for goods than passenger transport.

Referring back to the data on passenger x km shown before, the energy data reflect the preminence of car mode in countries like USA and the anomalous case of Japan.

a) Energy in passenger transport

The remarks on the different structure of the passenger transport system given in the preceding paragraph, based on the passenger x km data, are confirmed when analysing the percentage distribution of energy consumption in the various mode: car, bus, rail, air (see Fig. 2.13).

With the exception of the East area, it is the car that gets the largest share of energy used for passenger transport. This is due both to the higher share of passenger x km and to the higher energy intensiveness (kcal/pass x km) of car with respect to other modes.

	TOTAL	PASSENGER	FREIGHT	PASSENGER	FREIGHT
		(10^6 T.O.E.)		(%)	
WEST EUROPE	162.2	122.1	40.1	75.3	24.7
U.S.A.	381.4	318.0	63.4	83.4	16.6
JAPAN	38.2	15.8	22.4	41.4	58.6
CANADA-AUSTRALIA	36.3	26.3	10.0	72.4	27.6
WEST AREA	618.1	482.2	135.9	78.0	22.0
URSS	89.7	24.7	65.0	27.5	72.5
EAST EUROPE	21.1	9.7	11.4	46.0	54.0
CHINA	19.3	7.3	12.0	37.8	62.2
EAST AREA	130.1	41.7	88.4	32.0	68.0
ADVANCED DEVELOPING COUNTRIES	53.8	19.1	34.7	35.5	64.5
LESS ADVANCED COUNTRIES	18.1	8.4	9.7	46.4	53.6
DEVELOPING COUNTRIES AREA	71.9	27.5	44.4	38.2	61.8
WORLD	820.1	551.4	268.7	67.2	32.8

Fig. 2.11 - Share of Energy Consumption Between Passenger - and
Freight Transport Systems (1975)

Fig. 2.14 confirms this when looking at single countries and subareas. The case of Japan is a confirmation of the higher energy intensiveness of the car mode. In fact notwithstanding lower share of passenger x km (see Fig. 2.3) the energy share of this mode is higher than that of the collective modes.

Whereas on the average the East area presents an equal share of energy used in the various modes, in URSS the air energy use is higher than the others (40% of the total energy used in passenger transport), due to its relevance in passenger transport (12% of pass x km) and of the high energy intensiveness of the air mode. The air energy use share is almost everywhere higher than that of the other collective transport modes (bus and train).

L. LUCIO BUSINARO AND A. FEDRIGHINI

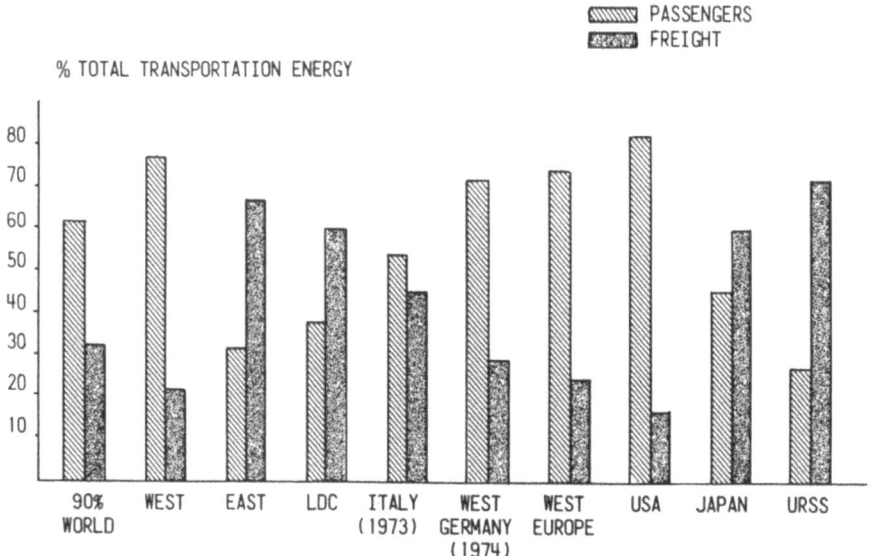

Fig. 2.12 - Energy Use Share Between Passenger and
Freight Transportation

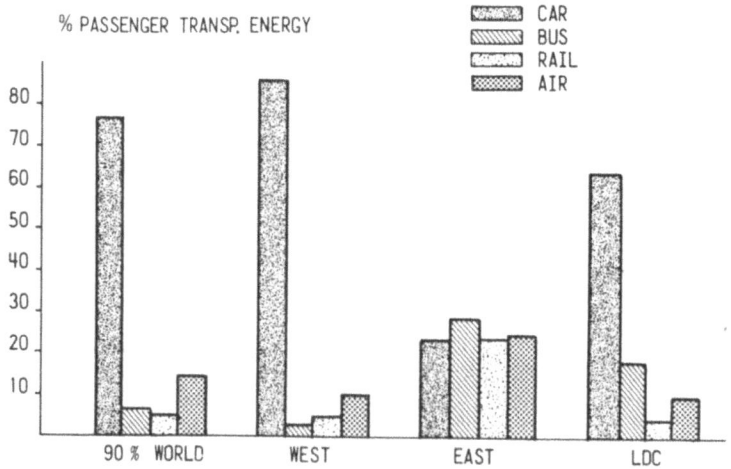

Fig. 2.13 - Percentage Passenger Transport Energy by Mode

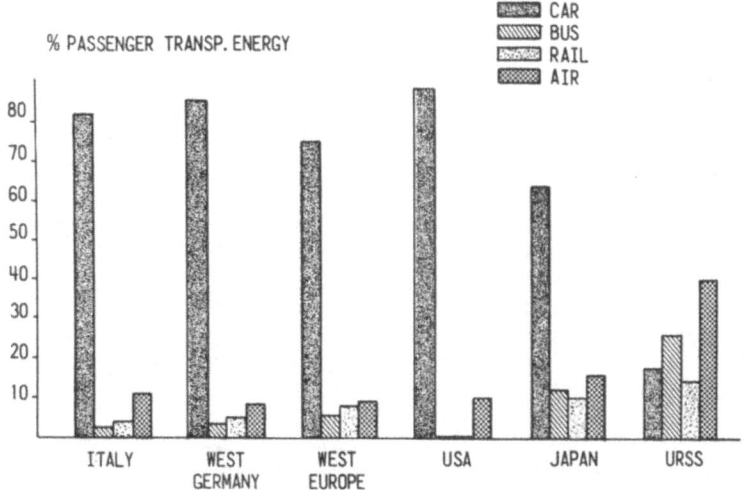

Fig. 2.14 - Percentage Passenger Transport Energy by Mode
in Selected Regions

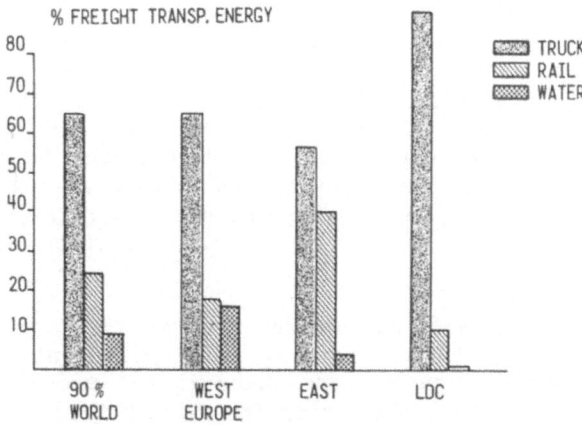

Fig. 2.15 - Percentage Freight Transport Energy by Mode

Train represents a negligible part of energy use in USA. Italy shares the typical structure of the western european countries.

b) Energy in goods transport

More than 65% of the energy used for goods transportation in the world is due to road transport (Fig. 2.15). This percentage increases to 90% in the LDC area, where more than 50% of the ton x km of goods is transported by trucks. The difference between the two precentage values is an indication of the higher energy intensiveness (kcal/ton x km) of the road transport with respect to the other surface modes (train, inner water). A further confirmation is obtained by comparing Fig. 2.4 with Fig. 2.15. The percentage distribution of road ton x km is 28% in the West area and 12% in the East area whereas the energy share is, respectively, 65% and 57%.

Rail is the second largest energy user in goods transport, with percentage values variable from 40% in the East area to 10% in the LDC. Inner water transport is comparable with rails only in the West area, and lower or negligible in the other areas.

In Fig. 2.16 the data are further disaggregated geographically. The general characteristics of each area are confirmed with the exception of URSS (with lower road energy percentage share than the average in the East area), and Japan where inner water transport uses much more energy than rail.

It is interesting to remark the case of USA which has a road mode energy use lower, percentage wise, than that of the average West area. This is in line with the structure of the goods transport shown in Fig. 2.8, indicating that rail and inner water are the prevailing goods transport modes.

In the analysis, the energy use in other modes of transportation were not included, wether because not available (air transport, oleoducts and gasducts) or because of the difficulties to subdivide the ton x km to each area as is the case for the internationsl sea transport, and its related energy use. For completeness's sake, we give some indication on the latter modes:

- in USA, in 1976, the energy used in the pipeline was 5.5% of the total (or 6.8% according to a different bibliographic source[10]). Such high percentage value is not probably reached by other countries, with the exception of URSS, due to the large extension of the pipeline network in USA. The results of our analysis should therefore not be invalidated by having neglected the pipeline modes.

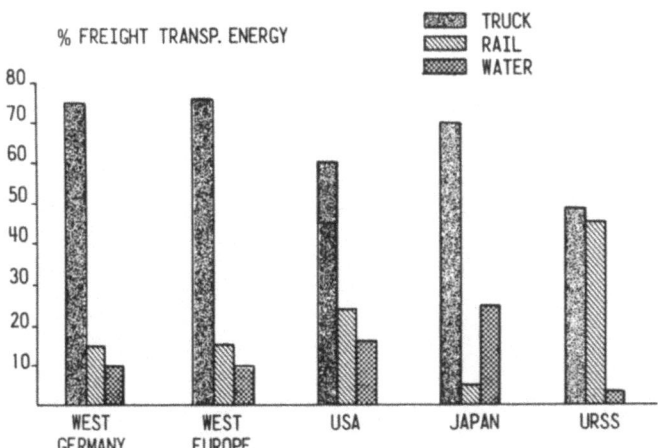

Fig. 2.16 - Percentage Freight Transport Energy by Mode
in Selected Regions

	% TOTAL TRANSPORT	% FREIGHT TRANSPORT
U.S.A.	2.8	14.8
WEST EUROPE	15.99	43.4
JAPAN	29.31	41.8
URSS	0.67	0.9

Fig. 2.17 - Energy Use in International Sea Transport - 1975

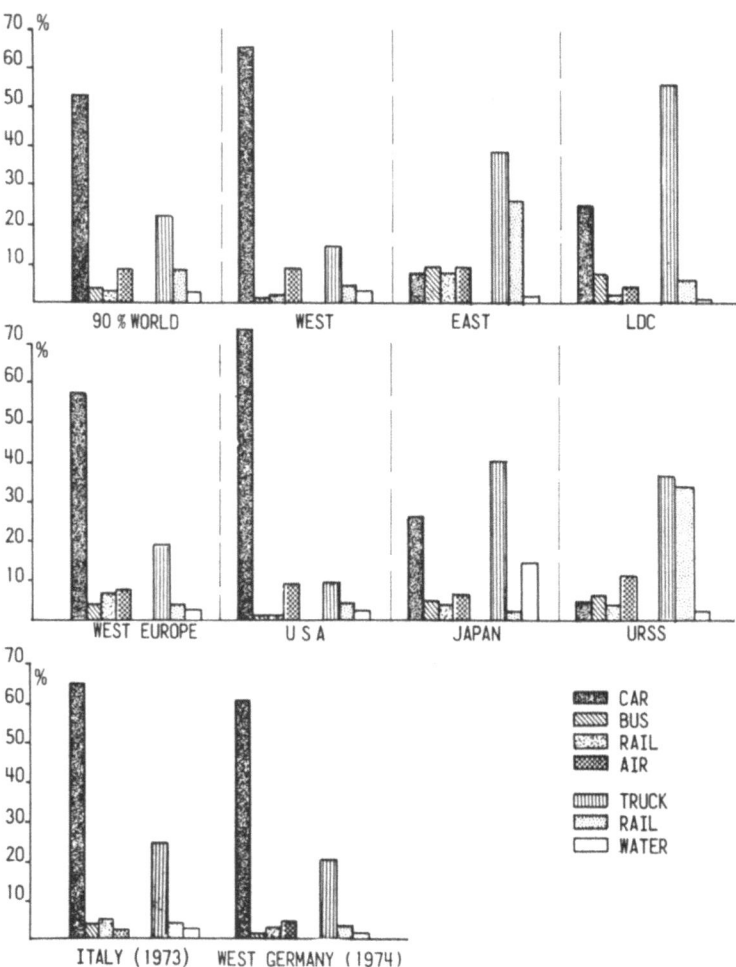

Fig. 2.18 - Percentage Passenger and Freight Transport Energy:
Modal and Regional Splitting

- it is different the case for the international sea
transport. It represents a large fraction of energy used in goods
transport especially for Europe and Japan[3] (see Fig. 2.17).
Notwithstanding this we consider that the conclusions on ways and

means to energy saving prospects will not be greatly affected. As
a matter of fact one such way is related to the possibility of
shifting from one mode to another more energy efficient. The
international sea transport could be considered, as a first
approximation, difficult to be substituded by, or be the
substitute for, other transportation modes (with the exception, in
special cases, by sea-bottom pipelines).

c) Comparison of energy use structure in passenger and goods
   transport

It might be of interest to summarise here the structure of
the energy use in the total transport system (passenger and
goods). In Fig. 2.12 the relative importance of passenger and
goods were presented adding together all modes.

In Fig. 2.18 the data are disaggregated among the different
modes. It is apparent the world-wide importance of car (more than
52%) in the use of energy in transport. Truck comes second (22%),
followed by train (11%), air mode (8%) bus (3.6%) and inner water
(3%). Road transport - adding car, truck, bus - accounts for 78%
of the total energy use.

The West area has an energy use structure similar to the
World average, where car reaches 65% and total road transport 82%
of the transport energy. Quite different are the structures of the
East and LDC areas.

In the East area truck and train are the larger energy users,
while car is marginal. Total road transport (car, truck, bus)
account for 56% of energy use. In the LDC area truck is by far the
larger energy user (55%), and together with car (25%) and bus the
total road transport uses 87% of the transport energy.

The similarity of West area structure to that of the World
average is depending on the high ratio of energy consumption in
West (70%) referred to World total. When disaggregating the West
area into single countries one find the same type of structure,
with the exception of Japan. On the other hand URSS, accounting
for 65% of the East area transport energy use, shares the same
structure of this area.
We might again be tempted here, as when discussing the passenger x
km and ton x km data, conclude that there are two typical
patterns, Western ad Eastern, for the transport system
development.

2.3 Energy intensity and energy efficiency in transport

In the two preceding paragraphs we separately analysed the
data on:
      - transport "work": pass x km, ton x km
      - transport energy: 10^6 TOE or kcal

	ENERGY INTENSITY		ENERGY EFFICIENCY	
	$\dfrac{Kcal}{Pass \cdot km}$	$\dfrac{Kcal}{Ton \cdot km}$	$\dfrac{Pass \cdot km}{1000\ Kcal}$	$\dfrac{Ton \cdot km}{1000\ Kcal}$
WEST EUROPE	437	467	2.29	2.14
U.S.A.	859	184	1.16	5.43
JAPAN	215	644	4.65	1.55
CANADA-AUSTRALIA	501	194	2.00	5.15
WEST AREA	621	263	1.61	3.80
URSS	241	171	4.15	5.85
EAST EUROPE	246	301	4.06	3.32
CHINA	94	203	10.64	4.93
EAST AREA	207	196	4.83	5.10
ADVANCED DEVELOP COUNTRIES	271	564	3.69	1.77
LESS ADVANCED COUNTRIES	151	295	6.62	3.39
DEVELOPING COUNTRIES AREA	218	470	4.59	2.13
WORLD	495	252	2.02	3.96

Fig. 2.19 - Total Transport Energy Intensities and Efficiencies

We should now analyse the ratio of the two above characteristic parameters, which can be defined as:

$$\text{- transport "efficiency"} = \frac{\text{transport "work"}}{\text{transport energy}} = \begin{cases} \text{pass} \times \text{km/kcal} \\ \text{ton} \times \text{km/kcal} \end{cases}$$

As a matter of fact we will refer to the reciprocal of the "efficiency":

$$\text{- transport energy intensity} = \frac{\text{transport energy}}{\text{transport work}} = \begin{cases} \text{kcal/pass} \times \text{km} \\ \text{kcal/ton} \times \text{km} \end{cases}$$

The data on energy intensity have been obtained simply, at each country or area aggregate level and for the considered modes, by dividing the two data on transport energy and transport work. The results obtained are reported for passenger and goods transport in Fig. 2.19 averaged over all modes.

Notwithstanding its simple definition, transport energy intensity is a complex parameter which is dependent on a multiplicity of factors, among which:

- operative aspects (load factor, occupation factor, empty backhaul, quality and lenght of the routes, speed ecc.).
- technological aspects (power plant efficiency, operation and traffic control, innovative aspects of infrastructures, etc).
- different mix of the modes and of the vehicles within a mode (including vehicle "age").
- type of load. Light, expensive or perishable goods are better transported by air or by truck, whereas heavy and cheap goods by train or by inner water.
- preferences due to psychological attitude or social acceptability (frequency and waiting time, safety, door-to-door capability, individual habit, etc).
- mode availability (in some zones only cars, or trains, or air are available to perform the transport).

Furthermore one should not forget the importance of indirect energy for manufacture and maintenance of vehicles and infrastructures and for modes interfacing. We remember that in this analysis we considered only the energy intensity related to the direct energy use.

In Fig. 2.20 the data of Fig. 2.19, related to energy intensity for passenger transport, are plotted as a bar-chart, in kcal/pass x km, for the world and the three main areas, and for specific countries.
The West area has the largest energy use for unit of transport "work": 3 times larger than the East area and LDC.
The high value of the world average energy intensity is determined by the high weight of the West area. USA has an impressively high energy intensity (see the right hand size of Fig. 2.20) when compared not only to URSS and Japan but also to Western Europe. One perceives here the influence of the larger car size and lower occupancy factor prevailing in USA, when compared to Europe. The prevailing use of collective mode explains the lower energy intensity of East and LDC areas as well as of Japan.

The energy intensity data of Fig. 2.19 for goods transport are illustrated as bar-chart in Fig. 2.21.
As a surprise, the West and East areas show energy intensities which are half the value for LDC area.
One should remember that, while the structure of the West area (see Fig. 2.6) is well balanced among all modes, and that of the

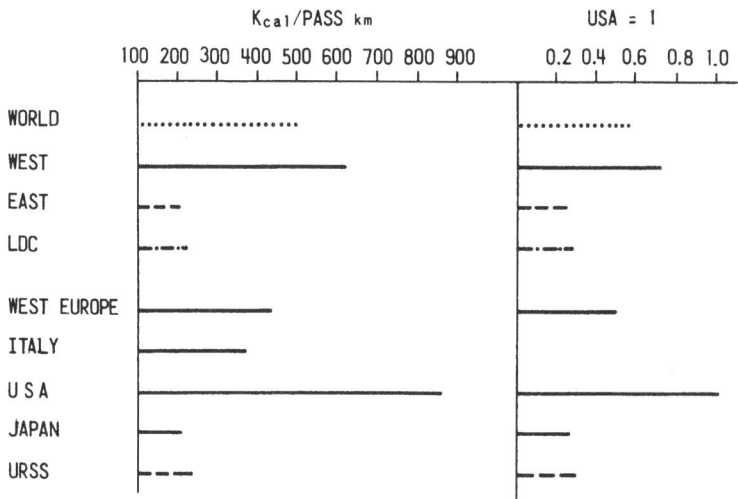

Fig. 2.20 - Passenger Transport Energy Intensity

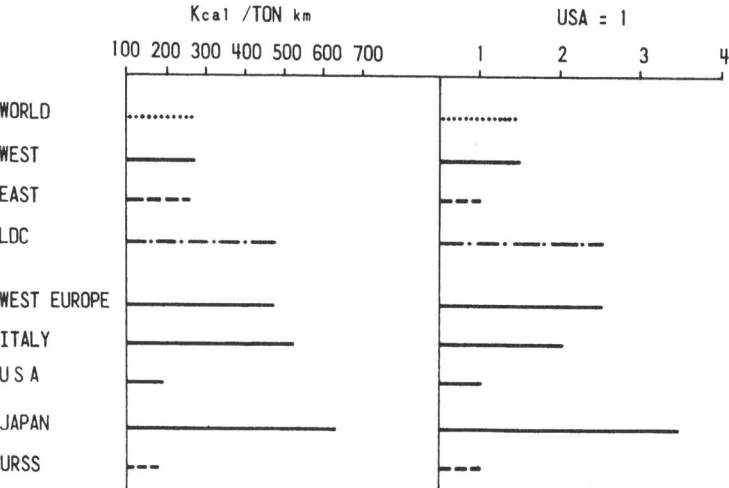

Fig. 2.21 - Freight Transport Energy Intensity

East area shows a predominance of the energy efficient train mode, the LDC goods transport structure is dominated by road transport. Because of the high weight of West and East area (91%), the world average energy intensity is aligned to their values.
Worthly to note is also the very low value of goods transport energy intensity for USA and, on the contrary, the very high value of Japan (higher than LDC). The case of USA, as that of URSS, is due to the predominance of rail in goods transport, while for Japan the truck mode is predominant.
West Europe has a high value of energy intensity similar to that of LDC, due to the importance of the road mode.

2.4 Transport system evolution patterns

The analysis of the preceding paragraph has shown that there are different degrees of energy efficiency in transportation and that they are correlated to different structures of the transport systems. One is therefore tempted to jump to the conclusion of developing a first receipe using the different cases of different geographical areas as the results of a "what-if" game: f.i. shifting from private to collective passenger transport and from road to rail for goods transport. If not on a world-wide basis, the "what-if" game could be tempted within macro-area with similar geographic and socio-economic characteristics. Before judging to which extent this is feasible, one should try to get better ideas on what is the "intrinsic" dynamics of the transport system because it represents the "inertial trend" of the system that should be taken into consideration in order to assess how realistic it is to shift the system from one state to another state thought-of.

To avoid wishful thinking in decision making, it is necessary to develope a cognitive base (both technical and cultural) on the system behaviour from which a political and strategic plan could be developed to optimize the transport system operation on a country (or larger area) level.

For instance, with reference to Fig. 2.20, we already indicated that the main responsibility of the lower energy efficiency in USA is to be looked for in the high percentage weight of the passenger x km transported by car mode. There are here two possible actions: one which is asking for technological innovation on the car itself to increase intrinsic efficiency, the other that relays on the possibility to perform a macrovariation on the society behaviour by shifting from private to collective transport.

While the first line of action is not at all easy to follow and might find its limitation on technological barriers and lack of human and financial resources and requires 10-20 years (because of the intrinsic slow process of innovation diffusion) the second line of action might even be more difficult because, to be effective, it requires corrective measures at global system level which might be vanified by first and higher order reactions at transport subsystems level and because of interactions with other societal systems. So one has to consider the interrelation between car, the other transport modes, the economic parameters (f.i. per capita GNP, degree of industrialization), the social factors (f.i. the degree of urbanization).

The difficulties to plan effective actions increase with the "density" and therefore the higher degree of interactions among the transport systems modes themselves and with external systems. The number of critical parameters with a threshold value, below which the system behaviour shows instabilities, increases, whereas the threshold values decrease. It is therefore high the possibility that actions tending to modify the intrinsic system development will excite instabilities (or be ineffective). Unfortunately the world area more responsible for the energy use in transportation is the West area, which is the more inefficient in transport energy, and presents a very dense and complex system.

All this is well known and complex mathematical models have been developed to simulate transport system "dynamics". These models, when well designed, will help the decision maker. In the next paragraph one such model will be presented and it will show, for the case of a complex system like USA, unexpected results on the impacts of actions which, based on intuition, should be highly effective.

These mathematical models, even when very complex, are almost limited to perform the accounting of effects of interactions simulated by simple equations, the complexity coming from the system dimension.

Reality is much more complex and one should try to validate the models using the cases of different regions and countries as "experiments". We think it might help very much to grasp from this set of "esperiments", even in a very qualitative way, the system behaviour, by developing some kind of correlations with macro-parameters characterizing the total society system of which the transport system is part. Such correlations have not to be interpreted as "laws of nature" on which to base cause-effect reasoning but simply as heuristic tools.

In order to try such correlations, we have considered as macro-variables external to the transport system:

. the per capita Gross Domestic Product (GDP, given in 1974$)
. the percentage of urban population

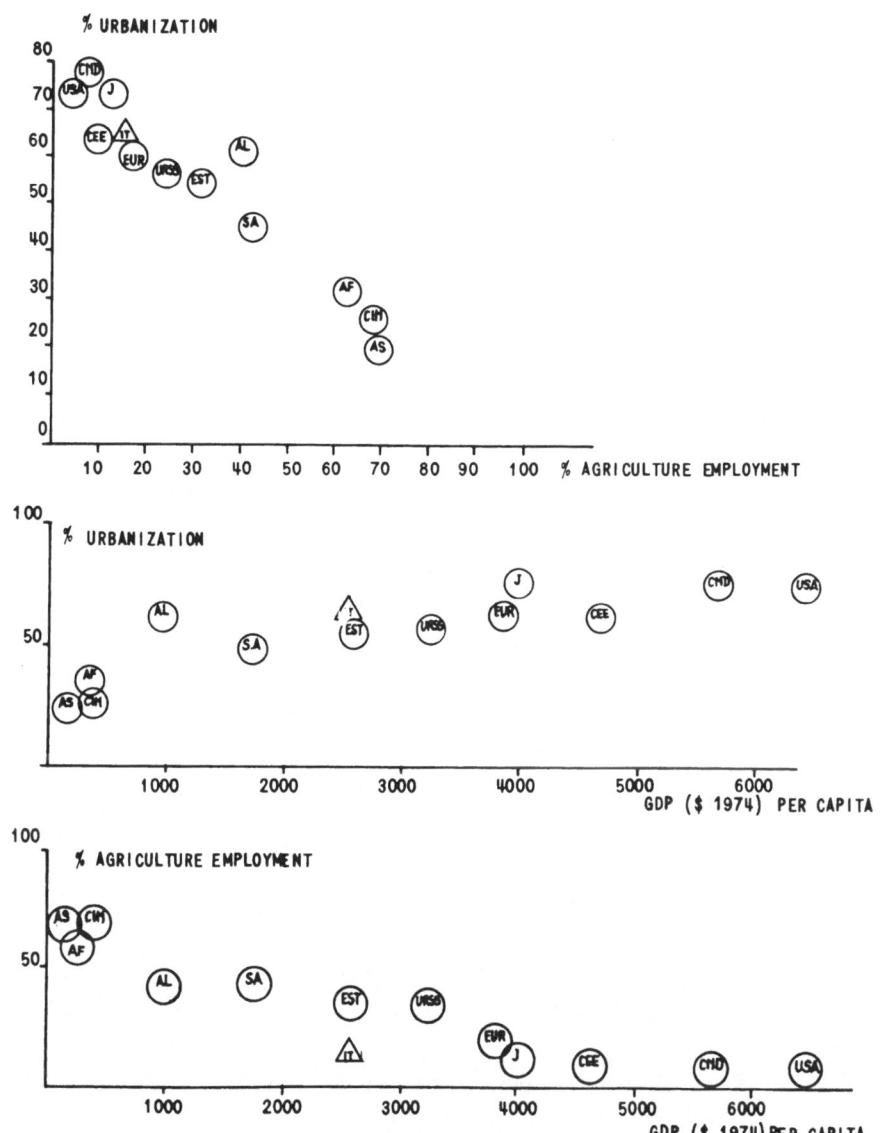

Fig. 2.22 - Macro-Economic Parameter Correlations

the percentage of the total working population in the agricultural sector. The complement to 100 of this number could be considered as the degree of industrial development (working force in industry and in service sectors) of a country.

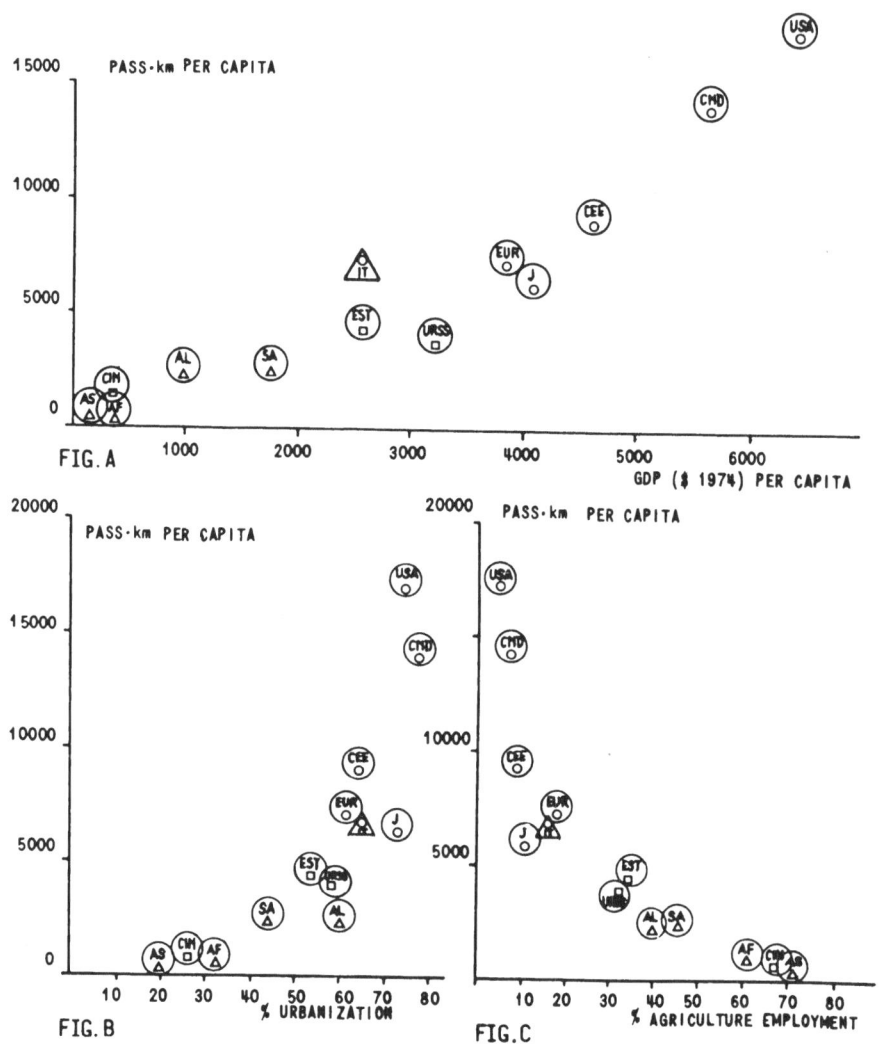

Fig. 2.23 - Total Passenger · Km per Capita Versus Socio -
Economic Macro Variables

For the urban population the data refer, for the greatest
part of the 34 countries considered, to the percentage of total
population living in towns with more than 10,000 inhabitants. For
few of the cases the bibliographic source was not precise on this
respect, but we are confident that the degree of incertitude so

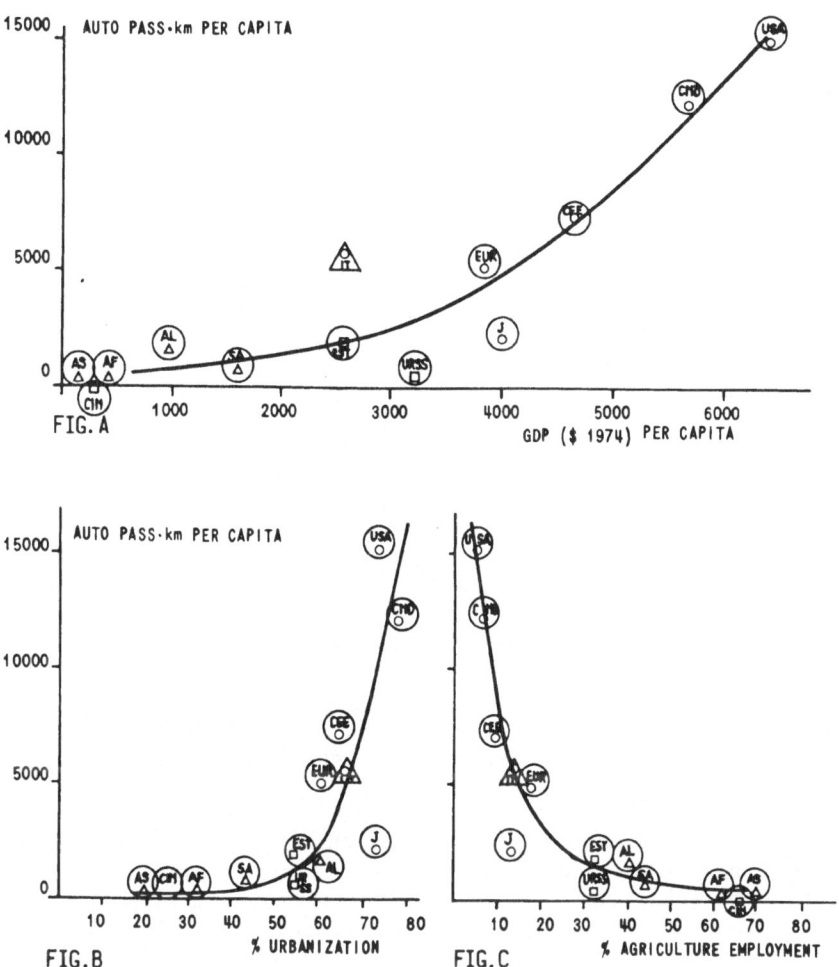

Fig. 2.24 - Auto Passenger · Km per Capita Versus
Socio - Economic Macro Variables

introduced is not such to change significantly the results of the
analysis. Furthermore not always the data refer to the same year
for all 34 countries. The time difference is not more than 5 years
with respect to the base year 1975, and therefore it is not such
to have seen major changes in the urbanization level.

The correlations two-by-two of the three parameters are shown in Fig. 2.22.

The correlation between the percentage of work force in agriculture and GDP/capita is clear even if not univoque. The same is apparent in the correlation between degree of urbanization and percentage of agriculture work force. These correlations indicate, even if in a non rigorous way, that the shift from a "closed" agricultural society to an "open" industrial society produces "wealth" and urbanization. Less neat it is instead the correlation between the degree of urbanization and GDP/capita: there are countries with the same GDP/capita with large differences in the degree of urbanization (Japan and West Europe). On the other hand countries like Italy have the same degree of urbanization of the average of CEE but with 60% of the GDP/capita.

a) Passenger transport correlation with macro-variables

To develope the "heuristic" correlations that relate the transport system to external macro-variables we start first plotting the average travelled distance (the transport "work" per inhabitant, passenger x km/inhabitant) respectively versus GDP/capita, % of urbanization, % of work force in agriculture. Fig. 2.23 shows a good degree of correlation. The "wealth", the degree of urbanization and of industrial development tend to increase mobility measured in term of average travelled distance per capita. The same type of correlation is obtained when looking at the average travelled distance in the car mode (see Fig. 2.24).

Among the West countries Japan is an exception, as was already noted when analysing in paragraph 2 the transport "work" structure. While from the point of view of the possession of the automobile per inhabitant, Japan is well in the West area trend of increasing number of car/capita with increasing GDP/capita, the average travelled distance is at variance with the world wide strongly correlated trend of increased mobility with increased number of car/capita (see Fig. 2.25).

The average distance travelled on collective mode of transportation (bus, train, air) shows a less clear correlation with the chosen macro-variables (see Fig. 2.26). In any case it appears clear that the growth in collective transportation vs. GDP/capita or urbanization is much less sharper than that of the car mode. If one separates the three main geo-political areas considered, West, East, LDC, one finds a better correlation at least between average travelled distance versus GDP/capita (with the exception of Japan which seems to follow the East and LDC correlation). To which extent the separated trends between East and West are an indication of different patterns of transport system evolution is matter of opinion. To support the idea it is

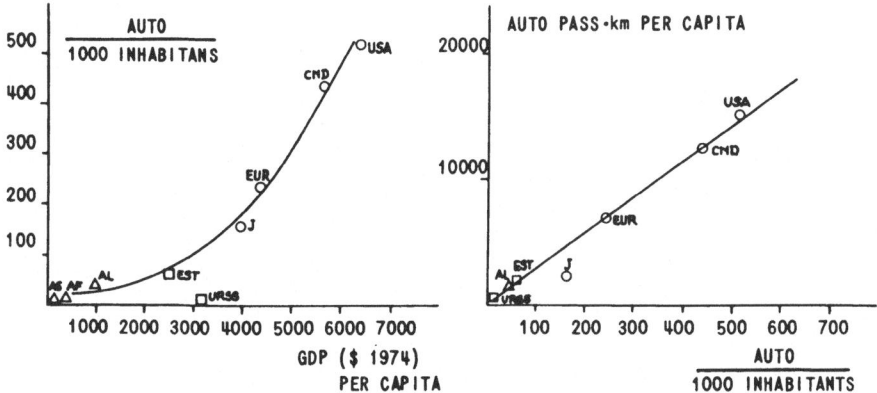

Fig. 2.25 - World Trends of Auto per Capita and Auto
Pass · Km per Capita

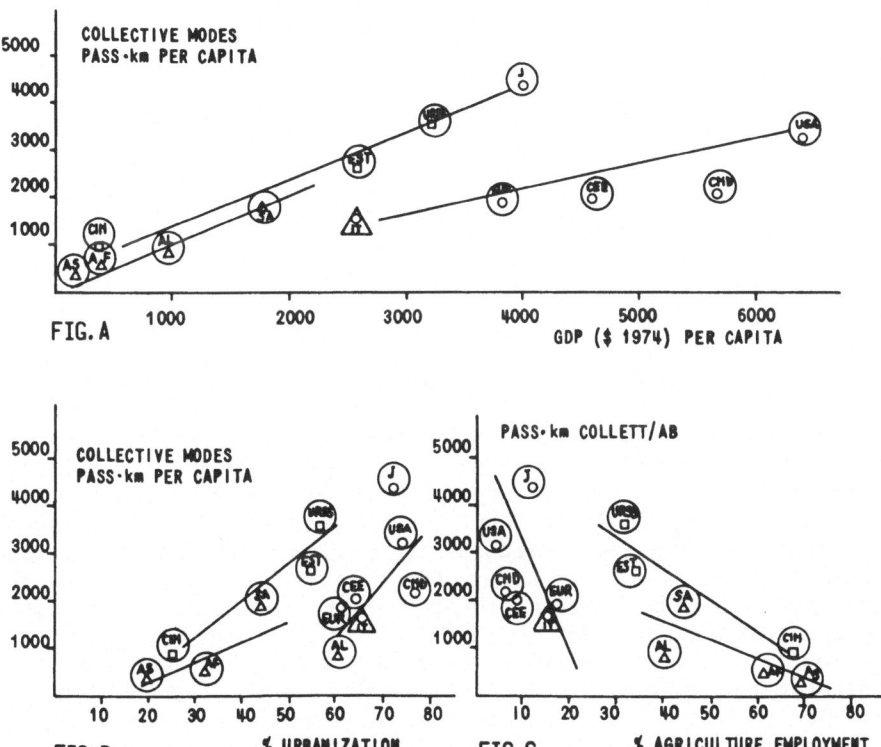

Fig. 2.26 - Collective Modes Passenger · Km per Capita Versus
Socio Economic Macro Variables

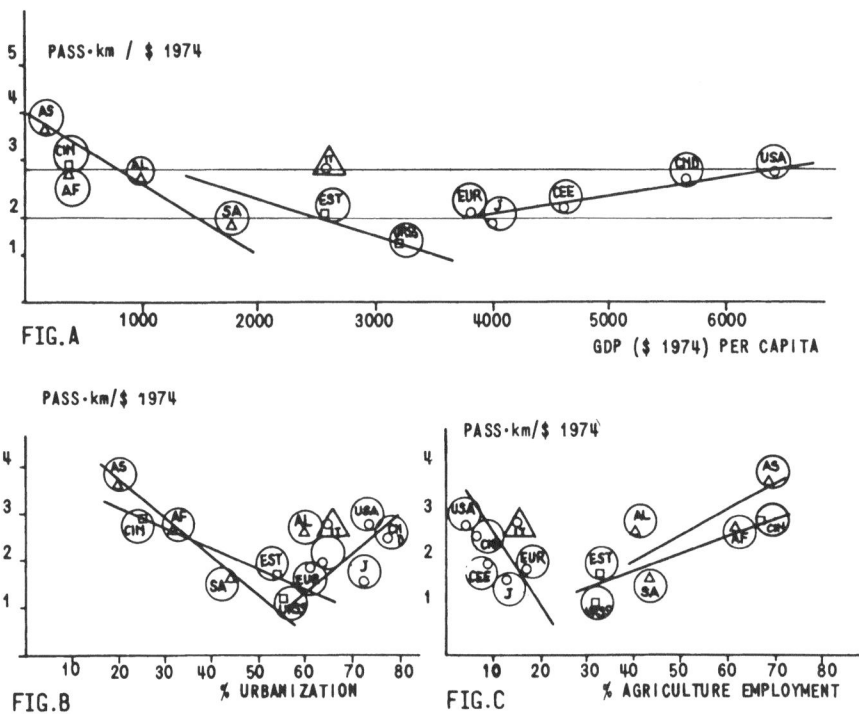

Fig. 2.27 - Total Passenger · Km per GDP Versus
Socio-Economic Macro Variables

interesting to note that the separate trends appear much more
clear in collective transportation for which central planning
should be more effective than with respect to the private mode.
    To look if other "heuristic" indications come to evidence
from changing the transport indicator to be correlated with the
external macrovariables, we have considered instead of the average
travelled distance, the transport "work" obtained per unit of GNP
(pass x km/$_{GDP}$), which could be considered either a rough
indicator of the "economic efficiency" of transport or the
tendency to dedicate a lower fraction of GNP to satisfy the
mobility needs. The results are plotted in Fig. 2.27 for the total
transport modes, and in Fig. 2.28 and 2.29 respectively for the
car and the collective mode. It is interesting to note that this
indicator has a very low variation with respect to the GDP/capita
(within a factor of 1.5, for a variation of a factor 6 of
GDP/capita). The plot as a function of degree of urbanization
clearly shows a V shaped correlation for total transport modes and
for the car mode (Figs. 2.27 - 2.28). The high value of the
indicator for an agricultural economy is a sign of reduced

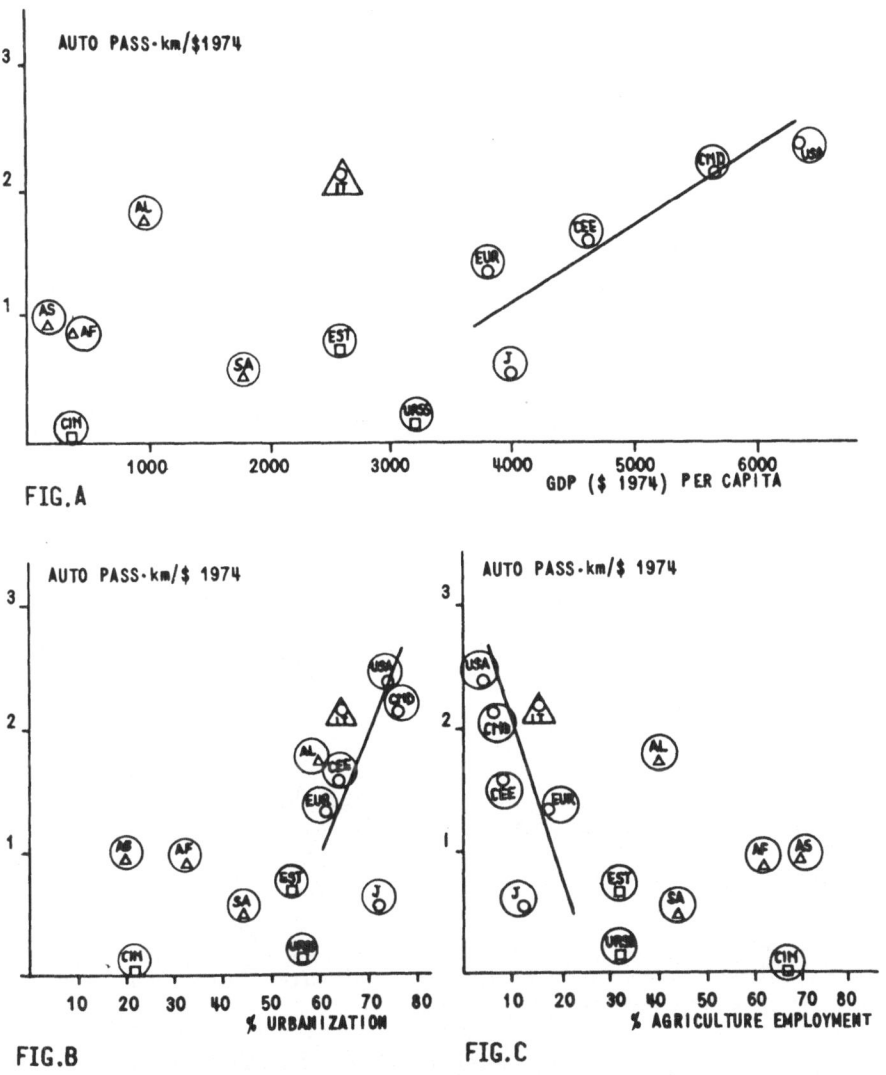

Fig. 2.28 - Auto Passenger·Km per GDP Versus
Socio-Economic Macro Variables

mobility and "economic" efficiency. The increased need of mobility
with increasing urbanization (refer to Fig. 2.24) is compensated
by an increasing transport economic "efficiency" (see Fig. 2.27).

The data on the collective modes (Fig. 2.29) if interpreted
as indicating "economic efficiency" show a lower efficiency for

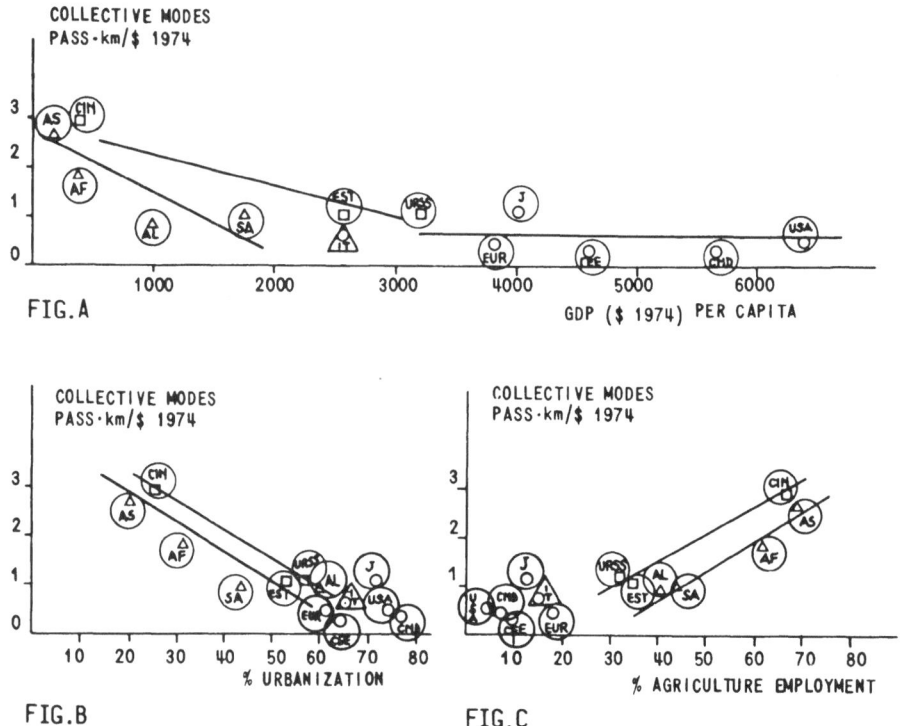

Fig. 2.29 - Collective Modes Passenger·Km per GDP Versus
Socio-Economic Macro Variables

the West area and with the increasing degree of urbanization. The
remarks above are not-conclusive and are here reported to
stimulate further research.

We will pass now to analyse the correlation taking the
energy intensity as an indicator of the transport system. Fig.
2.30 shows a clear correlation of kcal/pass x km with each one of
the three macro-variables. The plot suggests the consideration
that the "richer" is the society, with a higher degree of
urbanizazion and of industrialization, the higher is the tendency
to "waste" energy for transportation.

Less efficient transport is correlated to longer average
travelled distance (see Fig. 2.31 A). The car mode is the
principal responsible: on one side it provides the opportunity for
higher mobility (longer travelled distance), on the other side it

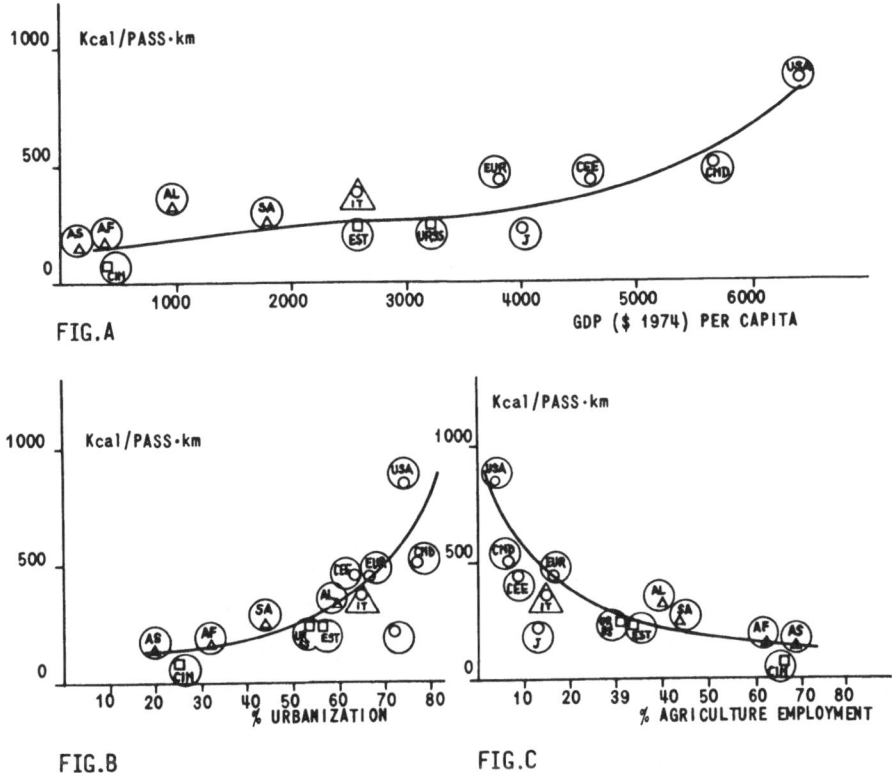

Fig. 2.30 - Energy Intensity for Passenger Transport Versus
Socio-Economic Macro Variables

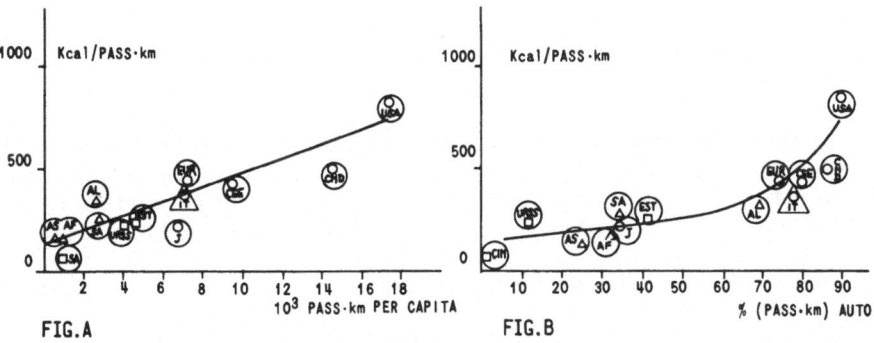

Fig. 2.31 - Energy Intensity for Passenger Transport Versus
Mobility and Auto USAGE

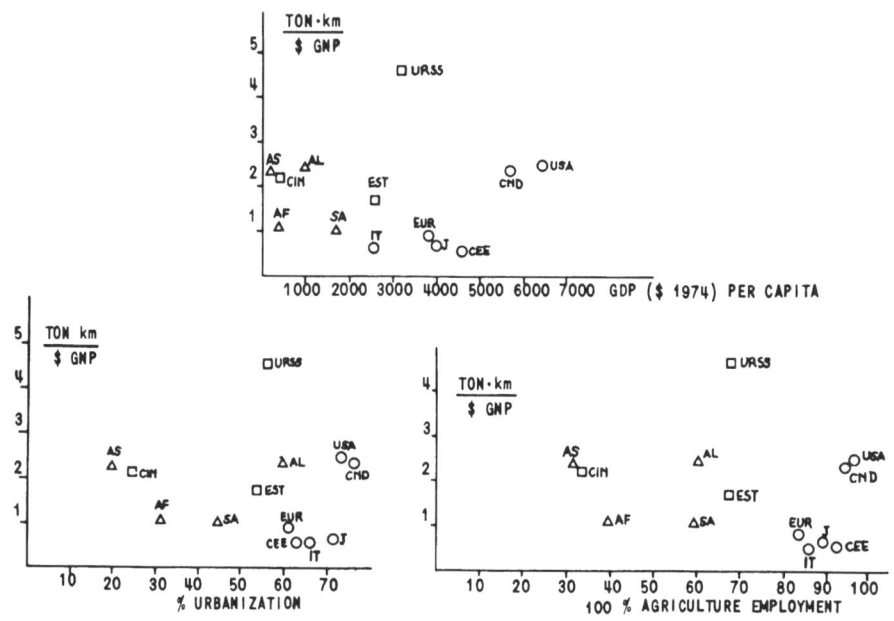

Fig. 2.32 - Total Freight Ton·Km per GDP Versus
Socio-Economic Macro Variables

reduces the average energy efficiency (see Fig. 2.31 B).

b) Goods transport correlation with macro-variables

We attempted to correlate the freight transport "work" and
energy intensity to the same macro-variables used for the case of
passenger transport. No clear correlations come out of the
analysis, possibly due to the reason that the chosen variables are
less adequate to take into consideration the goods transport
system's interrelations with external systems.

Other macro-variables have been preliminary investigated such
as the population density (inhabitants/km²), the ratio between the
lenght of road/rail networks, the density of road network per
inhabitant or per km², without finding clearer correlations.

We limit ourselves here to present some of the partial
results obtained without attempting interpretations. The analysis
has served to give indications on how to proceed further, and we
hope that the challenge will be taken in the future.

In   Fig. 2.32 the transport   "work" per   GDP ("economic

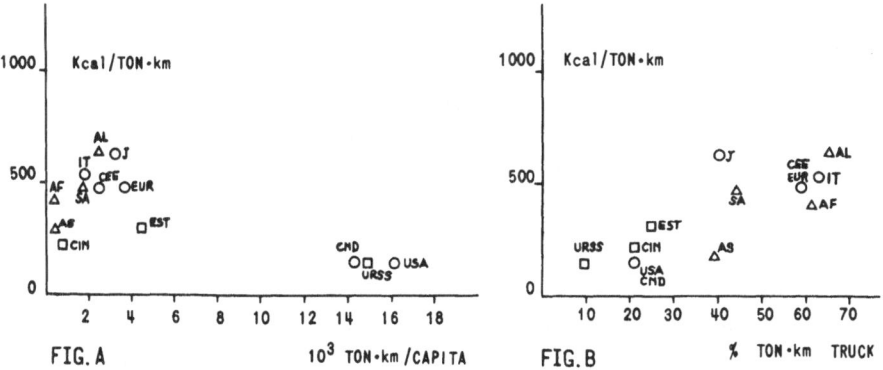

Fig. 2.33 - Energy Efficiency of Freight Transportation
Versus Ton·Km

efficiency") is plotted versus GDP/capita, % urbanization, % agricultural work force. By separating the West countries one can show a certain degree of correlation with improved "economic efficiency" (or increased need of transportation) with increased GDP. The case of URSS comes into evidence has having the largest value of goods transport "work"/GDP.

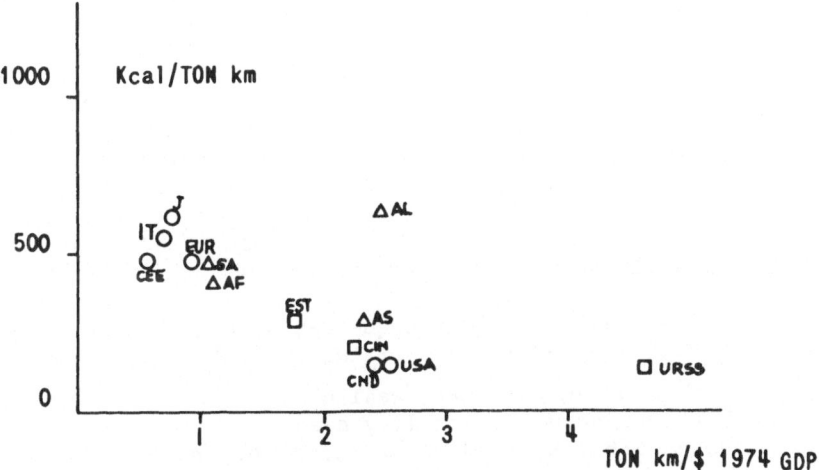

Fig. 2.34 - Energy Efficiency of Freight Transportation
Versus Ton· per GDP

When analysing transport energy intensity, one detects a
different behaviour with respect to passenger transportation:
while Fig. 2.31 A shows an increasing energy intensity with
increasing passenger transport "work" per capita, Fig. 2.33 A
shows a decreasing trend of goods transport energy intensity
versus goods transport "work"/capita. Nevertheless, when the goods
mobility need is satisfied by increasing the share of road
transport mode, the energy intensity increases (see Fig. 2.33 B).

Fig. 2.34 seems to suggest that transport energy efficiency
and "economic efficiency" tend to increase in a direct
proportional way.

c) A model for passenger transport evaluation

We will attempt to extract from the correlations illustrated
above a model of evolution of the transport system in parallel
with the evolution of the external society system only for the
case of passenger transport.

In the introduction to this paragraph we stated that
simulation models should be validated with experiments, and that
different region or country cases represent the equivalent of
different experiments, with varied internal parameters. The
correlations obtained were simple ways to reach a "perspective
view" of the set of experiments.

We should now try the much more ambitious and less soundly
based exercise to make use of the "astronomer paradigm" described
in the introduction. According to this paradigm we could consider
the different region and country cases as indicating different
stages of the evolution of the transport system. The ambition is
now here, not only to validate simulation models, but to grasp the
intrinsic dynamics of the system in order to predict future state
for the transport system in a region or country. Since the
prediction is based on the intrinsic dynamics of the system
derived by its historical development, changes with respect to the
predicted future state are possible, but the more the thought-of
state of the system is different from its "natural" development
the more difficult will be to obtain such a change and higher the
risk that the well thought and planned actions be ineffective.

The model is condensed in the sequences of Fig. 2.35 and Fig.
2.36. In Fig. 2.35 the graphs illustrates the following sequence:

- a) industrialization produces "wealth"
- b) "wealth" increases the mobility needs
- c) increased mobility need is satisfied by an increased use
  of the car mode
- d) the increasing use of the car mode worsens the transport
  energy efficiency.

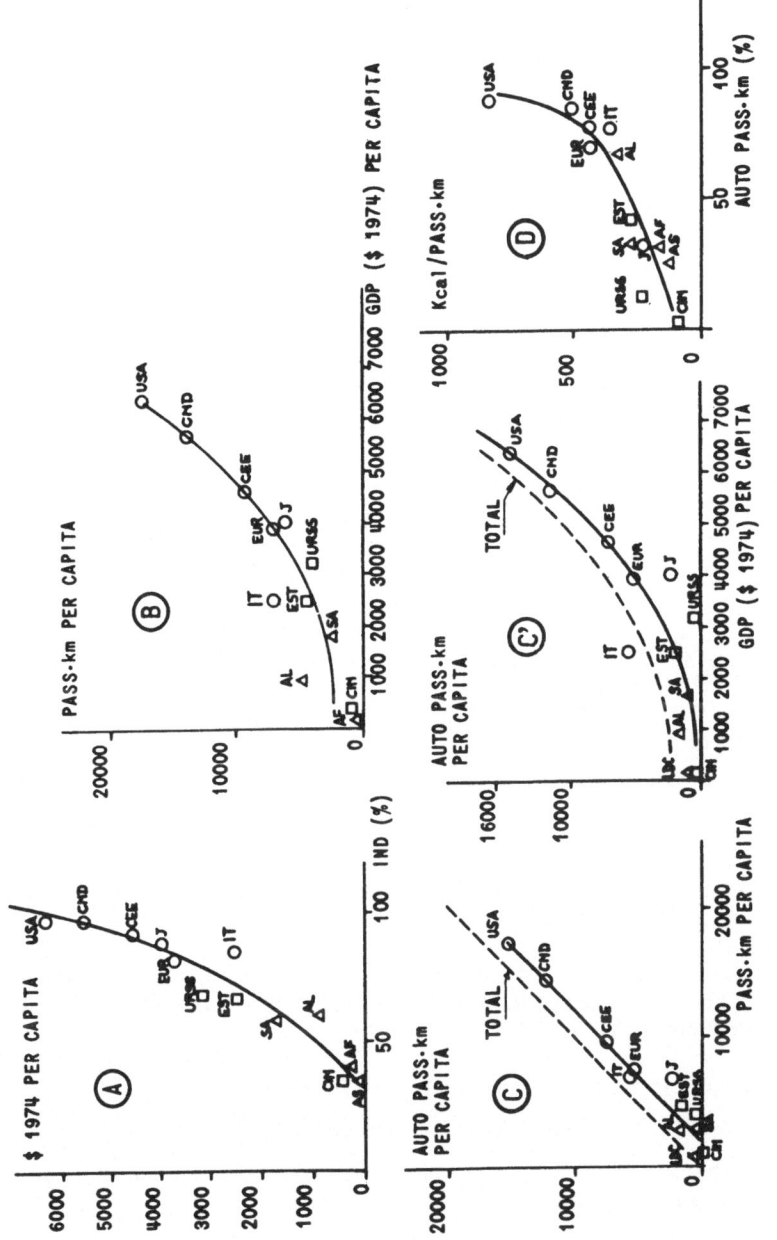

Fig. 2.35 - Passenger Transport: Evolution Patterns

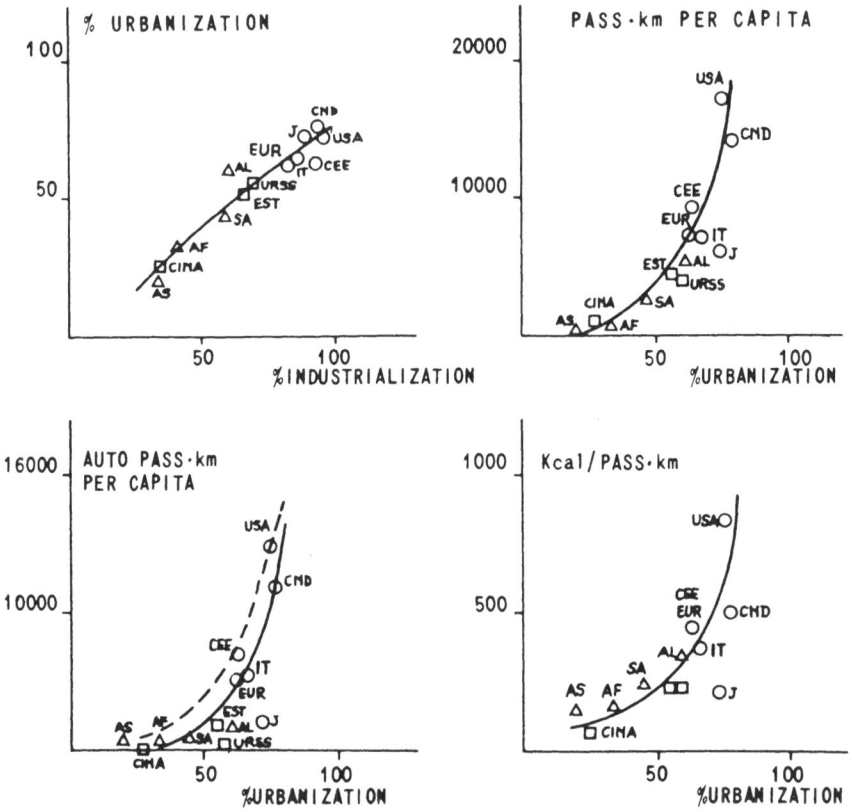

Fig. 2.36 - Passenger Transport: Alternative Evolution Patterns

The graph c') in Fig. 2.35 is a confirmation that to satisfy an increasing mobility need, the larger is the per capita income, the higher is the use of automobile.

In Fig. 2.36 an alternative, but not independent sequence, based on the human settlement characteristics more than on wealth, is illustrated:

a) industrialization induces urbanization

b) urbanization requires increased mobility needs

c) urbanization favours the use of car mode to satisfy the mobility needs

d) as a consequence the energy intensity in transportation increases.

Considering the degree of industrialization or of urbanization as exogeneus variables, whose future value can be predicted by analysing historical time series, it is easy to understand the use of the model to predict the corresponding future state of the transport system. With the help of the analysis of the transport system structure in terms both of transport "work" and transport energy illustrated in parag. 2.1 and 2.2, one could make further steps in detailing the model. This attempt is not yet done and is left for future work.

As already said the purpose of this presentation, with all its risks to be judged too superficial, is that of thoughts provoking. A trivial suggestion to the decision makers coming from the model is that actions leading to improve the efficiency within each mode have a higher probability of being effective with respect to mode shifting. According to the illustrated sequence, mode shifting towards reduced car mode use, should be interpreted as if the total system goes "backwards" in the industrialization or in the urbanization level. Without implying that the model illustrate a cause-effect relationship, the remarks are here given just for the purpose of thought provoking.

3. PROSPECTIVE FOR ENERGY SAVINGS IN TRANSPORTATION

The "structural" analysis of the transport system illustrated in the previous chapter has brought to evidence the variety of existing forms which we tried to reduce to a common basis with the help of hystorical, geographic, socio-political considerations. In particular there is evidence of a regular trend of changing in the percentage weight of the different modes of transport when passing from the West, to LDC, to East countries. For passenger transport the data show an inverse correlation between the transport energy efficiency and the diffusion of the car mode.

The prospective for energy saving in transportation passes through the potentiality of intervention for:

1 - improving the intrinsic efficiency in each transport mode, ranking the car at top priority
2 - shifting mobility towards more efficient modes, which for passenger transport means reducing the percentage weight of the car mode
3 - reducing total mobility.

The potential interventions have been listed in decreasing order of credibility, especially if one put the condition that the intervention on the transportation system should not have major effect on the socio-economic system.
Continuing the remarks sketched at the end of the preceding

chapter, the analysis so far done gives some indication on the policy for actions on energy conservation (refer to Fig. 2.35):

a) the non-linearity of the correlation between the energy intensity (kcal/pass x km) vs. the percentage of the share of the car mode indicates that higher conservation effects are obtainable in the regions and countries of higher mobility (with the exception of Japan).

b) a reduction of the energy intensity obtained by shifting to lower car mode use, has to be attentively analysed for its effects on the total mobility and the GDP/capita (remember the importance of the contribution of the car industries to the GDP and the need for huge investments for collective transport infrastructure which should be diverted from other uses).

Therefore the reduction of the average intensity by acting on the intrinsic energy efficiency of the car with technological innovations seems to have the highest potential on one side (remark a), and the least impact on the socio-economic structure (remark b). Later on, in this chapter, referring to the results of a study performed with the help of a simulation model for the USA case, we will find a confirmation of the higher potentiality, for energy saving, of technology innovation in car with respect to modal shifting. Before illustrating this case study, we will briefly comment on the potentiality and difficulties of some types of actions proposed to increase energy conservation in transportation. The discussion will confirm the somewhat trivial remarks that the potentiality for energy conservation is tied with the possibility of performing many small steps in different directions, even if the result to be expected is less than the sum of each of them (due to non-compatibility, secondary effects, etc.).

The types of intervention to conserve energy can be classified in five categories[6] (see Fig. 3.1). To the three types of action already indicated at the beginning of this paragraphs, it is added:

4 - increase the load factor (or occupancy factor): the intensity of use of vehicles and infrastructures is increased and therefore also the productivity of the transport system.

5 - improving operations and management: it is a matter of actions often amenable to quick interventions, leading to optimise, from an energy point of view, the management of the mobility demand/offer by improving the operation characteristics of the mode and of the use patterns.

Many of the actions listed in Fig. 3.1 have a direct impact

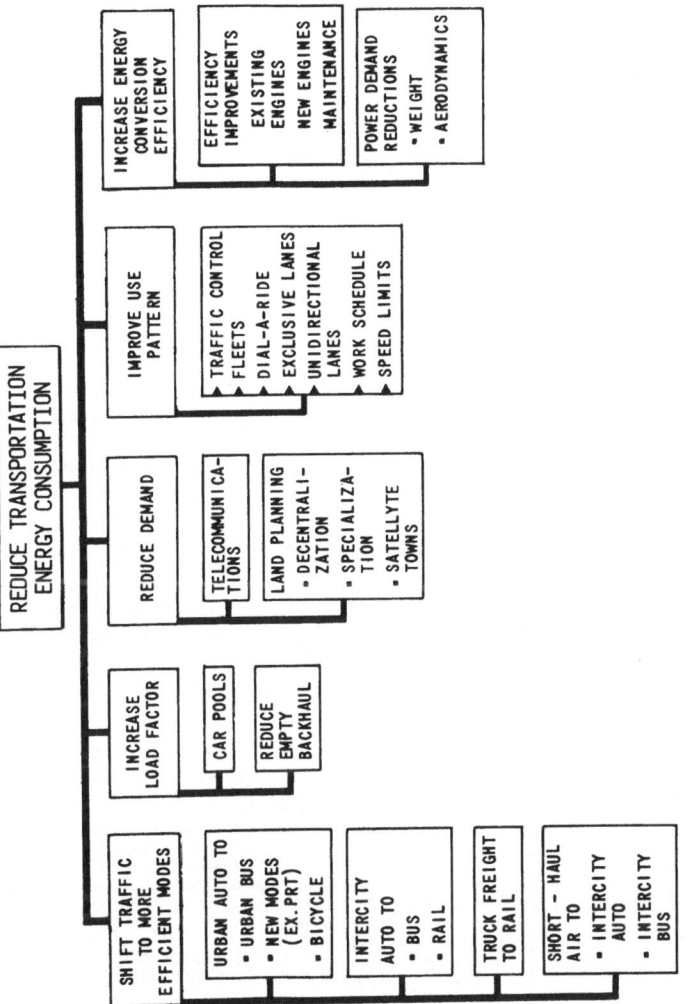

Fig. 3.1 - Alternatives for Reducing Transportation Energy Consumption

on the transport system. Others (f.i. land planning, staggered
working hours), by acting on external systems, have only an
indirect impact on the transport system. The data or estimated
values that we will present below are taken mainly from literature
dealing with the USA case, and therefore difficult to transfer to
other cases. Whenever possible, together with the potential energy
saving an estimate of time necessary to implement the proposed
actions and the impact on noxious emissions are also given.

3.1 Review of estimated energy savings in transport

a) Modal shifts

    In Fig. 3.2 the potential savings by specific modal shifts

	IMPROVED EFFICIENCY %	FUEL SAVINGS AS % OF TOTAL DIRECT TRANSIT ENERGY	YEAR TO OBTAIN MAXIMUM BENEFITS	EMISSIONS BENEFITS	COMMENTS	REF.
URBAN AUTO TO:						
• URBAN BUS		0.7 - 0.8	15	MINOR		6
		2.7 1.9			} 50% OF COMMUTERS TO DEDICATED BUS SERVICE	26
		(8.6-11.6)*			} *% OF TOTAL ENERGY USE IN SURFACE PASSENGER TRANSPORT IN U.K.	22
• BICYCLE		0.5 - 0.7	10	SIGNI-FICANT		6
		1.6			} 50% OF THE PEOPLE TO WALK OR BIKE UP TO 2-5 MILES, INSTEAD OF	27
		1.8			DRIVING	26
INTERCITY AUTO TO:						
• INTERCITY BUS		0.2	15	MINOR		6
		1.5 2.5			} 50% OF INTERCITY AUTO PASSENGERS	27 26
• INTERCITY RAIL		0.01	15	MINOR		6
		1.5 1.5			} 50% OF INTERCITY AUTO PASSENGERS	27 26
TRUCK FREIGHT TO RAIL.		0.4 - 1.4	15	MINOR		6
		3.4			} 50% OF INTERCITY TRUCKING	27
SHORT-HAUL AIR TO:						
• INTERCITY AUTO		0.2 - 0.4	5	MINOR		6
• INTERCITY BUS		0.2 - 0.4	15	MINOR		6
		0.29 0.7			} 50% OF SHORT-HAUL AIR PASSENGER	27 26

Fig. 3.2 - Modal Shifts

are indicated for the case of USA, with the exception of reference[22] which is referring to U.K. Concerning passenger shift from car to collective mode, besides the general remarks given above, the following ones are to be remembered:
. the increase of the ground area occupied by the infrastructures and their insertion in an already congestioned environment will require long time of implementation, possibly of the same order of that foreseen for a new generation of cars (10-15 years)
. the huge investement capital required could exceedingly increase the cost per unit of "transport work" with respect to the energy saving benefits
. taking into consideration that the indirect energy (at least in USA) used for transport reaches 42% of the direct energy use (see Fig. 1.9) with peak of 100% for bus and 117% for rails, every modal shifting suggested action requires to perform an attentive energy balance to assess its potential energy saving.

The data in Fig. 3.2 indicates that, apart from the variance in the estimated values, the global effect of modal shifting in energy conservation is not larger than few percent of the total energy used in transport. As a marginal remarks we note that one should have expected a higher beneficial effects when shifting to bicycles or walking. In practice however, even if 50% of the people uses bicycle, the contribution to the total travelled distance is very small, because each bicycle trip is within 2-5 miles.

b) <u>Increasing the load factor</u>

The load factor (or occupancy factor), together with the propulsion efficiency, the vehicle drag and rolling resistance, is one of the variables that more influences the specific energy consumption. In the passenger transport system its value varies from 0.2 to 0.8. In the car mode the occupancy is low in the commuter trips and high in the weekends extraurban trips or during holidays (see Fig. 3.3). In the collective modes the occupancy factor is rather high in peak hours, but it dives to very low levels in other periods of the day.

For the goods transport case the load factor remains within the same range as the case of passenger transport. In fact, considering also the empty return trip, the truck average load factor in USA has a value of approx. 0.5^{27}.

In Fig. 3.4 some estimates of energy saving are listed, which are obtainable through an increase of the load factor. Car-pooling shows good potentiality with low implementing time: it is already well diffused in USA, even if no statistical data are available to confirm the estimates given in Fig. 3.4.

EARNING A LIVING	
HOME-TO-WORK RELATED BUSINESS	1.4 PASSENGERS 1.6 PASSENGERS
FAMILY BUSINESS	

SHOPPING
    DENTAL/MEDICAL
    OTHER
	2.1 PASSENGERS
	2.0 PASSENGERS
	1.9 PASSENGERS

CIVIL, EDUCATIONAL, RELIGIOUS	2.5 PASSENGERS
SOCIAL AND RECREATIONAL	

VISITING FRIENDS , etc.
    PLEASURE DRIVING
    VACATIONS
    OTHER
	2.3 PASSENGERS
	2.7 PASSENGERS
	3.3 PASSENGERS
	2.6 PASSENGERS

AVERAGE, ALL PURPOSE	1.9 PASSENGERS

Fig. 3.3 - Passenger per Trip Statistics[6]

c) Reduction of transport mobility

Modern cities and life style are structured around road transport. This means that while only 25-35% of world population lives in urban areas, the related energy use in transport has a much higher percentage weight.

With a view to reduce in a significant way that energy use, one should restructure the town plans and the passenger and goods distribution system to reduce the need of diffusive transport, the distance of the average travel, the mobility demand in general.

That type of strategy even if pursued with determination requires half a century to be effective. It will, furthermore, be very difficult to implement without destroying the old urban plan with all the related implications on the historical, artistical, landscape assets.

	IMPROVED EFFICIENCY %	FUEL SAVINGS AS % OF TOTAL DIRECT TRANSP. ENERGY	YEAR TO OBTAIN MAXIMUM BENEFITS	EMISSIONS BENEFITS	COMMENTS	REF
CARPOOLS (WORK TRIPS ONLY)						
• at 47% PARTICIPATION		1.5 - 1.9	1	- 14 %		6
• at 70% PARTICIPATION		3.8 - 4.9	1	- 30 %		6
		>2.4			} 50% OF URBAN	26
		3.1			} COMMUTERS	27
AIR PASSENGER LOAD FACTOR IMPROVEMENT (FROM 50% to 70%)		2.3 - 3.7	4	PROPORTIO- NAL TO FUEL SAVING		6
DOMESTIC FREIGHT TRUCK:						
• LOAD FACTOR INCREMENT		1.8 - 2.1	5	MINOR		6
TRUCK CAPACITY (DOUBLE)	5.8					19

Fig. 3.4 - Increasing Load Factor

One suggested means to reduce the mobility need is the diffusion of telecommunications. On its direct impact in reducing some types of travel (business trips, meetings, etc.) opinions are in disagreement. As a matter of fact we have seen in the past both the increase in telecommunications (for instance intercity phone call by direct dialing) and mobility.

To substitute travel demand with telecommunications depends not only on cost but also on psychological attitude. Nevertheless the potential impact of telecommunications is large in energy conservation when effective in reducing mobility needs. In Fig. 3.5 the data estimated in a study for the U.K. case, are reported[22]. In the same figure are also listed the effects of energy conservation, by reducing, no matter why, the private mode mobility demand.

d) Improving transport operation and management

In Fig. 3.6 a list of the effect of energy saving by actions falling into this category is given. Each single action does not result in high saving. Neverthless many of the listed actions can

	FUEL SAVINGS AS % OF TOTAL ENERGY USE IN SURFACE PASSENGER TRANSPORT IN U.K.	REF.
SUBSTITUTION OF TELECOMMUNICATIONS FOR 50% OF BUSINESS TRAVEL BY PRIVATE VEHICLE	6.2 - 9.9	22
REDUCTION BY 50% OF SHOPPING AND PERSONAL BUSINESS TRAVEL BY PRIVATE VEHICLE	3.2 - 5.2	22
REDUCTION BY 50% OF OTHER PERSONAL AND SOCIAL TRAVEL BY PRIVATE VEHICLE	16.6 - 18.7	22

NET SAVINGS AS PERCENTAGE OF TOTAL SURFACE TRANSPORT
ENERGY USE IN U.K.

Fig. 3.5 - Reduce Demand

be added and therefore reach value of approximately 5-6% of the
total transport energy use (the data refer to USA). In chapter 4
we will illustrate a specific intervention planned for the traffic
control of the town of Torino, in Italy, with an innovative
system, and the projected energy saving.

Speed limits, at least at the level already implemented in
many countries, had a high impacts more on the safety aspects,
reducing number of accident and casualties, than in fuel savings.
This is so at least to the extent that the reduction of speed
limit will not have an impact in down-rating the power installed
on car, accepting also car's lower acceleration performances.

For some of the listed type of actions in Fig. 3.6 no
estimates are available. We reproduce here for the case of
"jitneys" ("public" car) an excerpt from ref. 30: "If buses
costing $ 40,000 each can carry an average of 1,000 passenger per
day, eight public cars costing $ 5,000 each can perform the same
task by carrying 125 persons per day. Compared to a private car,
which might carry 10 people per day (5 trips with an average of 2

	IMPROVED EFFICIENCY %	FUEL SAVINGS AS % OF TOTAL DIRECT TRANSP. ENERGY	YEAR TO OBTAIN MAXIMUM BENEFITS	EMISSIONS BENEFITS	COMMENTS	REF.
CONTROL OF URBAN TRAFFIC FLOW		0.4 - 0.7	10	MINOR		6
TRAFFIC RESTRAINT IN URBAN CENTRES		$(<1)^*$			* % OF TOTAL ENERGY USE IN SURFACE PASSENGER TRANSPORT IN U.K.	22
PROVISION OF CYCLE FACILITIES AND PEDESTRIANIZATION		$(<1)^*$				
ELIMINATE 50% OF URBAN CONGESTION		1.1 1.6				27 26
URBAN FLEETS OF PUBLIC CAR OR "JITNEYS"						
DIAL-A-RIDE BUS SYSTEM						
EXCLUSIVE LANES						
UNIDERECTIONAL LANES						
WORK SCHEDULE						
SPEED LIMITS:						
• 55 mph (HIGHWAY)		0.9 - 1.2	1	MINOR		6
• 50 mph (HIGHWAY)		2.9 2.8				27 26
• AIR PASSENGER: CRUISE SPEED REDUCTION		0.2 - 0.4	0	PROPORTIONAL TO FUEL SAVING		6
• DOMESTIC FREIGHT TRUCK		0.5 - 0.6	1	MINOR		6

Fig. 3.6 - Improve Use Patterns

people) the public car can do 15 times as much work. The service provided by 100,000 private cars, could be supplied by 6,600 public cars"

The potential energy saving of the "Dial a bus" system finds its optimum for distance which are intermediate between short trip with multiple destination (well covered by taxis) and longer door-to-door destination usually covered by private car[31] (see Fig. 3.7).

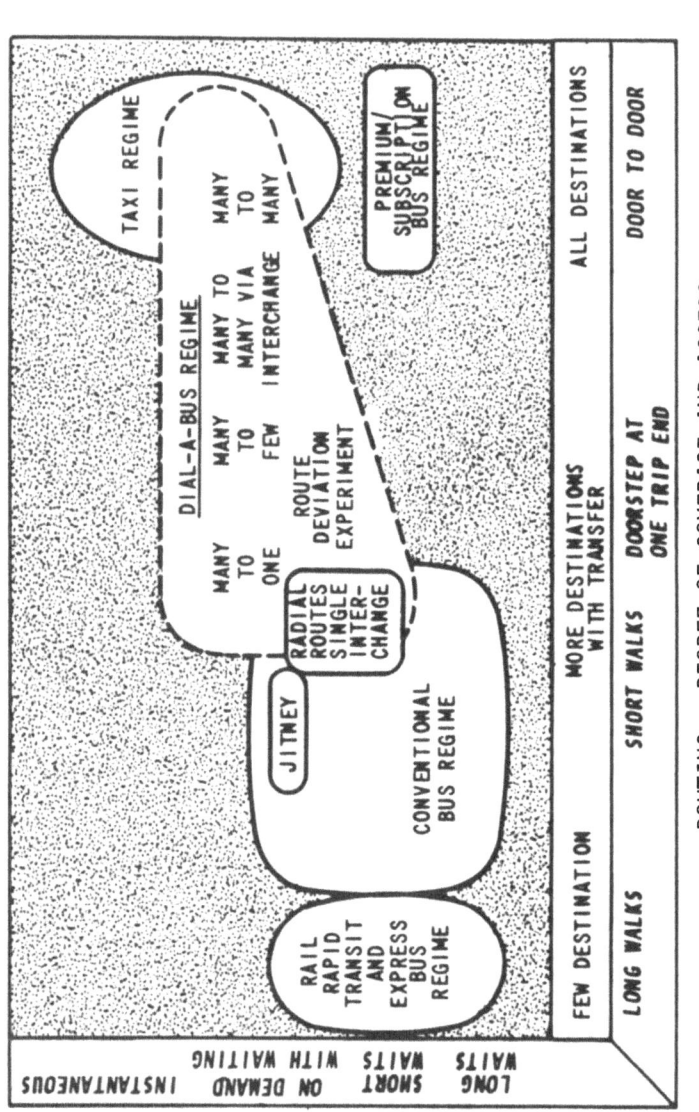

Fig. 3.7 - Conceptual Framework for Dial - A - Bus Services[31]

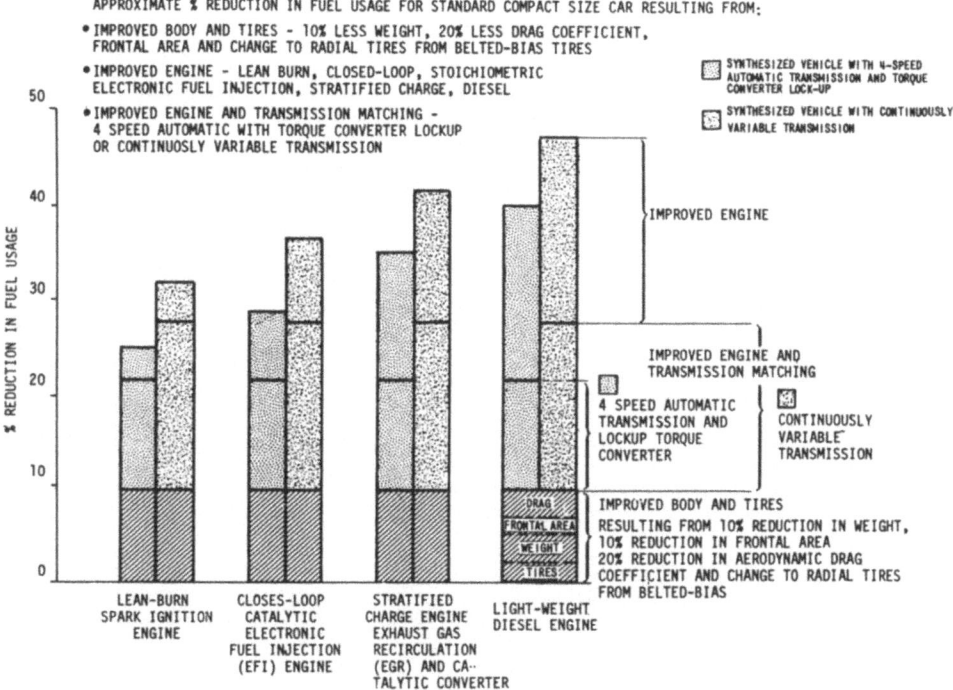

Fig. 3.8 - Reduction in Fuel Usage From Proposed Improvements[28]

Few cases of Dial-a-bus systems are already in operation in USA, Sweden, France, Germany, but no energy saving analysis is available. From the economic point of view it is known that the service is not rentable and requires substantial public financial aids.

e) Increasing vehicle energy conversion

The case will be dealt with at a greater length in the next chapter.

The savings potential of initiatives falling within this category is much larger than that of any of the other approaches, because motor vehicles - and notably the cars - now consume the major share of transportation energy and operate at an average efficiency below the state of the art.

It is important to point out that:

. gains in car efficiency can have relatively little impact on the "auto usage/GDP per capita" correlation (see Fig. 2.35);
. in general, improved vehicle efficiency has no significant

ENGINE	WASTE HEAT RECOVERY OPTION			
	REGENERATION	TURBO-CHARGING	TURBOCOMPOUNDING (100% EFFICIENT POWER TURBINE)	RANKINE ENGINE COMPOUNDING
DIESEL	8	6	11	26
ADIABATIC DIESEL	25	6	34	40
SPARK IGNITION	25	5	13	22
GAS TURBINE	--	--	--	37
STIRLING	--	--	--	7

Fig. 3.9 - Maximum Theoretical Fuel Economy Improvement in Truck Engines

impact on the performance quality, so that it is not required to change the behaviour of either consumers or institutions - except for vehicle manufacturers, of course;

. implementing improvements in vehicle efficiency can reduce total cost of transport; to the extent that it does, it is favored by market forces.

The major disadvantage of this measure is the relatively long implementation time - on the order of 15-20 years - required to realize its full benefit (time lag due to the time needed to industrialize a new technology and to fully change the car park).

Several studies have been made on the efficiency increase for all transportation modes, and many other are in course. Due to different ways of approaching the subject, the results are often spreading over a broad range of estimates, so that it is impossible to have a clear picture of what can be achievable in the future. However one result is evident and encouraging: the potentialities of this intervention are very large and could yield in a near future - within the end of the century - transportation energy savings on the order of about 30-40% on a world scale.

We show here only a few examples of proposed improvements leading to substantial reduction in fuel usage.

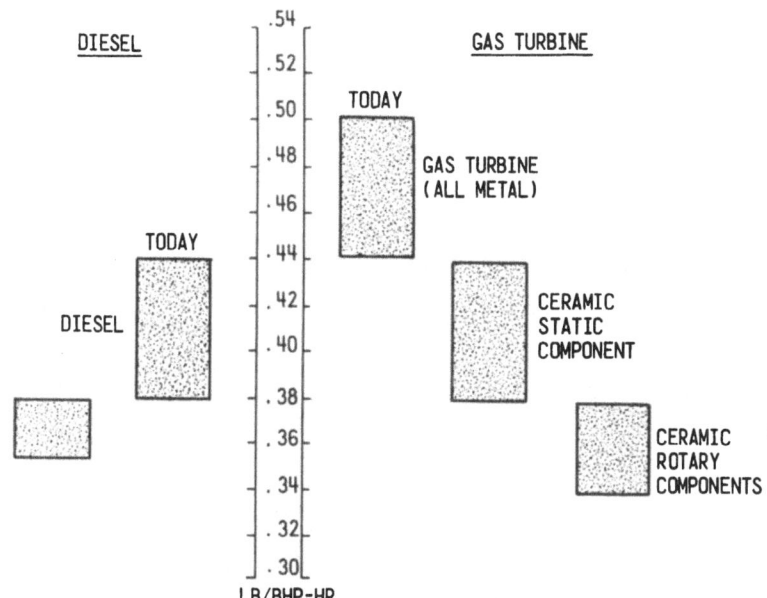

Fig. 3.10 - Comparison of Current and Projected Diesel and
          Gas Turbine Specific fuel Consumption
(G.M. Thur "Alternate fuel gas turbine engine head start
commercialization program," presented at Automotive Technology Development Contractor Coordination Meeting
October 23-25, 1978, Dearborn, Mich.)

Fig. 3.8 summarizes the fuel economy improvements for the
case of compact and standard size american cars[28]. Reductions in
weight, frontal area, aerodynamic drag coefficient and change to
radial tires account for a 10% fuel saving; improved engine and
transmission matching could yield another 10-15%; lastly, with
engine improvements as lean-burn, closed-loop, stoichiometric
electronic fuel injection, stratified charge, Diesel, one could
reach a total 50% reduction in fuel usage.

Fig. 3.9 summarizes the maximum theoretical fuel economy
coming from the waste heat recovery option for truck engines[32];
here too the expected results show noticeable benefits.

A comparison between Diesel and gas Turbine engines is
exhibited in Fig. 3.10
Before closing this paragraph let's report some results of

theoretical and experimental researches carried out in Italy by Fiat and other companies and institutions under a government project:[33]

. aerodynamic improvements could reduce the national car fuel consumption up to 5% and the national truck fuel consumption up to 6%;

. engine-and fuel partialization could reach energy savings for vehicle of about 20-30%;

. continuous transmission with optimal engine control could give up to 25% fuel consumption;

. stratified charge engine could improve efficiency between 13% - 16%.

Of course, the indicated saving should not be considered as completely additive.

3.2 A case study in transport energy conservation

To exemplify the complex mechanism of interaction among technology choices, transport system, economy in its complex, and the quality of the environment, we report here the results of a study performed under contract of the U.S. Dep. of Energy (see ref. 24-25). The study considers alternative scenarios for the period 1977-2025.

The complex system interactions have been studied by means of a set of mathematical models, denominated TECNET (Transportation Energies Conservation Network) and developed by R.M. Doggett, R. Meyers, M. Heller of Internation Research & Technology Corp.

The structure of the model is given in Fig. 3.11 The "core" of the system is a model which supply the forecasts of GNP, of its components, of the industry production (in $) for each of the 185 sectors in which the USA economy has been subdivided. The other modules of the TECNET system serve the purpose to forecast the transport "work" (passenger and goods) subdivided among the different transport modes, to evaluate the related energy use (both direct and indirect), and the noxious emissions.

a) TECNET general description

We will report here directly between quote marks excerpts from the paper of ref....

"   TECNET has been adapted from an extension of the "Environmental Protection Agency's Strategic Environmental

IMPACT FORECASTING OF INTERVENTIONS
ON TRANSPORT SYSTEMS
(1977 - 2000 - 2025)   (U.S.A.)

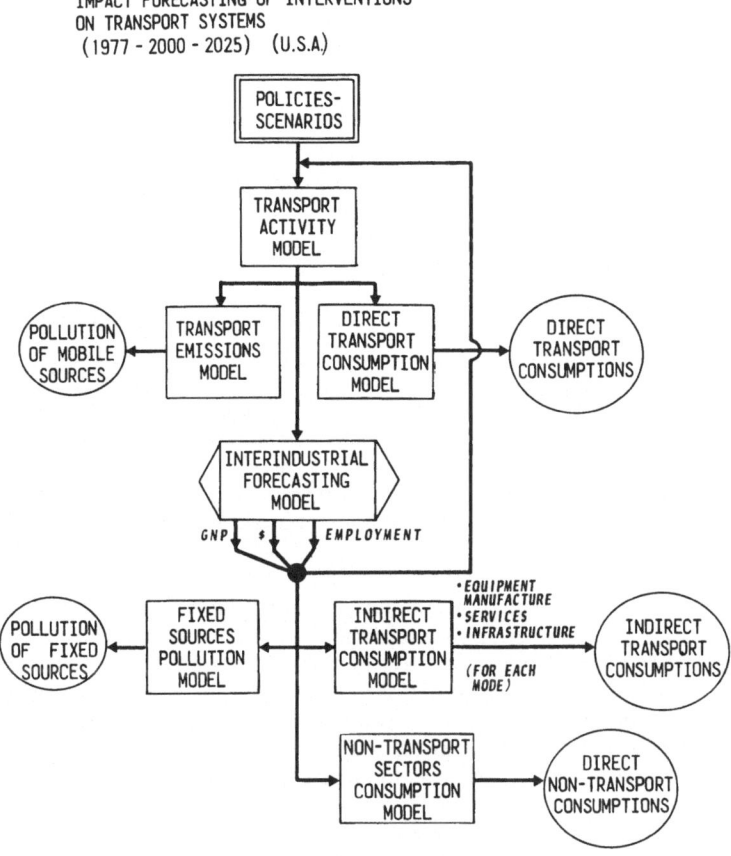

Fig. 3.11 - TECNET Model[24-25]

"Assessment System (SEAS)**. The SEAS model was originally
"designed to estimate the environmental impacts and pollution
"abatement costs that might result from EPA regulations out to the

----

"** For a complete description of SEAS, see Trading Off
"Environment, Economics, and Energy. A Case Study of EPA's
"Strategic Environmental Assessment System, by Peter House,
"Lexington Books, Lexington, Mass., 1977."

"year 1985. SEAS was later extended by Resources for the Future to
"estimate the health effects, resource consumption patterns and
"environmental degradation implied by several alternative futures
"out to the year 2025. Much of the Resource for the Future work
"has been incorporated into the structure of TECNET. However, a
"considerable amount of additional detail has been added to
"explicitly analyze the transportation sector. These additions
"include a preprocessor, through which the transportation
"researcher can specify a host of variables to simulate
"alternative transportation policies; a comprehensive
"transportation module that estimates activity levels (i.e.
"passenger miles, ton miles, and vehicle miles traveled), energy
"use, and pollution emissions by seven modes of travel; and a set
"of algorithms for estimating indirect energy consumption by the
"transportation sector.
"   The TECNET system provides some unique capabilities for
"analyzing transportation policy. In addition to providing
"estimates of direct energy use (i.e. energy consumed during the
"operation of vehicles), the model provides estimates of
"energy used indirectly in support of transportation activities as
"mentioned briefly above. Indirect energy use is estimated for
"each mode (auto, truck bus, air, water, rail, and pipeline) and
"is further disaggregated into three basic categories within each
"mode:

"   - Transportation equipment manufacture. This category
"includes all of the energy consumed in the process of producing
"finished vehicles from raw materials. It includes the coal used
"to make steel for use in automobiles, and the electricity used by
"metal stamping machines and along assembly lines. Also included
"in this category are estimates of energy embodied in the capital
"goods required to manufacture transportation equipment.

"   - Transportation services. This category includes energy
"consumed in support activities for transportation services, such
"as energy used by insurance and repair industries, and warehouse
"energy use. Also included in this category are estimates of
"energy embodied in the capital equipment used in support of
"transportation activities, such as airline baggage handling
"equipment.

"   - Transportation infrastructure. This category includes the
"energy used in the construction and maintenance of highways,
"railroads, airports, dock facilities, and pipelines.

"   TECNET also provides detailed estimates of the economic
"impacts of alternative policies affecting direct and indirect
"energy use by the transportation sector. These estimates are
"provided through manipulations of TECNET's macroeconomic

"forecasting system, INFORUM*. INFORUM generates forecasts of the
"dollar output and employment by each of 185 agricultural, mining
"manufacturing, and service industries. Estimates of aggregate
"economic statistics - Gross National Product and its components
"are also generated in INFORUM.

"   In addition, TECNET provides estimates of both mobile source
"and stationary source pollution emissions. Mobile source
"emissions are estimated for each of six transportation modes.
"Estimates are made for hydrocarbon, carbon monoxide, nitrogen
"oxides, sulfur oxides, lead, and particulate emissions.
Stationary "point source emissions of these pollutants plus
emissions of water "activity from INFORUM.

"   In summary, TECNET is a comprehensive transportation research
"tool that makes detailed estimates of total energy use by the
"transportation sector plus associated pollution emissions and
"economic activity. TECNET is now fully operational. A number of
"test scenarios have been run through the system to test its
"sensitivity to a wide array of policy alternatives.

b) TECNET system variables

    The principal variables of the system which are, in a
coherent way, assumed or forecasted are:

    * economy in general:

        - per capita revenue
        - population
        - number of families
        - working force
        - public expenditures

    * transport system:

        - transport "work":
          . passenger x miles traveled (PMT)
            PMT-urban

            PMT-extraurban
          . vehicles x miles traveled (VMT)
            VMT-urban
            VMT-extraurban
          . ton x miles traveled (TMT)

"* INFORUM stands for Interindustry Forecasting Model of the
"University of Mariland.

```
POPULATION          + 1. %        1977 - 1985
                    + 0.8 %       1985 - 2000
                    + 0.6 %       2000 - 2025

GNP PER CAPITA      + 3.7 %       1977 - 1985
                    + 2.6 %       1985 - 2000
                    + 1.5 %       2000 - 2025

CAR SIZES           17 %  LIGHT  ⎫
                    18 %  MEDIUM ⎬ 1977
                    65 %  LARGE  ⎭

                    30 %  LIGHT  ⎫
                    30 %  MEDIUM ⎬ 1985 - 2000
                    40 %  LARGE  ⎭

TRANSPORTATION
SPECIFIC            REF.TO FIG. 3.13
CONSUMPTION
```

Fig. 3.12 - TECNET Base Case: Internal Scenario Main Assumptions[24-25]

- transport model mix:
  . car (subdivided in small/medium/large)
  . bus
  . train
  . air
  . ship
  . truck
  . pipeline

* technologies:
  - materials
  - size
  - power plant

c) TECNET scenarios

   Among the analyzed scenarios we retained to report here the four we considered more significant. The four scenarios are compared to a base case which represents historical trends projections with the addition of legislative regulation already operative (for instance CAFE: Corporate Average Fuel Economy).

   As base year it is considered 1977, and the forecasts are for

	1985	2000	2025
AUTOMOBILE **	-37.5	-49.5	-48.8
TRUCK	-13.7	-26.2	-41.6
BUS	-10.9	-13.9	-14.9
RAIL	0.0	0.0	0.0
AIR	- 9.5	- 9.5	- 9.5
WATER	- 1.9	- 6.0	- 6.0
PIPELINE	0.0	0.0	0.0

* ENERGY INTENSITIES USED IN TECNET PROJECTIONS
ARE EXPRESSED AS BTU PER PASSENGER MILE TRAVELED,
TON MILE TRAVELED, OR VEHICLE MILE TRAVELED DEPEN
DING ON THE METHOD OF COMPUTATION FOR EACH MODE.

** THE SLIGHT INCREASE IN AUTOMOBILE ENERGY INTENSITY
BETWEEN 2000 an 2025 IS DUE TO AN INCREASED SHARE
OF URBAN (SMSA) DRIVING.

Fig. 3.13 - Changes in Transportation Energy Intensities*
        (Percent Change from 1975 Intensity)[24-25]

the years 1985-2000-2025. The following tables describe the
alternative scenarios and the results given in term of energy use
and general economy variables compared to the base scenario case
for the years 2000 e 2025.

Each scenario will be designated for short as:

. base case : inertial
. scenario 1: high conservation
. scenario 2: new engines
. scenario 3: electric car
. scenario 4: modal shift

The main assumptions characterizing the inertial scenario are
given in Figs. 3.12 and 3.13 The forecasted value for the

PASSENGER MILES TRAVELED (BILLIONS)

	1977	1985	2000	2025	AVERAGE ANNUAL PERCENT GROWTH		
					1977 1985	1985 2000	2000 2025
AUTOMOBILE	1765.6	2250.0	3308.7	5499.2	3.1	2.6	2.1
BUS	60.7	87.5	151.5	331.2	4.7	3.7	3.2
AIR	222.8	315.6	541.8	792.7	4.5	3.7	1.5
RAIL	22.0	30.2	57.1	149.9	4.1	4.3	3.9
TOTAL	2202.9	2882.0	4355.6	7277.3	3.4	2.8	2.1

TON MILES TRAVELED (BILLIONS)

	1977	1985	2000	2025	AVERAGE ANNUAL PERCENT GROWTH		
					1977 1985	1985 2000	2000 2025
TRUCK	762.5	1005.7	1704.7	4221.9	3.5	3.6	3.7
RAIL	828.0	1065.8	1604.1	3022.0	3.2	2.8	2.6
AIR	19.8	33.3	50.3	89.1	6.7	2.8	2.3
WATER	1405.9	1690.6	2349.5	4110.5	2.3	2.2	2.3
PIPE	788.1	792.2	1077.4	2020.8	0.1	2.1	2.6
TOTAL	3804.4	4587.5	6786.1	13464.2	2.4	2.7	2.8

Figure 3.14 – TECNET Base Case: Projections Transportation Activity
Levels[24-25]

	1977	1985	2000	2025	AVERAGE ANNUAL % GROWTH 1977 1985	2000	2025
					1985	2000	2025
DIRECT	20.0	20.7	27.4	45.5	0.4	1.9	2.1
INDIRECT	8.4	11.0	16.5	29.7	3.5	2.7	2.4
EQUIP.MFG.	4.1	5.7	8.8	16.3	4.3	2.9	2.5
SERVICES	3.2	4.0	6.2	11.2	2.9	2.9	2.4
INFRASTRUCTURE	1.1	1.3	1.6	2.2	1.7	1.3	1.3
TOTAL	28.5	31.7	43.9	75.3	1.4	2.2	2.2
INDIRECT AS A % OF DIRECT	41.9	53.3	60.3	65.3	3.1	0.8	0.3
DIRECT AS % OF TOTAL	70.0	65.3	62.4	60.4			

Fig. 3.15 - TECNET Base Case: Projections Direct and Indirect Energy Use by the Transportation Sector (Quadrillion BTU)

SCENARI

0	1	2	3	4
BASE CASE	HIGH ENERGY CONSERVATION	NEW CAR ENGINES	ELECTRIC CAR	MODAL SHIFTS
(SEE FIG. 3.12 and 3.13)	■ SOCIO -ECONOMIC SCENARIOS AS IN THE BASE CASE ■ DRASTIC ENERGY SAVING MEASURES IN INDUSTRY ■ NAT.GAS → COAL OIL → COAL ■ TRANSPORTATION EFFICIENCY INCREASE (SEE FIG.3.17) ■ NEW ENGINES (STIRLING AND GAS TURBINE) (SEE FIG.3.17, 3.19) ■ USES OF NEW MATERIALS FOR CARS (SEE FIG.3.18)	■ SOCIO-ECONOMIC SCENARIOS AS IN THE BASE CASE ■ NEW ENGINES (STIRLING AND GAS TURBINE) (SEE FIG.3.19) 1986 - 2000 100 % SUBSTI-TUTION ■ USE OF NEW MATERIALS FOR CARS (SEE FIG.3.18)	■ SOCIO-ECONOMIC SCENARIOS AS IN THE BASE CASE ■ ELECTRIC CARS= 1/3 FLEET IN YEAR 2025	■ SOCIO-ECONOMIC SCENARIOS AS IN THE BASE CASE ■ URBAN AUTO PMT - 10 % IN YEAR 2000 ↗ ↘ BUS RAIL ■ INTERCITY AUTO PMT - 10 % IN YEAR 2000 ↗ ↘ BUS RAIL ■ INTERCITY AIR PMT (DOMESTIC FLIGHTS) - 10 % IN YEAR 2000 ↗ ↘ BUS RAIL ■ EXPENDITURE FOR NEW CARS - 10 % ■ PURCHASE EXTRA-COST OF NEW BUSES AND TRAINS

Fig. 3.16 - TECNET Scenario Basic Assumption[24-25]

	1985		2000		2025	
	CHANGE FROM 1975	DIFF.FROM BASE CASE	CHANGE FROM 1975	DIFF.FROM BASE CASE	CHANGE FROM 1975	DIFF.FROM BASE CASE
AUTOMOBILE	-37.5	0.0	-58.4	-17.6	-65.8	-33.2
TRUCK	-17.6	-4.5	-51.4	-34.1	-62.5	-35.8
BUS	-10.9	0.0	-28.3	-16.7	-32.8	-21.0
RAIL	- 6.9	-6.9	-13.0	-13.0	-17.0	-17.0
AIR	-14.7	-5.7	-27.7	- 8.0	-36.6	-29.9
WATER	- 5.0	-3.2	-30.4	-26.0	-30.4	-26.0
PIPELINE	- 7.1	-7.1	-28.9	-28.9	-30.0	-30.0

NEW AUTOMOBILE MARKET SHARE (%)

	I.C.E.	BRAYTON	STIRLING	ELECTRIC
1985	100	--	--	--
2000	5	40	40	15
2025	5	40	40	15

Figure 3.17 - TECNET Conservation Scenario: Changes in Transportation Energy Intensities (Percent)

A) CONSERVATION SCENARIO, I.C.E. ENGINE

	BASE CASE	CONSERVATION CASE	DIFF. (%)
PLASTIC	107.8	266.4	147.1
STEEL	2211.3	1444.7	- 34.7
ALUMINUM	79.8	211.9	165.5
TOTAL, THREE MATERIALS	2398.9	1923.0	- 19.8

* WEIGHTS SHOWN ARE FOR COMPOSITE AUTOMOBILES, TAKING INTO ACCOUNT SIZECLASS AND ENGINE-TYPE DISTRIBUTIONS.

B) CONSERVATION AND NEW ENGINE SCENARIOS
   CAR WITH NEW ENGINES

	BASE CASE	BRAYTON	DIFF. (%)	STIRLING	DIFF. (%)
PLASTIC	107.8	102.6	-4.8	108.3	0.5
STEEL	2211.3	2005.5	-9.3	2250.2	1.8
ALUMINUM	79.8	60.4	-24.3	176.6	121.3
TOTAL, (THREE MATERIALS)	2398.9	2168.5	-9.6	2535.1	5.7

Fig. 3.18 - TECNET Alternative Scenarios: Differences in
           Materials Composition of new Automobiles
           (POUNDS)[24-25]

transport "work" (passengers and goods) is given in Fig. 3.14 The forecasted values for direct, indirect and total energy use are reported in Fig. 3.15.

For the alternative scenarios the following general

	I.C.E.	BRAYTON	DIFF. (%)	STIRLING	DIFF. (%)
1977*	18.9	18.9	0.0	18.9	0.0
1985*	27.5	27.5	0.0	27.5	0.0
2000	27.3	37.8	38.5	39.6	45.1
2025	26.9	45.8	70.3	42.3	57.2

*I.C.E. ONLY: ALTERNATIVE ENGINES PENETRATION BEGINS IN 1986

Fig. 3.19 - MPG Characteristics of new automobiles

	BASE CASE	HIGH ENERGY CONSERVATION (DIFF.%)	NEW ENGINES (DIFF.%)		ELECTRIC CAR (DIFF.%)	MODAL SHIFTS (DIFF.%)
			BRAYTON	STIRLING		
PASS x MILE (10^9)	7277	-1.3	--	--	--	+ 0.8
TON x MILE (10^9)	13464	-19.2	--	--	--	- 0.1
DIRECT ENERGY (10^{15} BTU)	46	-35.2	-10.5	-14.7	- 2.0	- 0.8
INDIRECT ENERGY	30	-31.5	- 2.2	+ 1.8	+ 2.8	+ 2.0
TOTAL ENERGY	76	-33.8	-10.8	- 8.2	+ 0.3	+ 0.3
GNP (10^9 $ 1977)	6187	- 1.3	--	--	+ 0.3	+ 0.6
EMPLOYMENT (10^6)	131	+ 0.5	--	--	+ 0.3	+ 0.5

Fig. 3.20 - TECNET Results for Year 2025[24-25]

assumptions were taken:
      . in the "high conversion" case a set of interventions
outside the transport system are included aiming at reducing the
energy use;
      . in the new engines and in the electric car cases, the
modifications, with respect to the base case, concern only the
automobile (including all indirect effects);
      . in the modal shift case the variation with respect to the
base case is limited to the shifthing of a part of the mobility
demand from car to collective modes.

      The basic assumptions of the four alternatives scenarios are
summarized in Fig. 3.16 The assumed technological changes are
summarized in Figs. 3.17, 3.18, 3.19.

d) TECNET outputs

      In Fig. 3.20 the results obtained for the alternatives
scenarios are summarized for the year 2025 as percentage variation
to the base case forecast. The data in Fig. 3.20 regard the
transport "work", the direct and indirect energy use, GNP and work
force.

      A first striking result concerns the electric car and the
modal shift scenarios, which, because of the increase in the
indirect energy use, result in a total increase in the use, well
contrary to the expectation.

      More specific remarks for each of the alternative scenario
follow.

      . High conservation.

            . The mobility decreases as a consequence of the decrease
in GDP;
            . the decrease in the direct energy use (see Fig. 3.21) is
a consequence of the mobility decrease and of the better energy
efficiency of vehicles;
            . the decrease in the indirect energy is related to
technological improvements and to the mobility reduction.

      . New engines
            . the mobility level shows practically no changes, because
there is no major GDP change;
            . the impact on direct energy (Fig. 3.22) concerns mainly
the automobile for which high increase in efficiency was assumed.
The small changes in the other modes are related to small
variation in GNP;
            . in the indirect energy use (see Fig. 3.23) one sees how
the heavy Stirling engine leads to an increase in energy for

	2000		2025	
	ENERGY USE	DIFF.(%)*	ENERGY USE	DIFF.(%)*
DIRECT	20.4	-25.4	29.5	-35.3
INDIRECT	12.5	-24.5	20.4	-31.5
EQUIPMENT MFG.	7.5	-14.7	13.4	-17.7
SERVICES	3.6	-41.6	5.1	-54.7
INFRASTRUCTURE	1.4	-11.6	1.8	-16.3
TOTAL	32.9	-25.0	49.8	-33.8
INDIRECT AS A % OF DIRECT	61.0	1.1	69.2	5.9

*DIFFERENCE FROM BASE CASE PROJECTIONS

Fig. 3.21 - TECNET Conservation Scenario: Direct and Indirect
Energy Use by the Transportation Sector
(Quadrillion BTU)

	2000					2025				
	BASE CASE	BRAYTON CASE	DIFF. (%)	STIRLING CASE	DIFF. (%)	BASE CASE	BRAYTON CASE	DIFF. (%)	STIRLING CASE	DIFF. (%)
AUTOMOBILE	10.216	8.003	-21.7	7.731	-24.3	17.234	10.062	-41.6	10.754	-37.6
HEAVY AND LIGHT TRUCK	5.741	5.712	- 0.5	5.721	- 0.3	10.979	10.836	- 1.3	10.874	- 1.0
PERSONAL TRUCK	1.681	1.679	- 0.1	1.679	- 0.1	2.207	2.203	- 0.2	2.203	- 0.2
BUS	.322	.321	- 0.3	.321	- 0.3	.707	.705	- 0.3	.705	- 0.3
TOTAL HIGHWAY	17.960	15.715	-12.5	15.452	-14.0	31.127	23.806	-23.5	24.536	-21.2
AIR	4.053	4.050	- 0.1	4.051	---	6.125	6.117	- 0.1	6.120	- 0.1
RAIL	1.253	1.248	- 0.4	1.255	0.2	2.473	2.462	- 0.4	2.474	---
WATER	1.060	1.050	- 0.9	1.053	- 0.7	1.854	1.826	- 1.5	1.834	- 1.1
PIPE	1.602	1.569	- 2.1	1.577	- 1.7	2.180	2.052	- 5.9	2.085	- 4.4
OTHER	1.439	1.439	---	1.439	---	1.770	1.770	---	1.770	---
TOTAL OFF-HIGHWAY	9.405	9.356	- 0.5	9.375	- 0.3	14.402	14.226	- 1.2	14.283	- 0.8
TOTAL SECTOR	27.365	25.071	- 8.4	24.827	- 9.3	45.529	38.032	-16.5	38.819	-14.7

NOTE: A DASH (----) INDICATES A CHANGE OF LESS THAN: .05 PERCENT

Fig. 3.22 - TECNET New Engine Scenario: Impact on Direct Energy
Use by the Transportation Sector (Quadrillion BTU)[24-25]

	2000					2025				
	BASE CASE	BRAYTON CASE	DIFF. (%)	STIRLING CASE	DIFF. (%)	BASE CASE	BRAYTON CASE	DIFF. (%)	STIRLING CASE	DIFF. (%)
AUTOMOBILES										
TOTAL INDIRECT	6.70	6.49	-3.2	6.94	3.6	11.49	11.11	- 3.3	11.93	3.8
EQUIPMENT MFG.	3.77	3.56	-5.5	4.00	6.1	6.91	6.53	- 5.4	7.32	6.0
SERVICES	1.95	1.95	-0.4	1.96	0.6	3.37	3.36	- 0.3	3.39	0.6
INFRASTRUCTURE	0.98	0.98	0.1	0.98	0.0	1.22	1.22	0.2	1.22	0.2
TOTAL DIRECT	10.22	8.00	-21.7	7.73	-24.3	17.23	10.06	-37.6	10.75	-37.6
DIRECT PLUS INDIRECT	16.92	14.49	-14.4	14.67	-13.3	28.72	21.17	-26.3	22.68	-21.
ALL MODES										
TOTAL INDIRECT	16.51	16.18	- 2.0	16.84	2.0	29.74	29.09	- 2.2	30.28	1.8
TOTAL DIRECT	27.37	25.07	- 8.4	24.83	- 9.3	45.53	38.03	-16.5	38.02	-14.7
DIRECT PLUS INDIRECT	43.88	41.26	- 6.0	41.67	- 5.0	75.27	67.12	-10.8	69.10	- 8.2

Fig. 3.23 - TECNET New Engine Scenario: Impacts on Indirect Energy
Use by the Transportation Sector (Quadrillion BTU)[24-25]

manufacture which is not the case for the Brayton engine.

- Electric car

. There are small mobility changes as a consequence of small variation in GDP;
. the decrease of 16.5% in the direct energy use (see Fig. 3.24), is reduced only to 5.6% when the electric utility conversion losses are included

	DIRECT			INDIRECT			TOTAL		
	REFERENCE	ELECTRIC	%	REFERENCE	ELECTRIC	%	REFERENCE	ELECTRIC	%
AUTOMOBILE	13.00	10.85	-16.5	11.63	12.21	5.0	24.63	23.06	-6.4
TRUCK	10.76	10.81	0.5	5.71	5.91	3.5	16.47	16.72	1.5
RAIL	1.98	2.01	1.5	2.77	2.80	1.1	4.75	4.81	1.3
PIPE	0.97	0.90	- 7.2	0.35	0.34	-2.9	1.32	1.24	-6.1
OTHER	9.38	9.39	0.1	10.65	10.73	0.8	20.03	20.12	0.4
ELEC. UTILITY LOSS. AUTO	0.00	1.42	---	---	---	---	0.0	1.42	---
GROSS AUTO	13.00	12.27	- 5.6	11.63	12.21	5.0	24.63	24.48	-0.6
SECTOR TOTAL	36.09	35.38	- 2.0	31.11	31.99	2.8	67.20	67.37	0.3

Fig. 3.24 - Electric car Scenario: Direct and Indirect Energy
Use by the Transportation Sector, 2025,
(Quadrillion BTU)[24-25]

	REFERENCE	ELECTRIC	%
INDUSTRIAL	35.61	35.70	0.3
COMMERCIAL	5.34	5.35	0.2
RESIDENTIAL	2.11	2.11	0.0
TRANSPORTATION	33.68	30.55	-9.3
ELECTRIC UTILITIES	0.07	0.08	12.9
TOTAL	76.81	73.79	-3.9

Fig. 3.25 - TECNET Electric Car Scenario: Comparison of Petroleum
Use, 2025 (Quadrillion BTU)[24-25]

. the indirect energy use (see Fig. 3.24) for the whole transport sector is larger then the base case, and it is responsible for a slight increase in the total energy use;
. if one refers only to the oil use, the electric car scenario (see Fig. 3.25) shows a decrease of 3.9% of oil needs by all the economy system;

	2000			2025		
	SHIFT CASE	BASE CASE	% DIFF.	SHIFT CASE	BASE CASE	% DIFF.
DIRECT	27.08	27.37	- 1.1	45.15	45.53	- 0.8
INDIRECT	16.73	16.51	1.3	30.32	29.74	2.0
EQUIPMENT MFG.	8.44	8.76	- 3.7	15.84	16.32	- 2.9
TRANS. SERVICE	6.72	6.18	8.7	12.29	11.22	9.5
INFRASTRUCTURE	1.57	1.57	0.0	2.19	2.19	0.0
TOTAL	43.81	43.88	- 0.1	75.47	75.27	0.3

Fig. 3.26 - TECNET Shift Scenario: Impacts on Total Energy Use by
the Transportation Sector (Quadrillion BTU)[24-25]

. the beneficial impact on noxious emissions reduction is limited to approx. 10% because of the importance of other mobile sources and of the stationary sources.

- Modal shift

. the global impact on GNP is positive but very small, partly due to an increase in the total passenger mobility (0.8% in 2025), even if differently distributed among the various modes with respect to the base case;
. the decrease in the direct energy use in car mode is almost entirely written off by the increase in bus and rail. The average energy efficiency (pass x km/kcal) slightly increases (2%);
. the indirect energy use worsens (see Fig. 3.26) because the indirect energy efficiency of the car is better than that of bus, even if it is worse than that of train;
. equipment manufacture requires less energy (see Fig. 3.26) which is more than compensated by increases in the energy use for the transport sources. The combined effects give an increase in the total energy use of the transport sector.

4. THE PROSPECTIVE FOR TECHNOLOGICAL INNOVATION IN TRANSPORT SYSTEM

As a conclusion of the analysis on transport system evolution patterns at the end of chapter 2, we put priority on the intervention to improve energy efficiency at single transport mode or vehicle level with respect to try to force variations in the transport system structure. The interesting case of TECNET model described in chapter 3, has shown that even with a model that simply make all the accounting, (rightly and extensively for direct and indirect energy use), modal shift does not seem to pay-off in term of energy savings.

In this chapter we will try, by analysing two "components" in the transport system - namely urban traffic control and car - to evaluate the prospective to improve, through technological innovation, the energy efficiency, without worsening system performance.

4.1 Energy conservation with a decentralized urban traffic control system

The behaviour of urban traffic, characterized by a great variability over short time (minutes) and by an average regularity over longer time, has originated, during the 60's, the implementation, in great urban areas, of control plans where traffic lights have a common cycle, and phases and synchronization are fixed. Such plans are based upon the average behaviour of traffic flow over a predetermined period of time[34,35,36,37].

At the beginning of the 70's such a scheme has been improved and automated plan selection has been introduced.

The plan selection is based upon continuous traffic measurements; anyway, as with previous systems, the number of plans is small and the entire system is conceived to "optimize" a stationary traffic. This hypothesis is far from being realistic in many actual situations and whenever public vehicles are present on the network. In order to overcome those difficulties, - to take into account the increasing role of mass transportation in urban environment, meeting new requirements in public transit journey time and in energy reduction for the overall transportation system - new traffic control systems have been conceived and designed.[38,39]

According to those new guidelines FIAT RESEARCH CENTER is developing a new system ("Progetto Torino") whose general specifications have been tested through a simulation program. System design and preliminary simulation results are presented in the following.

a) Underline System architecture

The system is conceived in order to meet some general specifications:
    . Adaptivity: the system can automatically adapt itself in real time to actual traffic characteristics. It is not tied to any predetermined control plan;
    . Decentralization: the control structure is of distributed type. Data processing tasks and control strategies elaboration are committed in part to peripheral processors.;
    . Reliability: the system is based upon specific criteria of reliability, diagnostics, efficiency, maintanability. It is able to work in a down-graded way, even when some specific equipments are off service;
    . Modularity: the system utilizes a large set of equal components and subsystems; in this way the system can reach good repairability and easy expandibility;
    . Public Transit Priority: the system gives effective priority to preassigned public transit lines. The priority can be "on the average" or "absolute".

In order to obtain the above performances the system must:

    . acquire and process on-line a high number of data both for public and private traffic;
    . be implemented in a cellular structure with modules able to

perform specific tasks at different hierarchical levels;
    . be able to forecast with high precision the movements of
every public transit vehicle.

    From the point of view of the hierarchy the system has two
levels:

    . First level : Area Control
    The Area Control is the higher hierarchical level. It
consists essentially of a multiprocessor minicomputer and its main
job is the treatment of information elaborating strategics plans
to be sent to a lower level as guidelines for the actual area-wide
control. The Area Control receives, through detectors, local
controllers, data transmission network, information about public
and private traffic over the whole controlled area. Strategies
elaborated by the Area Control cover an horizon time of some ten
minutes.

    . Second Level: Local Control.
    Local Control consists of a network of microprocessors, able
to receive data from sensors, to process them according to
specific control algorithms and guidelines from the higher level,
to trigger the traffic lights actuators. The actuated control is
of tactical type over un horizon time of few minutes.

b) Area traffic and energy use simulation models

    The potentiality in term of improved traffic flow, increasing
average vehicle speed and reducing energy consumption, was checked
by means of simulation models. The models relays on experimental
data of urban traffic characteristics and related energy use, in
particular for automobile. On this subject there are many papers,
all of them in agreement with the assumption that for urban
traffic, mainly for speed up to 60 km/h,

$$\Phi = A\ell + Bt \qquad (1)$$

    where: Φ = fuel consumption
             ℓ = travelled distance
             t = travel time
          A,B = coefficient to be evaluated esperimentally.

Relation (1) is equivalent to:

$$\Phi_\ell = A + \frac{B}{V} \qquad (2)$$

    where: Φ_ℓ = fuel consumption per unit length

             V = average speed on the link into consideration.

Following the authors opinion, the knowledge of the average speed on a travelled path length gives a sufficient estimate of the fuel consumption per unit length as accurate as more sophisticated formulas.

In order to account for the different traffic situations existing between USA cities - for which relations (1) and (2) were tested - and italian cities, the validity for our country of both relations was verified. Tests have been conducted in Torino on two urban links, central and peripheral, 9 km. each, by using three standard cars of the type FIAT 126, FIAT 128-1300 and LANCIA β-2000.

The experimental results confirmed the validity of formula (1) above and have permitted to obtain the proper coefficients. The validity of the formula (1) is confirmed by comparison with actual fuel consumption data .

The experimental coefficients obtained varies with cars. Average coefficients are obtained considering each of the experimented car as typical, respectively, of small, medium and large cars and averaging the results on the italian car stock population subdivided among the three classes.

The results show that:

. the coefficient A has to be interpreted as the fuel consumed per unit lenght for overcoming the rolling resistance;
. the coefficient B is related to fuel consumption due to mechanical losses and vehicle acceleration; moreover it depends upon the link parts travelled in neutral or by motoring over.

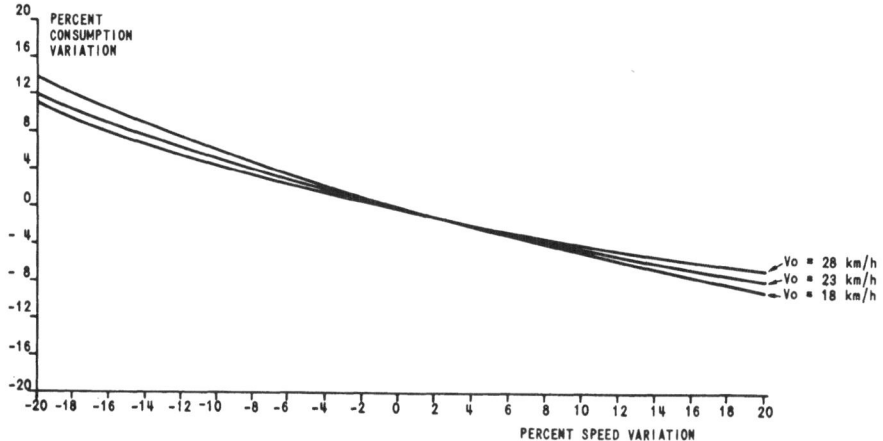

Fig. 4.1 - Vehicle Consumption Changes in Different Urban Traffic Situations

For variations around a given cruise speed, relation (2) gives the percentage of fuel consumption changes shown in Fig.

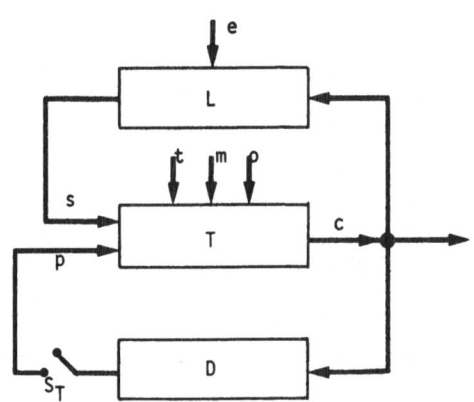

e : VECTOR OF REAL TRAFFIC-LIGHT CONDITION

p : "PERCENT OF TURNES" MATRIX: FOR EACH ROAD INTERSECTION AND ENTRACE BRANCH IT GIVES THE PERCENTAGE OF VEHICLES LEAVING A GIVEN OUTLET

c : REAL STATE VECTOR OF EACH VEHICLE IN THE NETWORK (POSITION, SPEED, ACCELERATION, etc.)

S : LIGHT CONTROL LAW

t : ROAD NETWORK TOPOGRAPHY

m : DESCRIPTION OF PUBLIC TRANSIT LINES

o : PRIVATE VEHICLE O/D MATRIX

S_T : FREQUENCY$_{-T}$ SAMPLER

L : LIGHT CONTROL MODEL

T : TRAFFIC SIMULATION MODEL

D : DRIVER HABIT MODEL

Fig. 4.2 - Traffic Microsimulation Model

4.1. In the case of a car fleet having the same average composition of the italian stock, from these curves it is possible to evaluate the fuel savings attained by cruise speeds

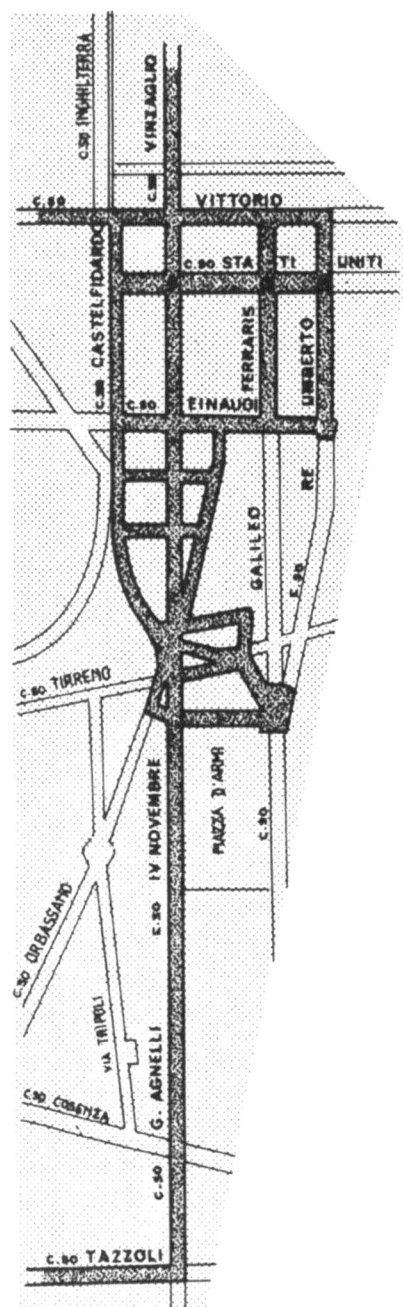

Fig. 4.3 - Progetto Torino: Map of the Light Controlled Crossroads

considerably different from the present one's.

In order to estimate cruise speed consistent with a hyerarchical-distributed traffic control system ("Progetto Torino"), a traffic simulation model was used, whose general structure is shown in Fig. 4.2.

The simulation is carried out dynamically through the three L, T, D models interaction. The T-model contains algorithms that displace the vehicles on the highways, in accordance with a "car following" law tested in Torino. The output gives details on position and speed of the vehicles at time intervals S_T (sampling frequency).

In the same time the D-model simulates the driver behaviour: at every S_T it provides with the percentage of turns at each cross road, in agreement with O/D matrix (o matrix in Fig. 4.4), the real traffic situation (c matrix) and the operating traffic light control (e vector).

The simulation model, coupled with strategies of traffic light control, appears to be a powerful tool, capable to simulate very complex networks where traffic flows on each branch can be distributed upon more lanes and the collective transport systems can travel either on common lanes (bus) or on exclusive lanes (streetcar).

From a functional point of view, at every time interval S_T the model:

. generates vehicles at the entrance branches
. solves problems arised at the road intersections
. simulates the vehicle displacements
. simulates the overtakings
. eliminates vehicles leaving the network
. updates statistical data on the network (flows, speed, number of stops etc.).

By means of such model the traffic on an area of the city has been simulated, with the goal to reproduce:

. the present situation (light controlled crossroads without co-ordination)
. the situation of a possible control system with fixed plains (Transyt model)
. the situation attainable by the "Progetto Torino" system (Fig. 4.3).

The results of Fig. 4.4 suggest that with the traffic lights control of "Progetto Torino" one could attain cruise speed for the

TOTAL MEAN SPEED OF SIMULATED CARS ON THE NETWORK

REAL SITUATION	TRANSIT*	%	PROG. TORINO CONTROL	%
27.75 km/h	31.68 km/h	+14.1	31.92 km/h	+15.0

TOTAL MEAN WAITING TIME OF SIMULATED CARS ON THE NETWORK

REAL SITUATION	TRANSIT*	%	PROG. TORINO CONTROL	%
17.91 min	14.82 min	-17.2	12.55 min	-29.9

(%) PERCENT IMPROVEMENT IN COMPARISON TO THE REAL SITUATION
(*) CONTROL SYSTEM WITH FIXED PLANS

Fig. 4.4 - Traffic Simulation Model

private traffic higher by 15% than the present speed.

From the comparison between this result and that one obtained with the consumption model (Fig. 4.1) one can conclude that a 15% speed increment corresponds to an energy saving of about 6%.

4.2 Prospective for technological innovation in automobile

Laymen consider car as a product which has been technologically innovated at a slow rate. The emission control problem earlier and now the energy crisis have put heavy requirements for radical technological innovations. It seems therefore important to have better judgement on the innovative performance of the automobile industries as measured by hystorical trends.

We report here the results of an analysis of design trends of European manufactured cars sold to the market from 1950 to today[47]. The analysis do not take into consideration the market share of the various models. The cars have been divided into three classes: small, medium, large cars.

FIGURE OF MERIT	INDICATORS
■ PAYLOAD / TIME TO ACCOMPLISH THE MISSION	■ INTERNAL VOLUME x MAX SPEED ■ INTERNAL VOLUME x MAX ACCELERATION

■ PAYLOAD / MISSION COST	■ INTERNAL VOLUME / EXTERNAL VOLUME
	■ INTERNAL VOLUME / TOTAL WEIGHT
	■ MAX PAYLOAD / TOTAL WEIGHT
	■ INTERNAL VOLUME / MAX POWER
	■ INTERNAL VOLUME /
	  ENERGY FOR TRACTION } FOR A
	■ INTERNAL VOLUME / } GIVEN
	  FUEL USE } MISSION

■ PAYLOAD / TIME TO ACCOMPLISH A MISSION ───────────── MISSION COST	■ (INT.VOL.)x(MAX SPEED)/MAX POWER ■ (INT.VOL.)x(MAX ACCEL.)/MAX POWER ■ (INT.VOL.)x(MAX SPEED) /EXT.VOL. ■ (MAX PAYLOAD x(MAX SPEED)/TOTAL WEIGHT

Fig. 4.5 - Indicators of Car Design Efficiency

To measure the innovative ability one should define some proper indicators, which should be related to the vehicle ability to perform its mission and the related costs.

Vehicle performance should be valued according to:
. payload
. time to accomplish a given mission
. safety and comfort.

The performance characteristics are not easily determined both due to the variety of missions and to the difficulties to define the payload (because of car flexibility in term of passenger and baggage transportation ability).

A first measurable set of "indicators" of vehicle performance are:
. vehicle internal volume (including baggage)
. vehicle acceleration (measured, f.i., as the time needed to cover a certain distance from idle)
. vehicle maximum speed.

The efficiency of a vehicle design should be judged on the

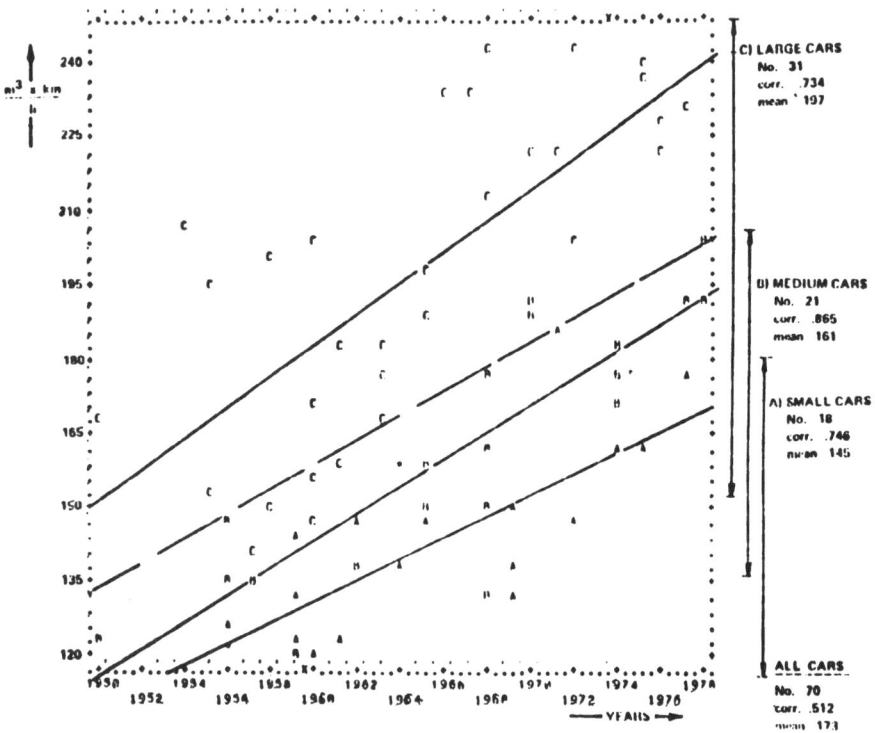

Fig. 4.6 - European Car Trends - Performance Characteristics
         Maximum Speed X Internal Volume

basis of the ratio of the quality of the service obtained by the
user (measured in terms of vehicle performance) to the cost of
that service. For the analysis we have selected some simple
indicators listed in Fig. 4.5.

    The performance characteristics of the vehicles have
constantly increased as can be seen in:
    . Fig. 4.6, which reports the data of (max speed) x (internal
volume)
    . Fig. 4.7, giving the value of the factor (max acceleration)
x (internal volume)

    Notwithstanding the increase in performance, purchasing price
has decreased (see Fig. 4.8), as well as fuel consumption (Fig.
4.9).

    Aerodynamic design has improved (see Fig. 4.10). An
indication of the constant technological development ability of

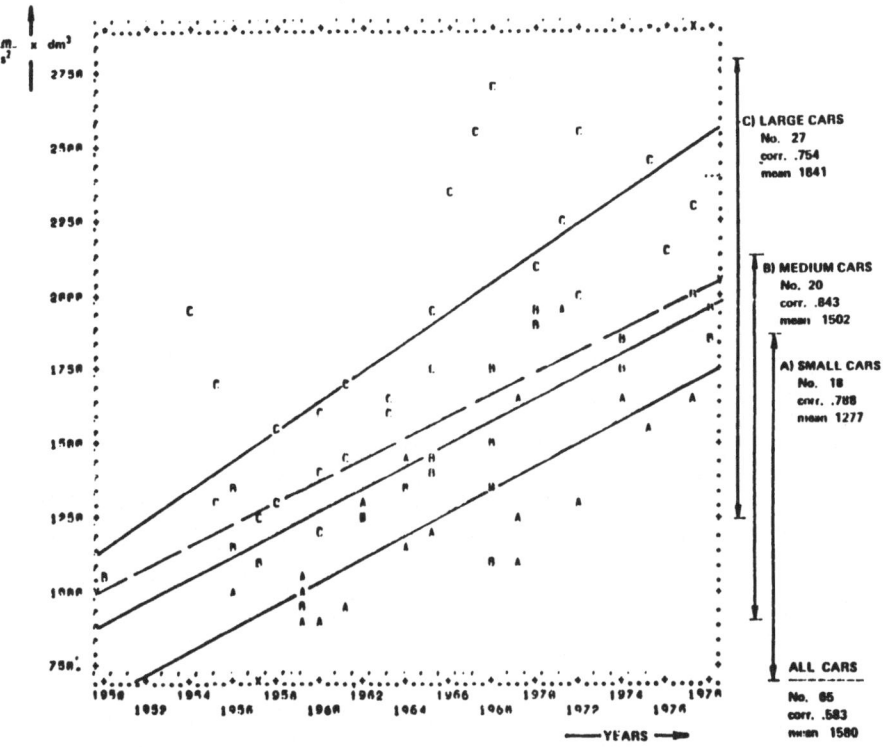

Fig. 4.7 - European Car Trends - Performance Characteristics
Acceleration X Internal Volume

the car industries is obtained by the trend of the engine specific
power (Fig. 4.11), which almost doubled in 30 years.

The continuous innovative process in the automotive
industries is not apparent to the laymen because it is the result
of several small improvements added together model after model and
not of sudden major technological changes.

The historical trends above illustrated set confidence on the
ability to meet the innovative target for energy conservation. The
question is to which extent the innovative pace could be
accelerated taking into consideration how a technologically
complex product the car is.

Taking fuel economy as the prime objective to be met in an
accelerated car innovation process, several routes should be
followed. We are in fact faced with a "global system problem"
which is difficult to divide into sub-problems (each with targets

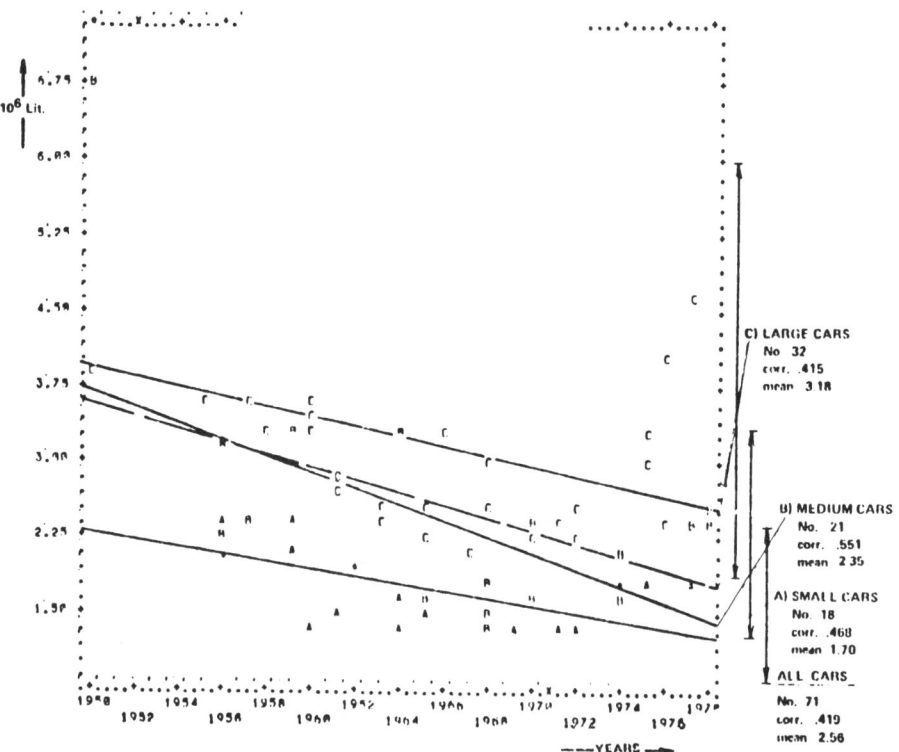

Fig. 4.8 - European Car Trends - Cost Index-Purchasing Price-at
           1974 constant money

well correlated to the global objectives); nevertheless one has to
simplify. A first large subdivision will be between:

. the problem of setting the performance specifications
. the technological problems to meet the energy saving
  objectives maintaining said specifications.

The first problem might be the more difficult one, if one
consider the risk in marketing vehicles with specifications that
might not meet the market requirements. Remember that the
analysis of the trends in the last decade shows an increase in
performance measured in terms of acceleration and speed.
Regulatory action will help in this direction, as well as a better
understanding of the car function to serve mobility needs.

Referring to the technological problems, there are several
possibility of actions which should be pursued in parallel. In
chapter 3 we have already showed the forecast of the gains in fuel
economy that might be contributed by improvements in the different

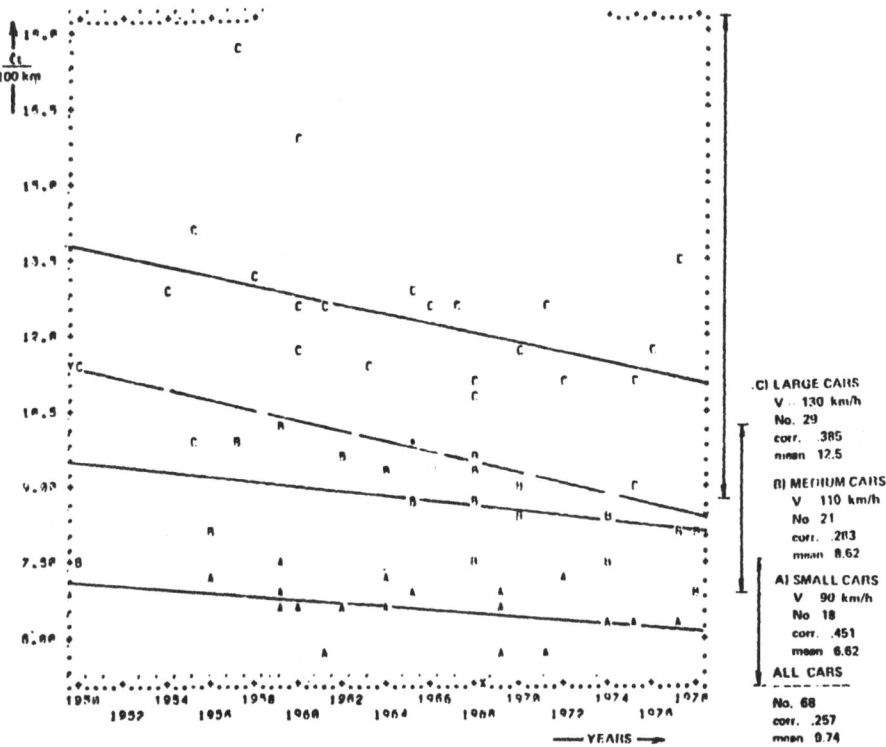

Fig. 4.9 - European Car Trends - Cost Index Fuel Consumption-at
constant speed

vehicle subsystems, for a gasoline engine vehicle. Unfortunately
the foreseen improvement cannot be considered as completely,
additives. Let us consider the various items one by one.

a) Vehicle design changes

The aerodynamic drag and rolling friction are the first
important item to be dealt with. The action to reduce
aerodynamic drag (Fig. 4.12) will reduce the freedom of the
stylist. Styling identity should be looked-for more acting on
surface decoration and interior arrangement than on the body
shape. The importance of total car weight on fuel economy is
remembered in Fig. 4.13. Building smaller vehicle is one way to
reduce car weight. Such possibility is still open for US cars, but
is much more questionable in Europe.

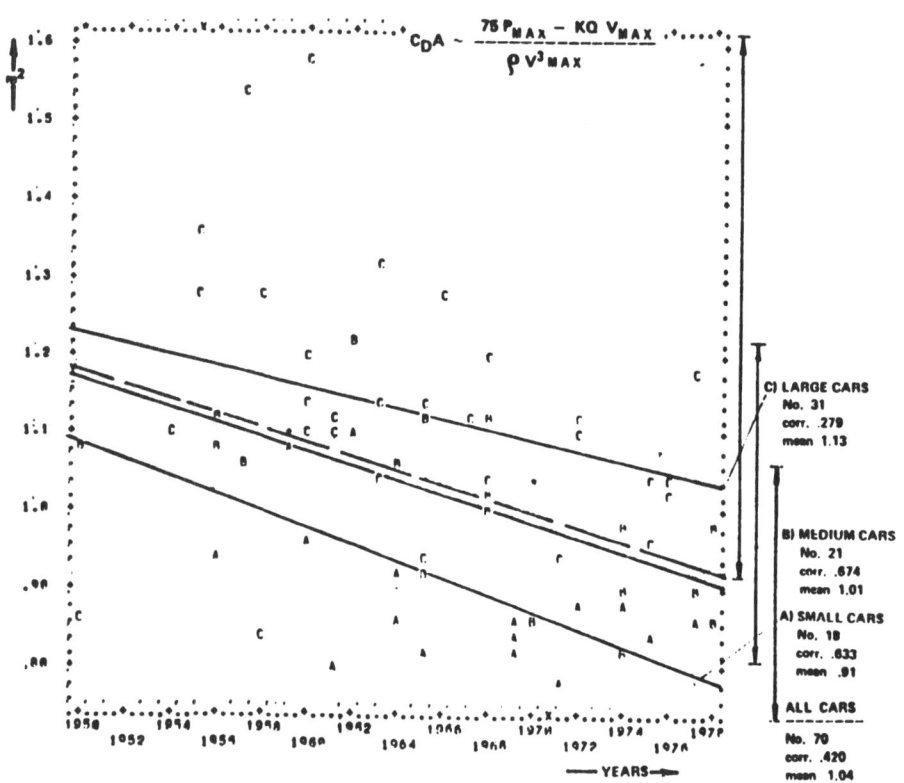

Fig. 4.10 - European Car Trends
Design Efficiency Factor-Aerodynamic
Resistance Index

b) Powerplant - vehicle matching

The matching gear ratio affect vehicle acceleration and fuel economy. As much as 50% improvement in fuel economy might results when an ideal variable gear ratio is compared to a constant gear ratio.

Gear lubricant might have a sizeable effect on fuel economy especially for short trip on cold starting.

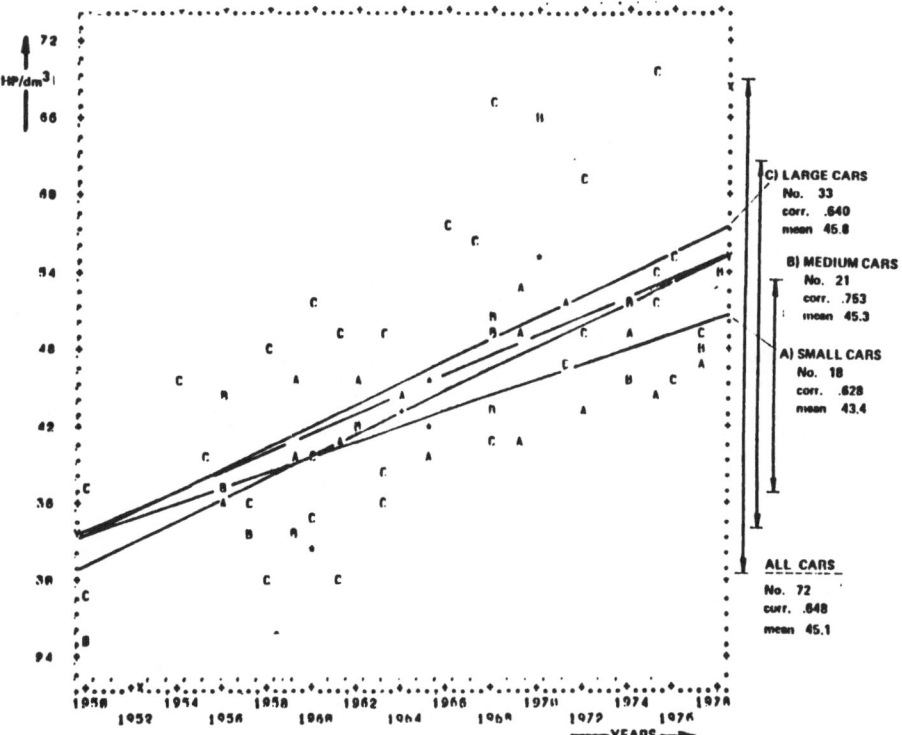

Fig. 4.11 - European Car Trends
Engine Specific Power

c) Engine efficiency improvement

As a result of the multiplicity of vehicle missions the
engine utilization pattern is for most of the time less than
20-30% of maximum power (see Fig. 4.14). This is the main reason
for the worse fuel economy of the gasoline engine with respect to
the diesel engine (because of the pumping loss at partially opened
air throttle).

Another big advantage of diesel is the lack of enrichment
needed in cold starting.

Penetration of diesel engines in the car market has already
reached sensible quotas especially in the medium and large car
classes.

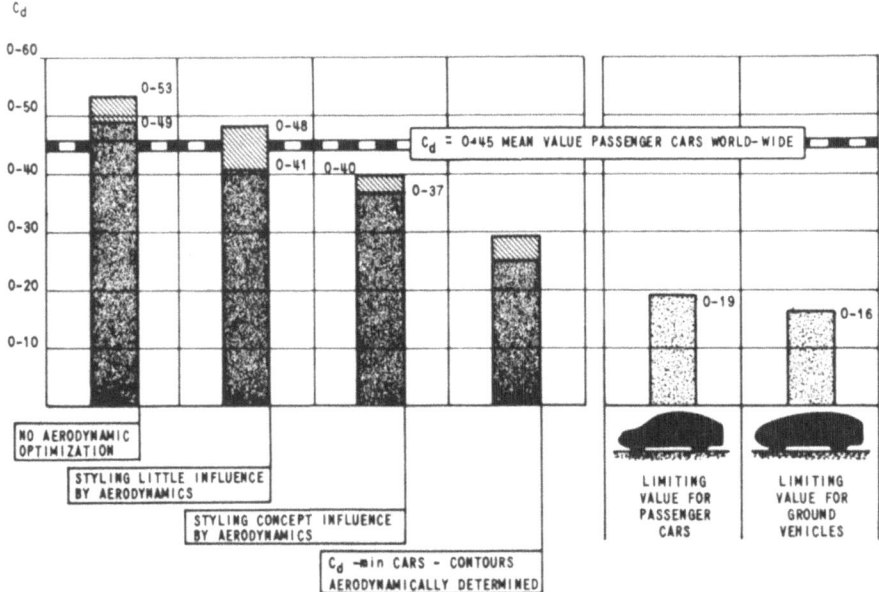

FIGURE FROM: FIALA - MECHANICAL ASPECTS - DESIGN FROM ECONOMY I. MECH. E., 1978 - C211

Fig. 4.12 - The Aerodynamic Coefficient C_d for Various Styling Concepts

To meet the diesel fuel economy challenge the gasoline engines should be able to operate at high compression ratio with very lean mixture. Electronic engine control might be the key of success for operating knock-free, in transient conditions such those existing in an engine.

Engine lubricant improvement might result in sensible fuel economy, taking into consideration that friction losses are responsible for more than 50% of the total engine mechanical losses .

d) Changing the material mixes

The reduction of the drag coefficient to a value of $C_D = 0.30$ (with respect to to-day European average of 0.46) and of 20-25% of car weight, represents a formidable task for the car designer.

The weight reduction target will also have revolutionary impact on the manufacturing technology because of the required

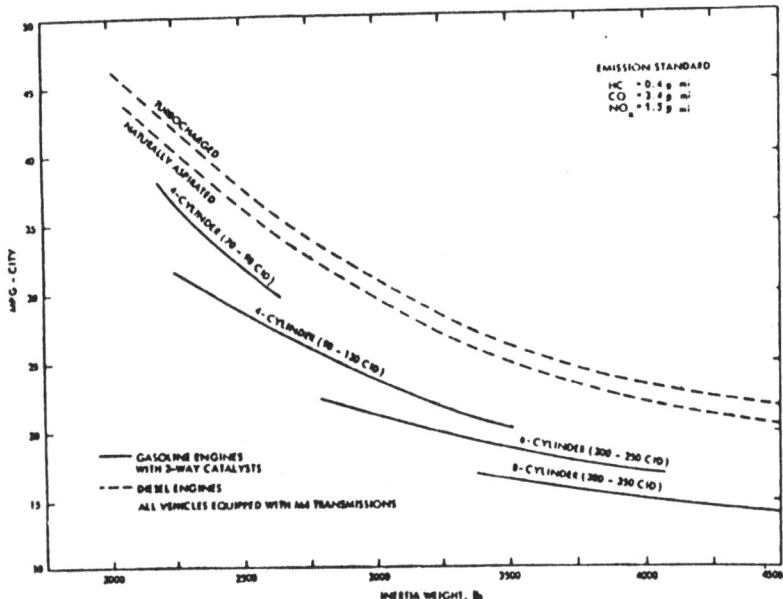

Figure from : SAE 780347 – A.F. Burke – The Moving Baseline of Conventional Powered passenger cars (1975 – 85)

Fig. 4.13 - Baseline Fuel Economy - 1978 Technology,
Urban Cycle[20]

change in material mix. All opportunities for material changes
should be looked for.

This will have quite an impact on the car cost. To give an
idea, the results of a study case for an American car on cost
increase by changing the material mix (for optimum material
substitution considering alternative material mixes such as Al,
HSLA, fiberglass and carbonfiber composites) is illustrated in Fig
4.15.

e) Integration of subsystem improvements

To have an indication on to which extent improvements in
subsystems can be integrated we will present the result of a study
where all the design choice are varied in a consistent way, to
have vehicles comparable to a base case.

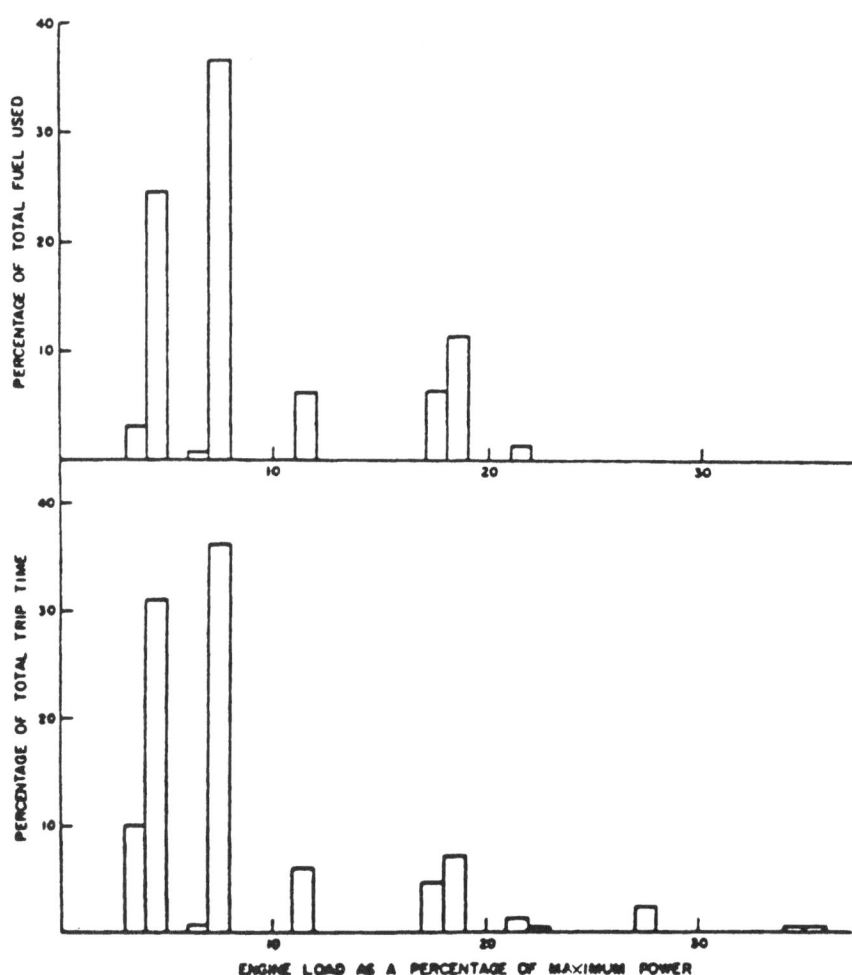

Figure from : D.R. Blackmore, A. Thomas, Fuel economy of the Gasoline engine, 1977, Mc Millan Press

Fig. 4.14 - Engine and Fuel Utilization Patterns Found for
Typical Road Service (1.8 ℓ Passenger Car)

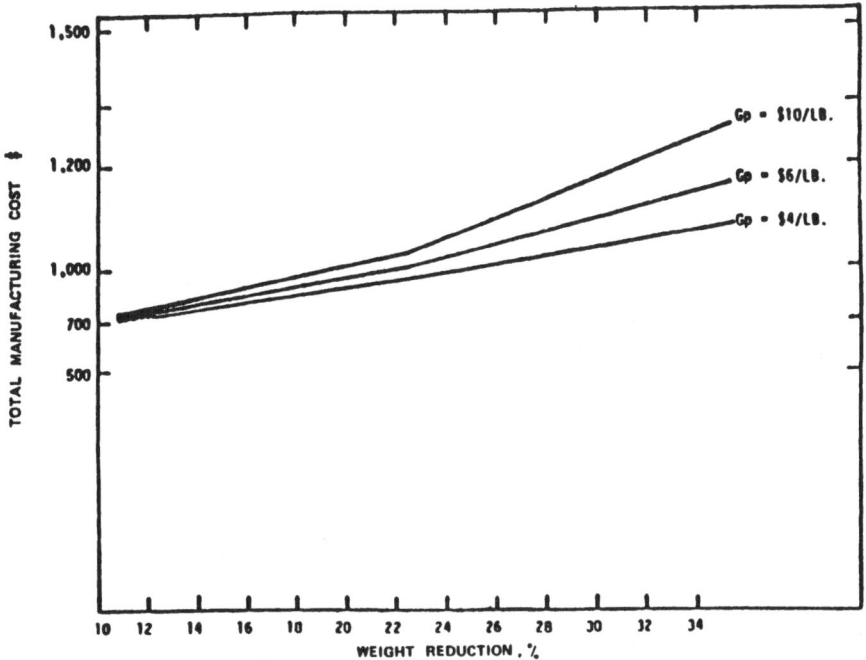

Table from : "The demand for advanced composite materials in the automotive industry,,
A proprietary study by Econ, Inc., February 1979

Fig. 4.15 - Weight Reduction Case Study - American Car
Total Manufacturing Costs for 26 Components
as Function on Percent Weight Reduction, 1977

Vehicle weight, drag coefficient and type of engines are
considered independent variables, while gear ratio and engine
displacement are chosen to meet the same acceleration. The base
vehicle is a 1150 Kg sedan, typical of European average design.

Fig. 4.16 shows the effect of drag coefficient reduction,
while Fig. 4.17 the effect of weight reduction.

The shaded area includes the effect of the different engine
types (fuel economy gain are normalized to that of the base car
with, respectively, the different engine considered). The fuel
economy gains shown in the two figures can be considered almost as
additives.

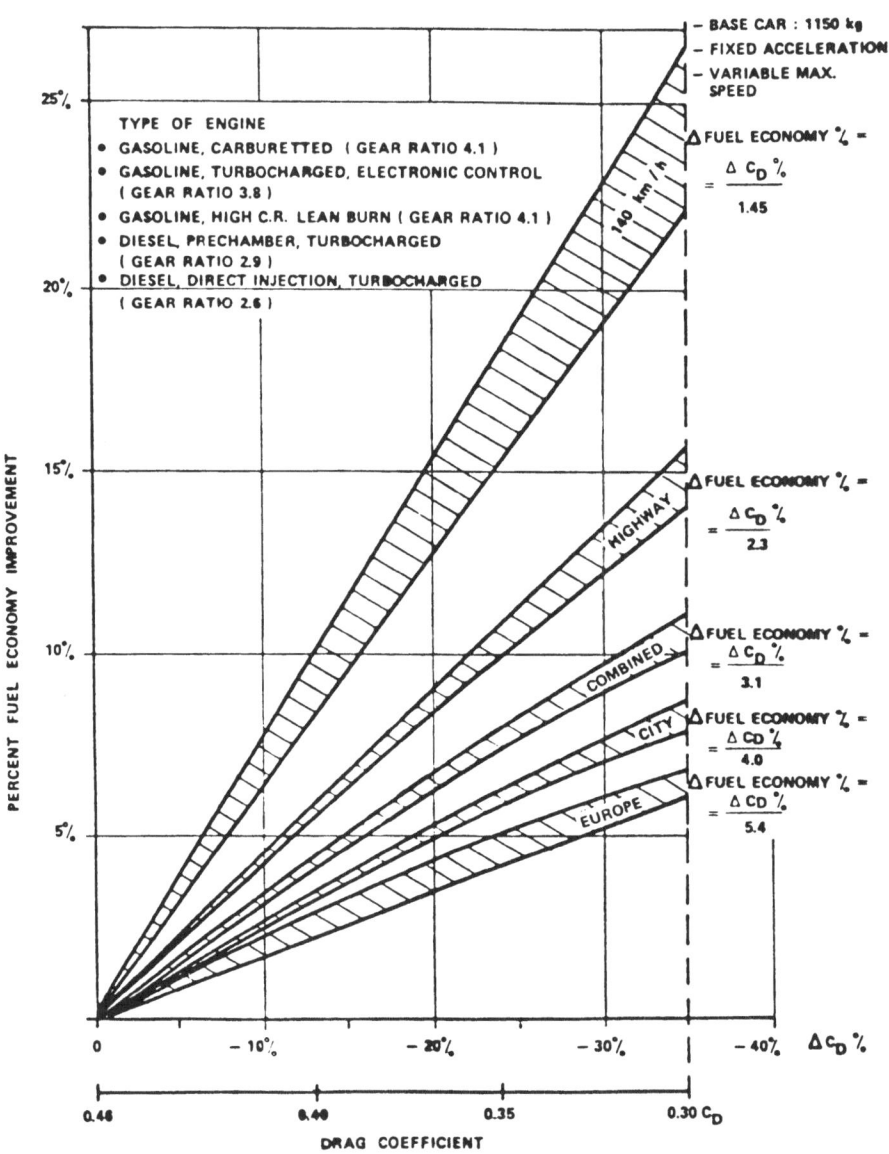

Figure from : FIAT Research Center internal report

Fig. 4.16 - Effect on Fuel Economy of Drag Coefficient Reduction

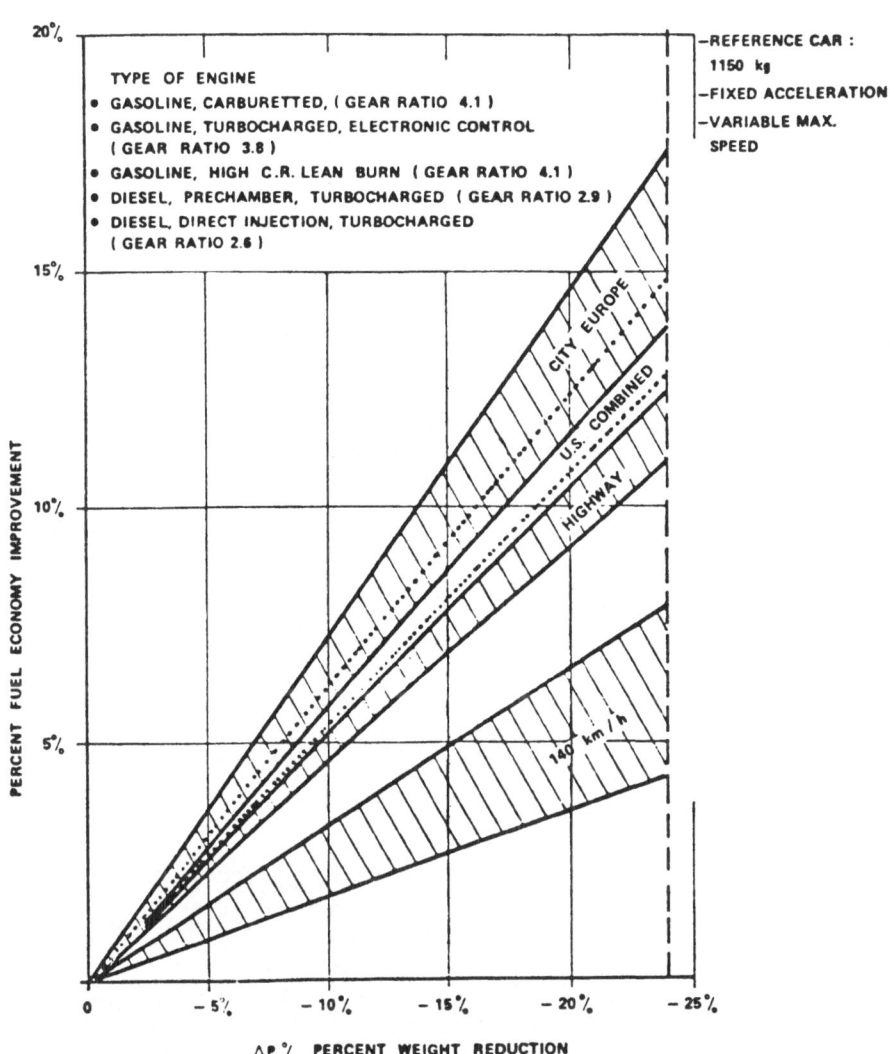

Figure from : FIAT Research Center internal report

Fig. 4.17 - Effect on Fuel Economy of Car Weight

4.3 The impact of car technological innovation on energy conservation

In the preceding paragraph we have shown that the potentiality for innovative change aimed at fuel saving in car is high and achievable. To put some weight of the effect on transport system energy conservation we will here report a case study for the Federal Republic of Germany[29].

In that country the energy use for highway transportation reaches about 18% of the total energy consumption and it is nearly 100% oil-derived, so that the highway transportation is strongly dependent on oil-price and availability.

Through a simulation model one tried to evaluate the possible energy savings attainable with technical interventions on the vehicles (automobiles, trucks and buses). For the auto case, according to the model assumptions, the total car energy request is given by the product of two main factors: the vehicle specific consumption or energy intensity (Wh/Pass.km) and the yearly amount of travelled passengers (Pass.km/year):

$$E\left(\frac{Wh}{year}\right) = f_{VEHICLE}\left(\frac{Wh}{pass.km}\right) \cdot f_{TRAV}\left(\frac{Pass.km}{year}\right) =$$

$$= [(f_\eta \cdot f_G) \cdot (f_{TR} \cdot f_W)] \cdot [f_L \cdot N_V \cdot L_y] \tag{1}$$

where:

f_η = engine specific energy consumption $\left(\frac{fuel.Wh}{mech.Wh}\right)$

f_G = transmission energy consumption $\left(\frac{mech.Wh\ in}{mech.Wh\ out}\right)$

f_{TR} = energy intensity for transportation work $\left(\frac{mech.Wh\ at\ wheels}{total\ weight\ (veh.+load)\ x\ km}\right)$

f_W = vehicle weight coefficient $\left(\frac{total\ weight}{max\ no.\ of\ passengers}\right)$

f_L = load coefficient per vehicle $\left(\frac{max\ no.\ of\ passengers}{avg.\ no.\ of\ passengers}\right)$

N_V = total number of vehicles

L_y = yearly travelled distance per vehicle

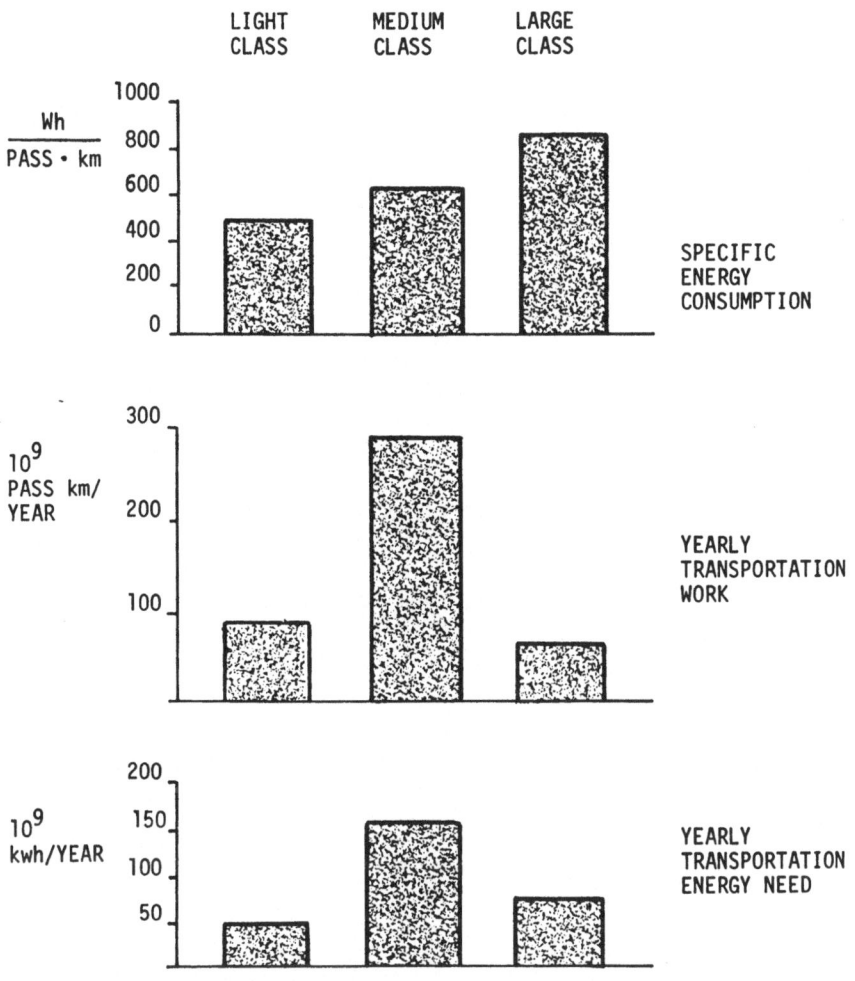

Fig. 4.18 - Federal Republic of Germany - 1978 Situation of
the Passenger Transportation by Car [29]

In other words one can assign to the first product
$(f_\eta \cdot f_G)$, the meaning of the reverse of the efficiency displayed by

| | | ENERGY SAVING FOR MEDIUM CLASS VEHICLE | |
		PARTIAL	COMPOUND
PROPULSION SYSTEM EFFICIENCY	• IMPROVED TRANSMISSION THROUGH ELECTRONIC CONTROL	22%	
	• PROPULSION SYSTEM WITH ENERGY STORAGE	10%	39%
	• INCREASING SHARE OF TURBOCHARGED DIESEL ENGINE	8%	
	• REDUCTION OF AUXILIARIES POWER CONSUMPTION	5%	
VEHICLE EFFICIENCY	• LOW FRICTION TIRES	20%	
	• AERODYNAMIC DRAG REDUCTION	25%	15%
	• FRONTAL AREA REDUCTION	3%	
	• RECUPERATIVE BRAKING	2%	
WEIGHT COEFFICIENT	• WEIGHT REDUCTION THROUGH LIGHTER MANUFACTURING	7%	7%
			52%

Fig. 4.19 - Technical Interventions Aimed at Improving the Car
        Efficiency [29]

the mechanical and propulsion systems, and to the second product
$(f_{TR} \cdot f_W)$ that of the reverse of the "efficiency" related to the
structural, aerodynamic, stylistic design of the vehicle.

    In order to simplify the calculations the model subdivides
the entire auto stock into three vehicle classes:

. light: up to 850 kg
. medium: 850-1200 kg
. large: beyond 1200 kg

    Fig. 4.18 shows the 1978 situation of both the above main
factors and of the yearly energy use in the Federal Republic of
Germany for each of the three vehicle classes. It is clearly seen
that the medium class largely exceeds the other two as far as the
transportation work (pass.km) and the yearly energy need are
concerned.

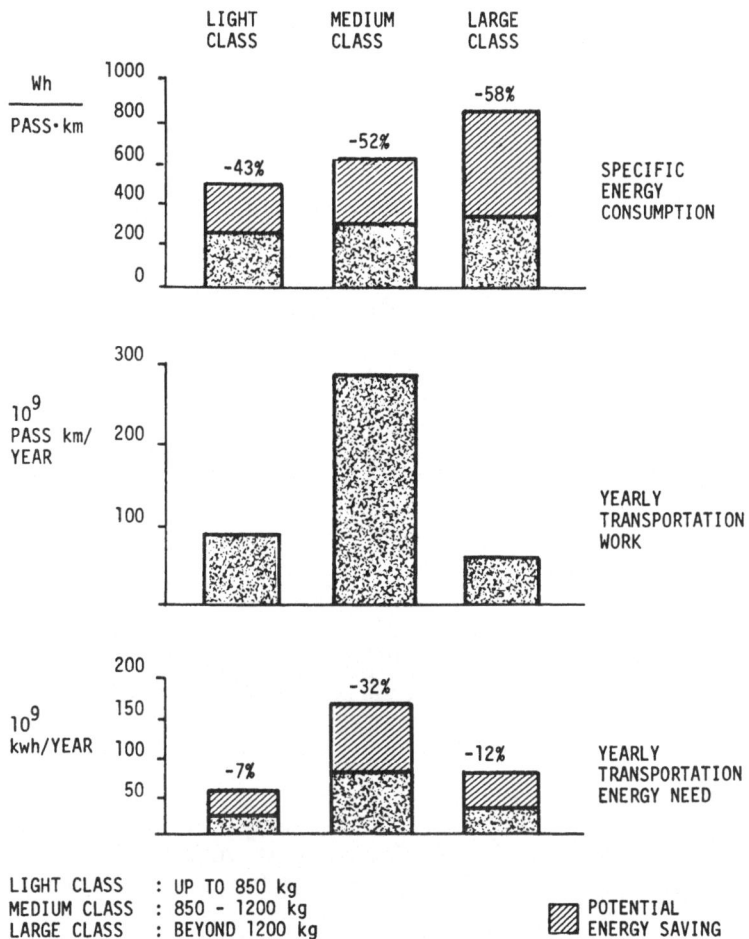

Fig. 4.20 - Federal Repubblic of Germany - 1978 Situation of the Passenger Transportation by Car and Potential Energy Savings Owing to Technical Interventions[29]

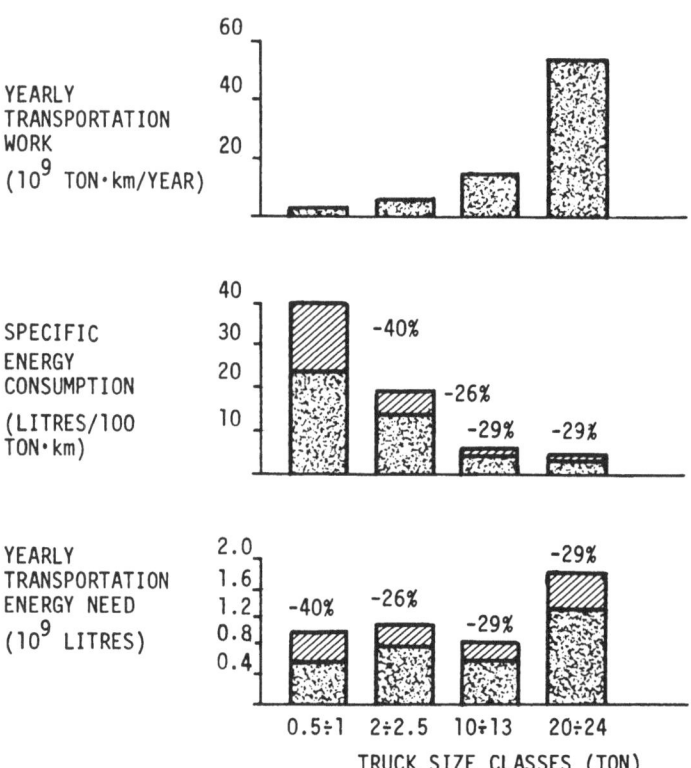

Fig. 4.21 - Federal Republic of Germany - 1977 Situation
          of the Freight Transportation by Truck and
          Potential Energy Savings Owing to Technical
          Interventions[29]

Specific consumption is calculated on the basis of a distance travelled by 1/3 in urban traffic conditions, by 1/3 with a constant speed of 90 km/h and by 1/3 with constant speed of 120 km/h, 1.6 pass/car being the occupancy coefficient.

The technological interventions aimed at improving the car efficiency are listed in Fig. 4.19 together with the percent potential energy saving for the medium class case. The total

	INTERIOR VOLUME	CURB WEIGHT	FRONT AREA	C_D	$A_F \times C_D$ (Sq Ft)	ENERGY SPENT ON CYCLE (HP hrs)			ENERGY PER INT. VOLUME Hp hrs / CU.FT
						PROP.TO WEIGHT	PROP.TO AIR DRAG	TOTAL	
	FT^3	(lb.)	(Sq Ft)			$f=29.10^{-3} xb$	$g=1.02 xe$	$h=f-g$	$i=h/a$
	a	b	c	d	e=cxd				
CLASS A									
AVERAGE	42.38	1559	18.15	0.46	8.35	5.75	8.38	14.12	0.33
BEST GASOL.	44.5	1576	18.3	0.43	7.86	5.8	8.0	13.8	0.31
DIESEL	--	--	--	--	--	--	--	--	--
CLASS B									
AVERAGE	46.4	1874	19.7	0.42	8.29	6.7	8.4	15.1	0.33
BEST GASOL.	48.4	1720	19.7	0.42	8.29	6.2	8.4	14.6	0.30
DIESEL	48.4	1808	19.7	0.42	8.29	6.5	8.4	14.9	0.31
CLASS C									
AVERAGE	48.8	2658	20.8	0.45	9.41	8.9	9.6	18.5	0.38
BEST GASOL.	48.0	2237	19.7	0.46	9.04	7.7	9.2	16.9	0.36
DIESEL	48.0	2557	19.7	0.46	9.04	8.6	9.2	17.8	0.37

Fig. 4.22 - Energy Spent Per Cub. FT of Interior Volume
(on European Cycle)

	FUEL ECONOMY MPG (COMPOSITE EUROPEAN CYCLE)	lb.OF FUEL (ON EUROPEAN CYCLE) $b=\dfrac{62 \text{ miles}}{a} \times \rho$	HP hr SPENT ON EUROPEAN CYCLE	LBS OF FUEL HP hr	ACCELERAT. SECONDS FOR 1000mt (0.62 miles)
	a		c	b/c	
CLASS A					
AVERAGE	30.35	12.67	14.1	0.90	40
BEST GASOLINE	31.0	12.4	14.4	0.86	40
DIESEL	--	--	--	--	--
CLASS B					
AVERAGE	28.9	13.49	15.1	0.89	37.5
BEST GASOLINE	28.9	13.2	14.6	0.83	38
DIESEL	35	12.4	14.9	0.83	38
CLASS C					
AVERAGE	22.8	17.8	18.5	0.96	39
BEST GASOLINE	23.8	16.1	18.3	0.88	34
DIESEL	27.0	18.1	20.3	0.89	44

Fig. 4.23 - Pounds of Fuel per Horsepower Required by the Vehicle
on European Cycle

specific energy consumption can be reduced by 43%, 52% and 58% for
the light, medium and large classes respectively.

With reference to the 1978 situation depicted in Fig. 4.18,
the potential yearly energy savings resulting from the mentioned
technical interventions for the three vehicle classes are shown in
Fig. 4.20. Clearly the bulk of the potential energy saved could be
obtained by technical improvements in the medium class alone. Of
course, non-technical interventions as carpooling, increased
intensity of car use (i.e. occupancy factor) due to increased
leisure time, shift to lighter or superlighter vehicles, walking
and biking up to 2-3 km at least, could be as well beneficial in
improving the situation shown in Fig. 4.20.

The model analogously run for the truck case gave the results
illustrated in Fig. 4.21.

As to the car, it is interesting to visualize[47] in Figs. 4.22
and 4.23 some data on potential benefits presented by recent
successful European cars and related in some way to both products
$(f_\eta . f_G)$, $(f_{TR} . f_W)$ of formula (1). For each class of car considered
in Fig. 4.22 and 4.23 (this subdivision differs from that one
assumed in the case study for Federal Republic of Germany) data
for best gasoline and best Diesel car show already sensible energy
efficiency improvements in comparison with average car.

From these results it is possible to define targets based
upon the best experience and/or future technologies.

REFERENCES

1. A. B. Rose, "Energy-Intensity and Related Parameters of Selected Transportation Modes: Freight Movements", ORNL-5554 (June 1979).

2. A. B. Rose, "Energy-Intensity and Related Parameters of Selected Transportation Modes: Passenger Movements", ORNL-5506 (January 1979).

3. R. U. Ayres, "Worlwide Transportation/Energy Demand Forecast: 1975-2000", ORNL/Sub-78/13536 (October 1978).

4. R. U. Ayres, L. Ayres, P. Patterson, "Uncertain Transportation Energy Futures - U.S. Compared with other Regions of the World: 1970-2000", The International Conference on Energy Use Management, Los Angeles (1979).

5. J. Hooker, "Some Problems of Definition raised by a Transportation Energy Data Base", The International Conference on Energy Use Management, Los Angeles (1979).

6. National Cooperative Highway Research Program, "Energy Effects, Efficiencies and Prospects for various Modes of Transportation", Transportation Research Board National Research Council, Washington D.C. (1977).

7. L. R. Brown, C. Flavin, C. Norman, "Running on Empty: the Future of the Automobile in an Oil-Short World", W.W. Norton & Company, New York (1979).

8. A. Frondaroli, P. Patrucco, "Transport Energy Consumption in two EEC Countries - Part I: Italy", FIAT Centro Ricerche, Orbassano (1977).

9. A. Frondaroli, P. Patrucco, "Transport Energy Consumption in two EEC Countries - Part II: Great Britain", FIAT Centro Ricerche, Orbassano (1977).

10. D. B. Shonka editor., "Transportation Energy Transportation Data Book: Edition 3", ORNL-5493 Special (February 1979).

11. N. N. "L'autotrasporto e le Ferrovie dello Stato di fronte al problema dell'energia", TRANS - Autocarri e Autobus No. 1 (January/February 1978).

12. K. M. Meyer - Abich, "Energieeinsparung als Neue Energiequelle", Carl Hanser Verlag, München (1979).

13. Ministero dei Trasporti, "Libro bianco: I trasporti in Italia", Istituto Poligrafico dello Stato, Roma (1977).

14. Enciclopedia Universale De Agostini (1974-1977).

15. "Energy efficiency in passenger and freight transportation. A comparative study of the Swedish situation", Transport Research Delegation, ISBN 91-85562-19-X (1979).

16. Colombo et al., "Il rapporto WAES-ITALIA: le alternative strategiche per una politica energetica", Franco Angeli Editore, Milano (1978).

17. F. Taylor, "International Economic Indicators - 1974 Edition", Long Range Planning Service Stanford Research Institute, Report Nr. 529 (November 1974).

18. R. Short, "International Economic Indicators", Business Intelligence Program, Stanford Research Institute, Nr. 592 (1977).

19. O. M. Bevilacqua, "Overview of the Potential for Energy Conservation for Intercity Truck", Purdue University, West Lafayette, Indiana.

20. A. F. Burke, "The Moving Baseline of Conventional Engine. Powered Passenger Cars (1975 - 1985)", S.A.E. Technical Paper Series Nr. 780347 (1978).

21. Authors Various: "Transportation Programming, Economic Analysis, and Evaluation of Energy Constraints", Transportation Research Board, National Academy of Sciences, Washington D.C. (1976).

22. D. Maltby et al., "The U.K. surface passenger transport sector", Energy Policy (December 1978).

23. D. Downs, "Tendances dans l'évolution du véhicule automobile de l'avenir". Proceedings IVA '79, Hamburg.

24. R. M. Doggett et al., "Ten Scenarios of Transportation Energy Conservation Using TECNET", Final Report. U.S. Department of Energy, HCP/M2101-1 (March 1979).

25. R. M. Doggett et al., "Further Development and Use of the Transportation Energy Conservation Network (TECNET)", U.S. Department of Energy, HCP/M2101-2 (March 1979).

26. R. E. Carrier, "Energy Conservation through Urban Transportation Planning", Transportation Review (1974).

27. Authors Various, "Energy and Transportation", S.A.E. SP-406 (February 1976).

28. D. A. Hurter, W.D. Lee, "A Study of Technological Improvements in Automobile Fuel Consumption", Automotive Fuel Economy, S.A.E. (1976).

29. J. Helling, G. Lerner, "Der Künftige Erdölbedarf des Strassenverkehrs - Eine Untersuchung Alternativer Szenarien", XVII FISITA International Congress, Hamburg (May 5-8, 1980)

30. W. Owen, "Transport, Energy, and Community Design", Futures (April 1976).

31. "Recent developments in urban transportation", NATO - CCMS, Report Nr. 82 (June 1978).

32. C. J. Leising et al, "Waste Heat Recovery in Truck Engines", S.A.E. Technical Paper Series Nr. 780686 (August 1978).

33. Atti del 1° Seminario Informativo su "Conservazione di energia nel campo della trazione", C.N.R. - Progetto Finalizzato Energetica.

34. J. Holroyd, D. Robertson, "Strategies for Area Traffic Control Systems: Present and Future", TRRL Report LR-569 (1973).

35. D. Owens, J. Holroyd, "The Glasgow Experiment: Assessments under Light and Very Low Flow Conditions", TRRL Report LR 522 (1973).

36. N. Gartner, J. Little, H. Gabbay, "Steady State Traffic Optimization by MITROP", Proc. IFAC/IFIP/FORES 3rd Int. Symp., Columbus, Ohio (1976).

37. D. Robertson, "TRANSIT: a Traffic Network Study Tool", TRRL Report LR-253 (1969).

38. G. Menga, P. Patrucco, "Area Traffic Control: a Decentralized Approach", presented to EURO-IFIP CONFERENCE (1979).

39. M. Inose, "Road Traffic Control with the Particular Reference to Tokyo Traffic Control and Surveillance", IEEE Proceedings vol. 64 Nr. 7, pp. 1028-1039 (1976).

40. D. Inaudi, G. Tagliaferri, "Urban Traffic Dynamic Simulation: a Microscopic Approach", First World Congress on Matematics at the Service of Man., Barcellona (1977).

41. H. Strobel, "Traffic Control Systems Analysis by Means of Dynamic State and Input - Output Model", IIASA-RR-77-12 (1977).

42. M. Ternullo, "Analysis and Interpretation of Fuel Consumption in Urban Traffic", FIAT Centro Ricerche, Internal Report, Orbassano (1978).

43. Evans, Herman, Lam, "Gasoline Consumption in Urban Traffic", Research Laboratories General Motors Corporation, Warren, Michigan 48090, Research Publication GMR - 1949 (February 1976).

44. Chang, Evans, Herman, Wasielewski, "The Influence of Vehicle Characteristics, Driver Behaviour, and Ambient Temperature on Gasoline Consumption in Urban Traffic", Research Laboratories General Motors Corporation, Warren, Michigan 48090, Research Publication GMR - 1959 (January 1976).

45. Evans, Herman, Lam, "Multivariate Analysis of Traffic Factors related to Fuel Consumption in Urban Driving", Research Laboratories General Motors Corporation, Warren, Michigan 48090, Research Laboratories Publication GMR - 1710 (October 1974).

46. "Energy balances of O.E.C.D. countries", O.E.C.D. (1976).

47. U. L. Businaro, G. Bolognesi, "Trends in European Car design", International Conference on Energy Use Management, Los Angeles, California (October 22-26, 1979).

THE APPLICATION OF TECHNOLOGY FOR ENERGY CONSERVATION

IN INDUSTRY

W. M. Currie

Energy Technology Support Unit
of the UK Department of Energy
at AERE Harwell, England

INTRODUCTION

In these lectures I shall be concerned with the achievement of more efficient industrial energy utilisation through the development and application of technology. My approach will be essentially pragmatic, that is to say, I am less concerned with developing theories, writing papers and giving lectures than with getting results. Therefore, not everything will be rigorous. I shall be particularly concerned with RDD&D, i.e. research, development, demonstration and deployment. Once upon a time people spoke only of research; then they began to speak about R&D; then it was appreciated that more attention had to be given to the further phase of demonstration; more recently, particularly in the United States, it has become fashionable to add deployment; and so we have this elaborate concept of RDD&D, essentially lengthening the innovation process, and I shall be discussing the relevance of this to energy conservation.

I should also make it clear that the views expressed in these lectures are personal views; they do not necessarily reflect the views of the British Government or the UKAEA. I will, however, be drawing heavily on UK data and cases for the purposes of illustration. This is simply because it is the easiest thing for me to do, but much of what I have to say will be representative of the situation in similar countries such as the USA, Italy, France, Germany etc. Indeed, my argument could be regarded as an explanation of the rationale underlying the broad approach adopted by several countries to energy conservation RD&D in industry.

Right at the outset I would like to emphasise that there is

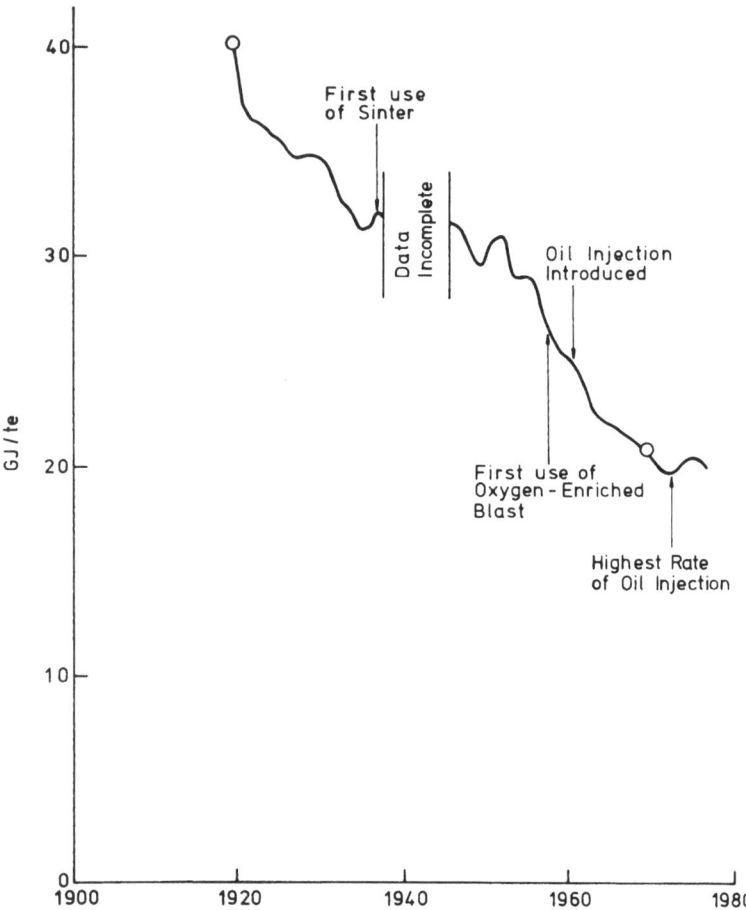

Fig. 1.   Coke and Oil Input to Blast Furnaces and Sinter Plants,
          Per Unit of Pig Iron Produced.

Source: Energy Research Group, U. of Cambridge.

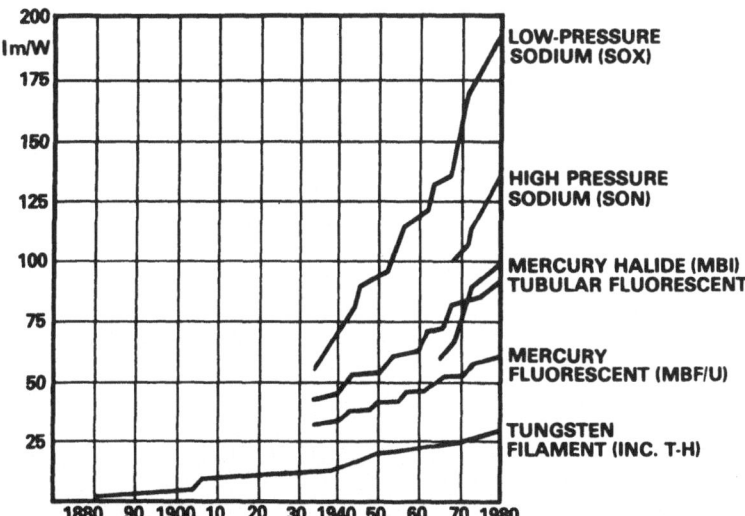

Historical chart of lamp development by Philips. Note: The efficacy figures
(vertical scale) are the maxima obtainable in each lamp type, and do not include
gear losses where applicable.

Fig. 2.   Improved Efficacy for Various Lamp Types.

Source: Energy Management, Jan. 1980, Dept. of Energy.

really nothing new about energy conservation. Over two thousand
years ago Cicero said that, "Economy is of itself a great revenue",
which would make a very good slogan for one of today's energy
managers. More recently, since the Industrial Revolution,
manufacturers have always been concerned with producing more
product and more profit from fewer resources. Energy is one of
those resources and so over the years there has been a great deal
of technological innovation to achieve higher efficiency of energy
use. This is particularly true in the energy intensive industries,
no more so than in those industries which themselves produce
energy.

For example, the average thermal efficiency of electricity
generation using steam, in England and Wales, has been improved
nearly fourfold over the fifty year period 1930/80. Some of this
achievement comes from the application of heat recovery, a topic I
shall be emphasising later. Indeed, the biggest heat recovery heat
wheel in the world is installed at a CEGB power station in the UK.
Over the same 50 year period the energy requirement in gigajoules
per tonne of output from blast furnaces in the UK has been reduced
by about half, as illustrated in Fig. 1, and in cement production
the energy requirement per tonne of clinker has been reduced by
40%. On a completely different tack, the maximum efficiency
achievable in lighting appliances has been improved more than
tenfold over the past 100 years, as illustrated in Fig. 2. These

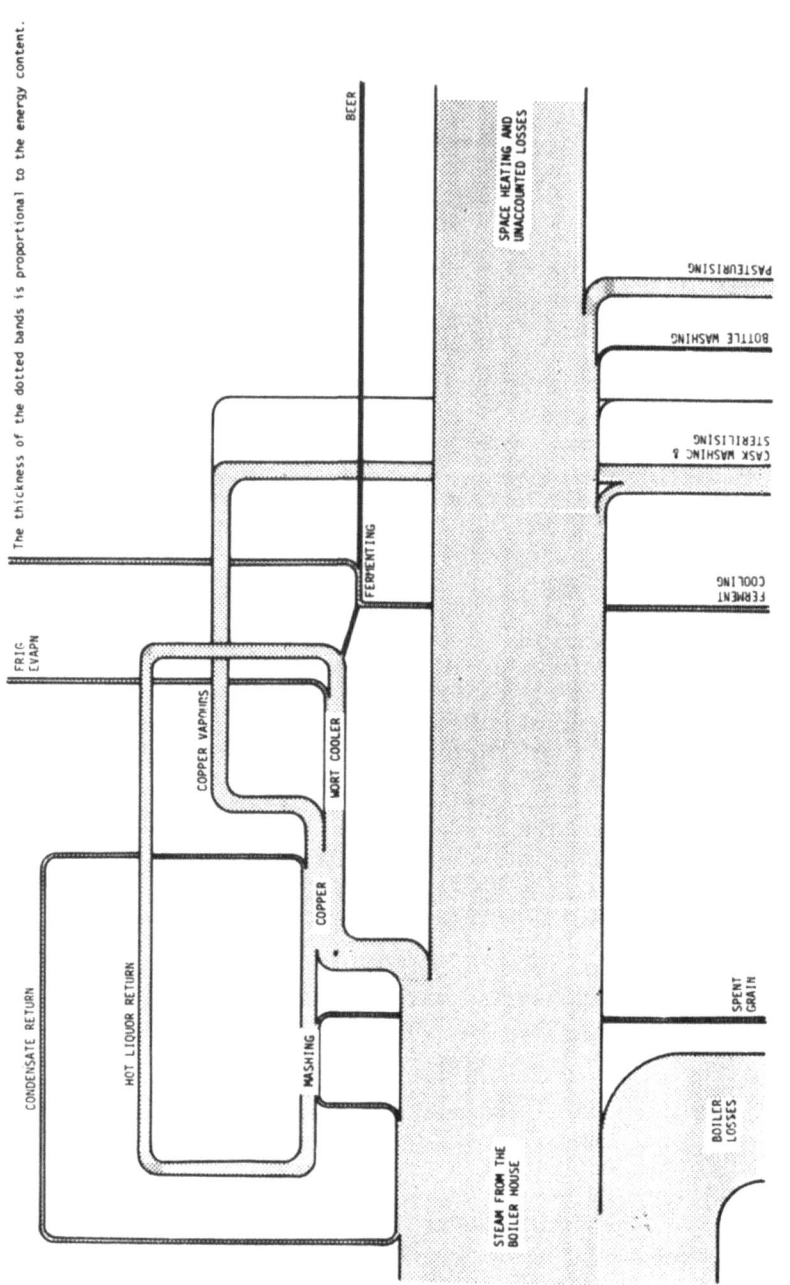

Fig. 3. Heat Balance of a Brewery Studies by Lyle in the 1930's.

Source: Ref. 1.

are dramatic reductions and it is against such figures that we have
to consider the prospects for future improvement.

Even in the non-energy intensive sectors of industry there has
for long been a concern with energy efficiency, certainly not a
dominating concern, but a concern nevertheless. For example, there
is a book which is now out of print in the UK, called "The Efficient Use of Steam" by Sir Oliver Lyle[1]. During the 1930's Lyle
was carrying out what we today would refer to as "energy audits" in
a variety of industries, including the brewing industry, where
energy is a fairly minor consideration in the overall economics.
Fig. 3 shows the results of one of Lyle's energy audits. You can
see that a very large part of the heat input from the boiler comes
out at the right hand side as "space heating and unaccounted
losses." In his book Lyle concluded that, "In this brewery 55% of
the steam is going about its private business unknown to the
management. It is up to the management to find out what these
private affairs are, always provided it is worthwhile." That was
before the so-called energy crisis. So what is new? Why all the
fuss? Is there really an energy crisis?

My own view is that there isn't a crisis as such. There were
two crises in the 1970's and there will certainly be more. But
crises are things which pass. The world is not going to run out of
energy. There is plenty of energy around. What is new is that the
cost of energy is likely to increase relative to other costs, and
there is now a widespread perception of this. Moreover the cost
movements may include sudden changes. Hence I would say there has
been a change of climate. Instead of energy being a cheap and
reliable resource which need not concern management greatly unless
it constitutes a very substantial proportion of the production
costs, it is likely to become an increasingly expensive and
possibly unreliable resource, and so we have to invest other
resources to finding ways of using it efficiently. And just as the
increasing cost of labour drives manufacturers towards automation,
so the increasing cost of energy must drive them towards energy
conservation through capital/energy substitution and technological
development. If there is a crisis at all it arises from the
incompatibility of long technological lead times with the short
term and fickle variations of the international oil market. It is
my task to consider how technology might contribute to greater
industrial energy efficiency and the ways and means of trying to
ensure that it does this.

My approach is first of all to examine the scope for conservation industry. This will take us through the subject of energy
analysis to the potential for conservation and the broader considerations that must be understood if one is to reach sensible conclusions about cost effective RD&D. I will then consider the
general difficulties of technological innovation and attempt to

draw out a few lessons of relevance to conservation. After that I
will discuss the particular difficulties of conservation - what I
refer to as the "barriers". Given the potential and the
difficulties I will then sketch out in broad terms the strategic
approaches that might be adopted, e.g. by governments, if limited
resources are to be deployed effectively in promoting conservation
innovation in industry. Finally I shall present a few case studies
which I hope, will show that the general methods and conclusions
discussed earlier really are relevant to the achievement of
practical results.

ENERGY ANALYSIS AND TECHNOLOGY ASSESSMENT

Faced with increasing energy costs the first step for a householder, company, country, or indeed the entire Western World, is to
find out how energy is used and evaluate the possibilities for
economy. Despite what I have said about the past there has never
been any systematic effort on a national scale to measure energy
use and assess the technological possibilities for conservation.
Hence, since 1973, in several countries there have been programmes
to do this. It is sometimes called energy analysis, sometimes
energy auditing and sometimes energy balances, by analogy with
financial auditing and balance sheets. Whatever term is used the
aims are broadly the same. Most countries also generate statistics
which can provide some sort of guide as to where energy is used.
However, statistics cannot really guide one towards technological
developments. One needs a lot more information than that. Nevertheless, let us spend a few minutes on statistics.

Energy Statistics

Most countries produce statistics from which information about
energy use in different industries can be extracted. I have only
very limited knowledge of other countries' approaches and so I can
tell you only about the UK. First of all, what do we mean by different industries? Most countries use some form of industrial
classification and in many cases this is based on the International
Standard Industrial Classification of all Economic Activities which
is issued by the United Nations. In the UK they maintain a
"Standard Industrial Classification" which takes account of the
United Nations classification. It breaks industry down into 27
"orders" and each order is further subdivided into what are called
"minimum list headings" or MLH's. There are about 140 minimum list
headings in the UK. That may seem a lot of detail but for our
purposes it is barely sufficient. For example, MLH 461 is
concerned with "bricks, fireclay and refractory goods." It is
further divided into two sub-groups, refractory goods and non-
refractory goods. These are both energy intensive sectors, but
they are rather different in the types of raw materials and kilns
which they use and also in the products which they produce. So

Industry Group	bought-in energy cost £M	energy* intensiveness by output	energy‡ intensiveness by value added
Iron and steel	207.03	.129	.411
Iron castings	38.34	.112	.230
Motor vehicles	36.14	.014	.038
Paper and board	33.74	.074	.198
Gen. chemicals (inorganic)	32.85	.097	.215
Cement	27.76	.263	.571
Gen. chemicals (organic)	27.32	.059	.189
Bricks and refractories	21.36	.116	.206
Glass	19.54	.078	.141
Steel tubes	18.70	.065	.160
Misc. building materials, etc	16.52	.037	.084
Resins, plastics materials and synthetic rubber	16.17	.034	.082
Rubber	14.92	.029	.055
Aerospace equipment manufacturing and repairing	14.91	.017	.031
Manmade fibres	14.02	.042	.086
Misc. metal goods	13.66	.017	.033
Bread and flour confectionery	13.50	.029	.065
Fertilizers	13.42	.059	.189
Aluminium and alloys	11.58	.038	.111
Brewing and malting	11.23	.011	.038
Gen. chemicals	10.67	.040	.088
Woollen and worsted	10.71	.019	.054
Gen. mechanical engineering	9.71	.018	.030
Milk and milk products	9.64	.012	.067
Textile finishing	9.42	.076	.119
Electrical machinery	8.83	.015	.028
Spinning and doubling	8.70	.028	.077
Shipbuilding and engineering	8.53	.019	.031
Misc. base metals	8.27	.022	.122
Copper brass and alloys	8.07	.014	.074

* i.e. energy purchased as a fraction of total output

‡ i.e. energy purchased as a fraction of (total output minus material costs)

Table 1.   Top 30 UK Industries by Energy Use.

Source: ETSU Internal Paper.

although they have a lot in common it is necessary to understand
the differences and, unfortunately, one can't get any of these
differences from the available statistics.

The principle source of published energy statistics in the UK
is the "Digest of Energy Statistics" published annually by the
Government Statistical Service. This details the production,
stocks and sales of the major fuels, it summarises prices, aggregate sectoral demand and overseas trade in energy products and
provides miscellaneous statistics such as GDP growth rates, mean
air temperatures and wholesale price indices. There is also a
shorter term publication by the Department of Energy called "Energy
Trends" which comes out monthly and there are additional publications by the nationalised fuel industries, the Institute of
Petroleum and the British Steel Corporation. The chief shortcoming
of these various statistics is that they are insufficiently disaggregated for our purposes. Another point is that they consist
entirely of averages, while on the technological front one wishes
to know also about the variations in energy use, the variations
between good practice and bad practice for example.

Even if we can get complete information about energy consumption this is still of little use. It is energy consumption per
unit of output that matters. In fact, my entire concern in these
lectures is with how to produce more goods, cost effectively from
less energy. Whether absolute energy consumption goes up or down
does not matter so long as the energy utilisation is increased.
Therefore one needs to be able to compare energy consumption with
some measure of industrial output, and that brings in other statistics. In the UK the only place where production and energy consumption statistics are gathered together is the "Census of Production" a major statistical review which is carried out every five
years or so. This entails a survey of all establishments employing
over 25 people. A great deal of detailed information is collected
and the big problem is how to analyse it all.

I would like to show you briefly what came out of such a
statistical analysis carried out by ETSU in 1974. The analysis was
based on the 1968 Census of Production. Table 1 is a tabulation of
the results for the top thirty energy consuming sectors of British
industry. It shows the value of energy bought in by each sector,
then the energy intensiveness by output, i.e. the ratio of the
energy purchases to the sales output for the industry, then the
energy intensiveness by value added, i.e. the ratio of sales
purchases to the value added. This last quantity indicates those
industries which consumed the highest amount of energy (in terms of
value) to achieve their purpose. This analysis was used to select
sectors for more detailed investigation. Let us turn now to what I
mean by detailed investigation.

Energy Analysis

The man who is running a factory is interested only in the energy that he purchases, what it costs him and how efficiently he can use it. This is the so called "heat supplied." The fact that one unit of electricity may require three units of fossil fuel to produce it is of little concern to him so long as the price is right. But if we are looking for technological developments that will achieve overall energy economy for the world then we have to think in terms of primary energy. Not only that, we have to understand the total energy usage throughout the entire sequence of operations in a production and use system. As a simple example, one can make motor cars lighter by constructing them of aluminium, so that they use less fuel. But aluminium requires more energy to produce than steel, weight for weight. So it may be a false economy to do this. Hence we need to examine the detailed ways in which the choice of material, process, system, etc. affects the total energy consumption. This is what we mean by energy analysis and after the 1973 energy crisis there was a great deal of excitement and interest in the effects of this.

The subject was considered so important that a workshop was convened in 1974 by the International Federation of Institutes for Advanced Study (IFIAS) and this workshop agreed certain conventions for international use, the so-called IFIAS Conventions[2]. These are widely used throughout the world but even so it is often still difficult to make comparisons between a particular industry in one country and in another. Several of the collaborative projects being developed under the auspices of the IEA include, as an initial step, an investigation of energy auditing methodology, following on from another report, the NATO/CCMS data base study published by the US Department of Energy[3]. This report is a particularly useful one for anyone starting an energy analysis exercise.

The basic ideas are shown in Fig. 4. First of all there are different levels of energy analysis. The first level considers the direct energy input to the manufacturing process plus any direct use of transport. The second level considers the energy sequestered in the materials that are fed into this process and the transportation energy used to make those materials. The third level considers the energy to extract the raw materials used to make the intermediate materials, plus transportation energy, plus the energy used to make the machines used in the final production process. Finally, there is the energy used to make the machines to make the machinery, plus the transportation energy in that. Obviously one could go on for ever in this way, so the guiding principle is that one probes down into these levels as deeply as is necessary to establish any energy components that are comparable in value to the uncertainties in the final figures or to the variations or potential savings that one is looking for.

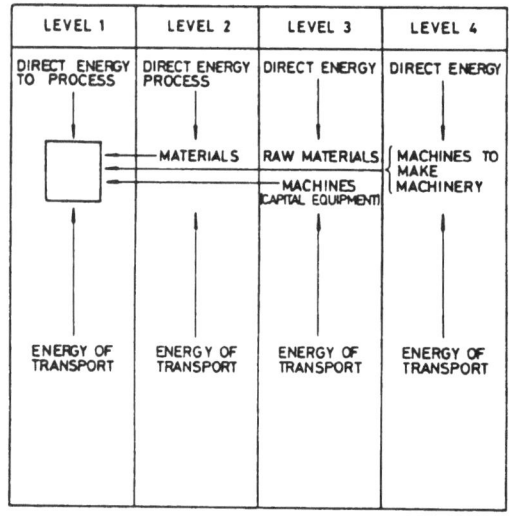

Fig. 4.   The Levels of Energy Analysis.

Source: Ref. 2.

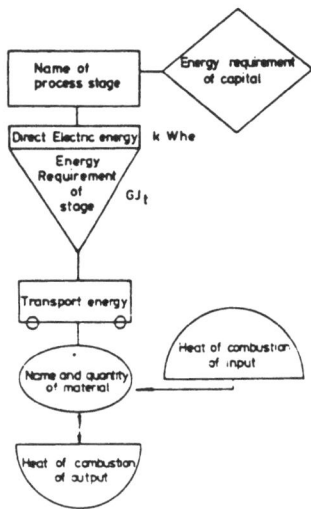

Fig. 5.   Energy Analysis Flow Chart Symbols.

Source: Ref. 2.

This can actually be of some significance in reaching conclusions. I'll give you two examples. In our study of the
refractories industry we came to the conclusion that certain
measures which might reduce energy consumption in refractories
production would not be worthwhile overall because they would
lead to refractories with a shorter life and since these
refractories are subsequently used in steel production, the
effect could be a greater overall energy consumption in the
production of steel. Also, in our study of the dairy industry,
because of the very large amounts of energy already sequestered
in milk before it ever goes into the dairy, we concluded that
any measures to save energy in dairies which incidentally
increased milk losses would be counter-productive because the
amount of process energy saved would be less than the amount of
sequestered energy lost in the milk.

The IFIAS workshop also agreed various standard symbols
for indicating the energy consumption in a flow chart, as shown
in Fig. 5. Fig. 6 shows how these symbols, along with the
actual numbers measured, can be combined into a flow chart for
a production process. In this case it happens to be a
brickworks. Such a chart helps an investigator in his
discussions with an industry to make sure that he really has
understood the way bricks are made and the various energy
inputs. You will also see at the bottom of Fig. 5 an equation,
GER = 3.49 GJ$_t$/tonne, which brings me to another set of
concepts that were agreed by the IFIAS workshop.

In global terms the most important quantity is the "Gross
Energy Requirement" (GER) which is the total amount of energy
sequestered in a product. Thus the gross energy requirement to
produce a pint of milk is $\sim$5.2 MJ. This includes the energy
to produce the fertilisers, to grow the grass, to feed the
cows, to produce the milk to be processed in the dairy, as well
as the energy of transportation of the delivered milk, of the
raw milk and so on. However, if one is looking only at energy
conservation in the dairy then one is more interested in the
Process Energy Requirement (PER), which is the energy requirement
for processing the product, within the boundary of the dairy
itself. In the case of UK milk the process energy requirement is
0.38 MJ/pint. As a result of a great deal of work by industry,
governments, research bodies, universities and consultants there
is now a great deal of published information on GER's and PER's
for a wide range of products. But once again these numbers are
usually averages and one still needs to know rather more than that
if one is going to formulate conclusions and recommendations about
technological change.

We have often discussed in ETSU the value of energy analysis.
A few years ago some people were arguing in the literature that

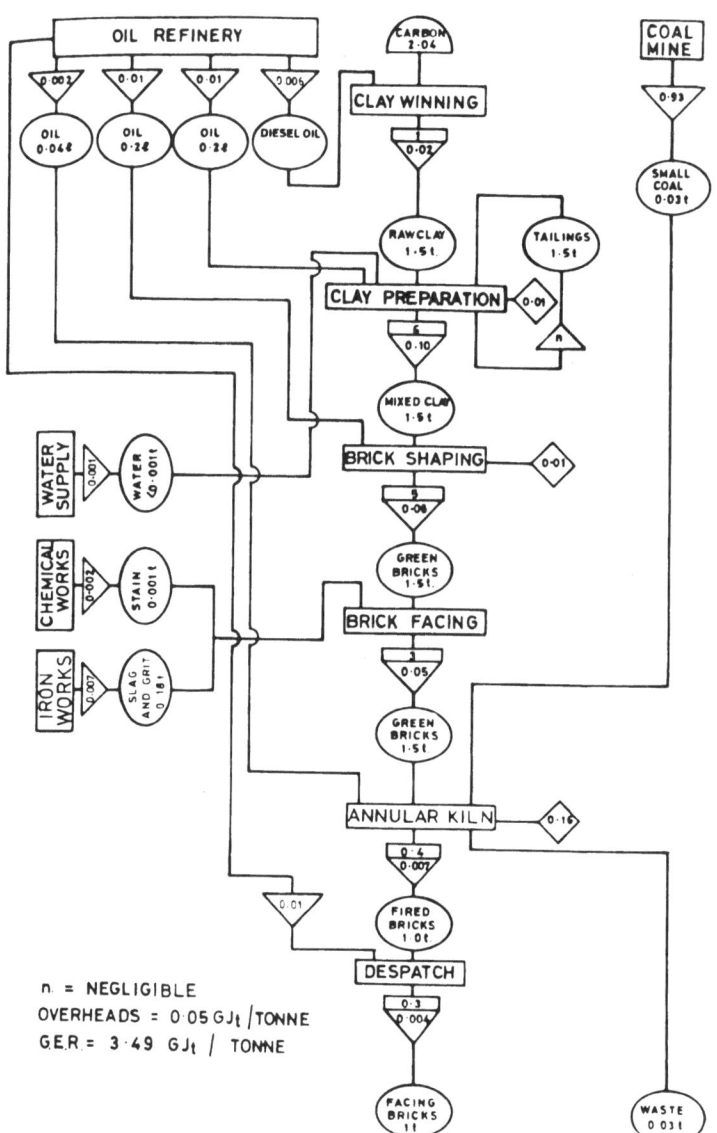

Fig. 6. Level 3 Analysis of Fletton Brickworks Making
Facing Bricks. Source: UK Energy Audit Report No. 2.

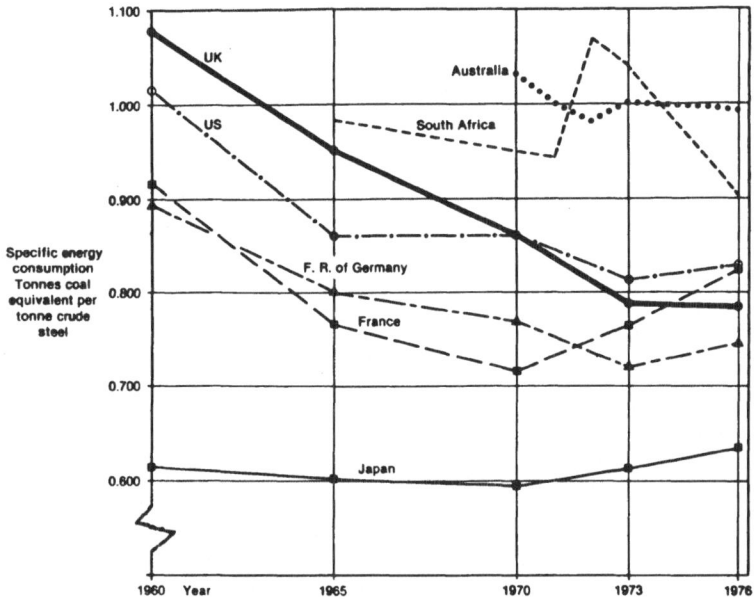

Fig. 7.   International Comparisons – Energy Per Tonne
          Crude Steel Production.   Source: Ref. 21.

energy analysis pointed up different conclusions and decisions from
those reached by more conventional means. We have considered this
but we have not been able to identify any such specific decisions,
eg concerning investment or R&D, that are different from what
economics would dictate. This is because in energy intensive
activities the energy inputs are reflected in the costs while in
non-energy intensive activities the energy inputs are of minor
significance. Nevertheless I believe it is a good discipline to go
through the sort of exercise just described. It clarifies the
energy aspects of a process, it prevents the pursuit of possibly
harebrained schemes and it sometimes reveals surprises, e.g. that
milk is energy intensive, requiring one third of its weight in oil
to produce.

     As an illustration of how careful one has to be, Fig 7, shows
data for the energy efficiency of steel production in several
countries. It suggests, for example, that the UK is very
inefficient and Japan highly efficient. A great deal has been
published in the press and technical literature about this and I
have heard disturbing comparisons drawn at IEA meetings. However,
when one looks into the data one finds that the information from
the different countries is not comparable. For example, one has to
make allowance for the proportion of ore which is imported, one has
to know the grade of ore used (a low grade ore requires more energy
than a high grade ore), the extent of recycling of scrap iron and
steel, the efficiency of electricity generation and the final

product mix. The curves given in Fig. 7 are not consistent with
respect to these factors and are therefore meaningless. In fact,
ETSU has concluded that one can only make such international
comparisons between particular process operations if one has the
sort of additional information I have just indicated. Energy
analysis can therefore help us to avoid spurious conclusions and
irrelevant debate. Let us therefore look at what is actually
involved in carrying out an energy audit.

Doing an Energy Audit

In many of the energy intensive industries it is traditional
practice to carry out what is called an energy balance. For
example, in the UK cement industry they have used a standardised
procedure since the early 1940's for carrying out an energy balance
on a cement kiln. This entails measuring the energy inputs to the
kiln and the energy outputs and making sure that they balance. If
they don't balance the engineer knows that something is wrong. If
they do balance then in all probability he knows where his energy
is going, and if he is consuming too much energy he can begin to
investigate ways and means of reducing it. Thus, an energy audit
is a study of the energy inputs and the outputs, including losses,
and both of these should balance if one has done the job properly.

The main things which one has to investigate in the energy
inputs to a process are as follows:

(a) Steam, heat transfer fluid, gas, oil or coal used for
     direct plant heating.

(b) Electric power to fans, compressors, pumps etc. and for
     motivation of plant.

(c) Heat energy already present in ambient air supplies and
     the humidity condition of the air.

(d) Energy present in the product entering the process
     plant.

(e) Also required is a record of the Energy Source costs, the
     product cycle e.g. drying time and the fractional use
     time of the plant per annum.

In the case of intermittent plant usage, it is also necessary to
obtain an estimate of the total metal weight of the plant so as to
assess the heat required to bring it up to operating temperature.
This can be quite important. For example, in the refractories
industries it has become increasingly popular in recent years to
use intermittent kilns rather than continuously burning kilns.
Provided the kilns have lightweight structures this can achieve

substantial energy savings as well as achieving higher productivity
and other benefits. The point is that you really have to
understand in detail what energy is being put into the process, and
what it is being put in for.

Equally, you have to understand very thoroughly where the
energy is going to, i.e. the energy output. The main things which
will have to be measured or investigated are as follows:

(a)   Housekeeping losses - steam leaks, hot air or gas escape
      from plant.

(b)   Radiation losses from the machine casing.

(c)   In the case of non-continuous operation, transient
      machine heating requirements.

(d)   Gas or air outflow losses (sensible and latent) from the
      process to atmosphere and, where direct gas or oil firing
      is used, the composition of the waste gases.

(e)   Cooling air losses.

(f)   The condition of the product leaving the plant (e.g. its
      temperature).

Some of these are not trivial. For example, it is not sufficient
to measure just the temperature of the gas or air outflow. It may
also be necessary to measure humidity so as to quantity latent heat
as well as sensible heat. It is also necessary to record details
of the production operation and output if one is to arrive at the
specific process energy requirement.

When one has made all these measurements the results should be
set out in a systematic way, perhaps as follows:

(i)   Details of product throughput.

(ii)   Details of heat source, moisture contents of the air in
       and product in and out. Also required is total moisture
       removed and total energy input to the plant.

(iii) Details of the exhaust air or gases - velocity, humid
       volume, equivalent weight of cold dry air at ambient
       temperature and the total moisture in the exhaust air or
       gases.

(iv)   A heat balance - the total energy output which should,
       within reasonable limits, agree with the total energy
       input given above.

(v)   An assessment of possible heat recovery, on whatever
        basis, may be chosen for calculating the potential heat
        recovery in relation to the type of heat exchanger to be
        used.

(vi)   Also, based on cycle time and the fractional use time of
        the plant, the amount of heat that could be recovered per
        annum.

(vii) At the quoted energy source cost, equivalent cost savings
        per annum.

It is quite amazing how many so-called professional energy consultants fail to incorporate all these points when making an energy
audit for a company. Indeed, although this is all really very
elementary it is often rather badly done and unless it is done
thoroughly one doesn't have a satisfactory basis on which to arrive
at recommendations.

By way of example I have prepared as an Appendix to the
lectures some sheets which contain details of an actual energy
audit carried out on a loose stock fibre drier in the woollen
industry. On sheet 3 of the Appendix you can see that the sum of
the energy inputs comes to 685.3kW while the energy outputs add up
to 678.9kW. There are therefore only 6.4kW of unaccounted losses
or, put another way, an error of less than 1%. This audit was
carried out to see if it would be cost-effective and technically
reliable to recover heat from the drier using a newly developed
prototype heat pipe heat exchanger unit. It is a case to which I
shall return later.

Summing up then, given some broad statistics which one can
acquire through desk research, together with field research
according to the principles of energy analysis which I have
outlined, one ought to be able to go into a particular industry and
carry out a series of measurements and investigations so as to
arrive at a picture of how energy is used in that industry. That's
all very well, but what we would really like to know is by how much
the specific energy consumption in the industry could be reduced
and what technological means need to be deployed or developed to
achieve that reduction. That takes us into a much more difficult
area. One needs to know what technology is available; how much it
costs; its performance and reliability; and one needs to know
additional things about the process. For example, if one is thinking about heat recovery what are the possibilities of fouling and
corrosion of the heat exchanger surfaces? Most heat exchangers get
fouled up and as they do their effectiveness is reduced so that
they don't achieve what the sales literature says they are capable
of. This may necessitate stopping production, or expensive
maintenance, or frequent replacement of heat exchangers. It may be

so bad that the manufacturer throws the heat recovery equipment out.

In Sweden they have found that for such studies it is a good idea to put two people on the job. The first person is an energy specialist who knows all about energy measurements, heat balances etc; the second person is a specialist in the particular industry concerned who knows all about the products and the processes, the managerial attitudes and the workforce and the peculiarities of the business. These two people together can usually come to a satisfactory evaluation of energy use and the possibilities for improvement. Ideally one would like to have an equipment specialist available also who knows about the capabilities of available equipment and the developments which are in the pipeline. In practice one man usually has to do it all consulting with other experts as best he can.

I would now like to tell you briefly about two of the approaches adopted in the UK for investigating industrial energy use and the prospects for energy conservation. The UK Departments of Energy and Industry have since early 1975 sponsored two major programmes of investigation into industrial energy use. These are "The Industrial Energy Thrift Scheme" (IETS) which has been sponsored by the Department of Industry, and the "Energy Audit Studies" (EAS) which have been sponsored jointly by the two Departments. My main involvement is in the EAS but I would like also to say something about the IETS.

UK Industrial Energy Thrift Scheme

This Scheme was launched in 1975 and had three main aims, as follows:

i)   To gather information on existing patterns of energy use in manufacturing industry;

ii)   To gather information on energy saving opportunities and the need for further research and development directed towards the improvement of energy utilisation;

iii)   To encourage the more efficient use of energy in manufacturing industry through improvements, process efficiency and the adoption of good housekeeping practices.

The Scheme has involved a campaign of one-day confidential visits to industrial concerns. Nearly 3,000 such visits have been made. In each case a confidential report is produced for the company visited, with recommendations, and the results are also aggregated statistically to arrive at broad conclusions for each particular industry sector and also for industry as a whole. For each sector

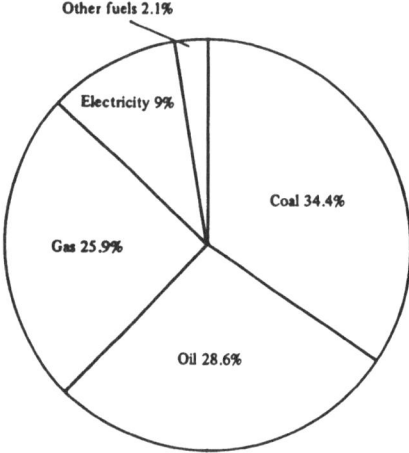

8a. Whole Sector
     (based on 61 sites)

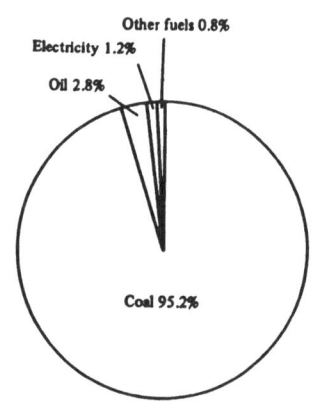

8b. Explosives and
     Fireworks (4 sites)

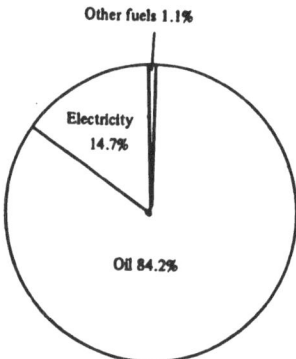

8c. Formulated Pesticides
     (6 sites)

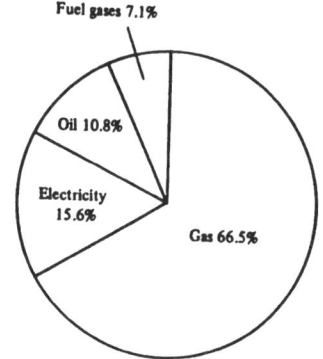

8d. Photographic Chemical
     Materials (8 sites)

Fig. 8. Energy Use in the Miscellaneous Chemical
         Manufacturing Industries.

         Source: UK IETS Report No. 18.

the appropriate research association has been contracted to make
the visits and produce the reports. These research associations
are institutions somewhat peculiar to the UK. Each one caters for
a particular sector and is largely supported by it through
subscriptions from members and also through R&D contracts both from
companies and from government. Some of the RA's also include
foreign members.

These RA's carry out the visits, put confidential reports to
the companies, prepare sectoral reports for the Department of
Industry, organise seminars to disseminate the results and generally promote interest in energy conservation. Also, they carry out
follow-up visits to see to what extent the recommendations have
been implemented. As a result, a great deal of information has
been gathered together and analysed and it is also claimed that the
Scheme has stimulated action in industry which is saving energy for
the nation valued at many times the cost of the Scheme.

The Scheme covers 125 of the minimum list heatings which I
explained earlier. So far 18 reports have been published and a
further 12 are in preparation. I can illustrate the results from
the latest report in the series, which is concerned with MLH 279,
the Miscellaneous Chemical Manufacturing Industries. The list of
the sub-sectors concerned is as follows:-

279.1   Polishes for wood, leather and metals

279.2   Formulated adhesives and gelatine

          (a)   Formulated adhesives, rubber adhesives,
                  glues

          (b)   Gelatine

279.3   Explosives and fireworks

          (a)   Explosives

          (b)   Fireworks

279.4   Formulated pesticides

279.5   Printing inks

279.6   Surgical bandages

279.7   Photographic materials

Fig. 8a shows the breakdown in fuels used by the industry as a
whole. However, this conceals very wide variations. For example,
the explosives industry uses more than 90% coal, as shown in Fig.
8b, while the pesticides industry uses 84% oil, as shown in Fig 8c,
and the photographic materials industry uses 66% gas, as shown in
Fig. 8d. This illustrates the very wide variations that can occur

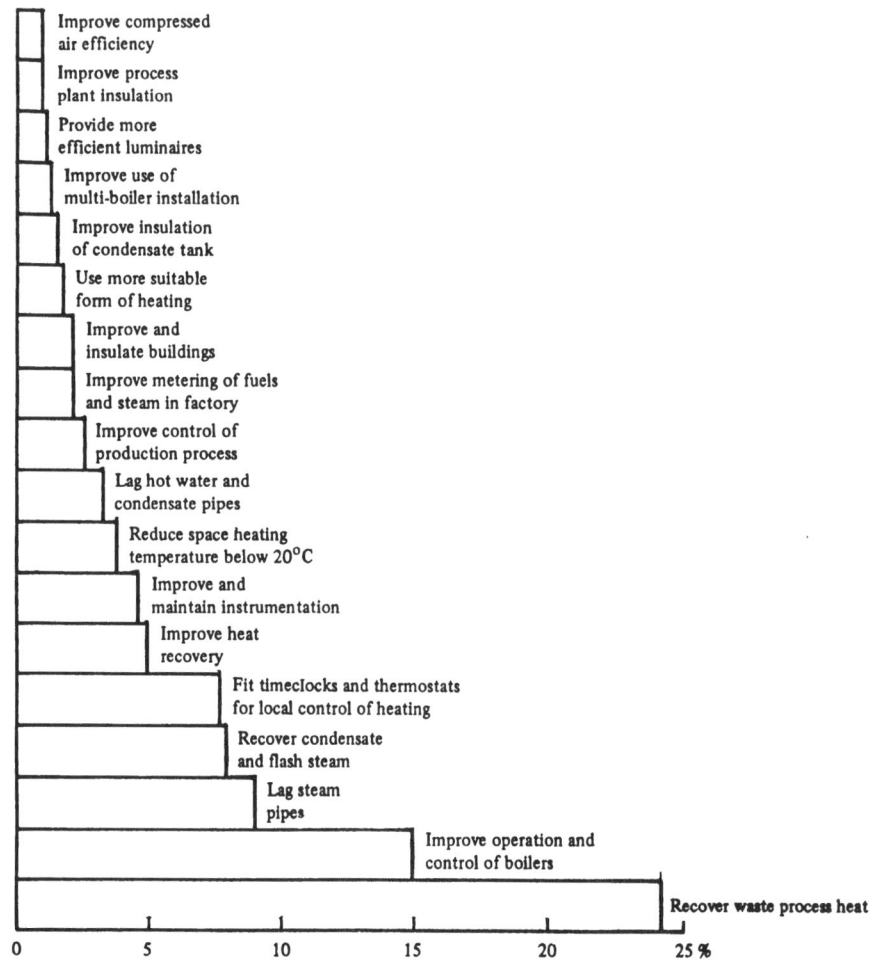

Percentage of potential energy savings

Fig. 9.   Possible Energy Savings from Various Measures
           in Establishments Visited.

           Source: UK IETS Report No. 18.

at this level of disaggregation. Finally, Fig. 9 gives a histogram
of possible energy savings identified in the visits. You can see
that waste heat recovery stands out a mile followed fairly closely
by improved operation and control of boilers. In fact, these two
measures, coupled with better insulation of pipes, account for 50%
of the potential energy savings identified, and the total potential
was assessed to be 20% of present energy consumption.

The chief limitation of this Scheme is that one cannot really
go into things in much technical depth in the course of a one day
visit. As I have already indicated, the carrying out of a reliable
energy audit is a non-trivial task, it requires careful preparation
and it certainly would require more than one day for a process of
any complexity. Nevertheless, the IETS has proved valuable in
identifying the opportunities for good housekeeping and also for
savings through short term, low cost, capital investment. These
are the means by which fuel costs can be cut more or less at a
stroke, by fairly simple and cheap means. So they are not to be
disparaged, but my interest is in discovering what technological
developments ought to be pursued and for that the Energy Audit
Studies are more appropriate.

UK Energy Audit Studies

The Energy Audit Studies were sponsored by two different
government departments and the approach varied somewhat between
departments and also from case to case. In some cases research
associations or other bodies were used to carry out the field work;
in other cases our own staff got heavily involved in the task.
Broadly speaking, however, the objectives of the studies were as
follows:

(i) to estimate the total energy, in primary equivalent terms,
used by the industry

(ii) to identify and quantify the energy used for different
products and/or different stages of the manufacturing
process

(iii) to assess the technological potential for energy conservation

(iv) to identify optimum technology and determine research,
development and demonstration (RD&D) priorities for energy
saving

(v) to provide background information for the determination of
energy conservation policy.

These are quite ambitious objectives and in some cases, for a

	Responsibility	Fieldwork
Published or in Press		
Iron Castings (1)*	ETSU	ETSU
Bricks (2)*	ETSU	British Ceramics RA
Dairy Industry (3)*	ETSU	NIFES, ETSU
Refractories (4)*	ETSU	British Ceramics RA
Glass (5)*	NEL	BGIRA
Aluminium (6)*	NEL	BNFMTC
Potteries (7)*	NEL	British Ceramics RA
Brewing (8)*	ETSU	NIFES, Brewers' Society
Coke (9)*	ETSU	Manufacturers' information
Lead & Zinc (10)*	NEL	BNFMTC
Cement	ETSU	Blue Circle Industries
Petroleum	ETSU	Oil Companies
Fertilisers	ETSU	PE Consulting Group
In Draft		
Chlorine-Alkali	ETSU	PE Consulting Group
Paint Finishing	NEL	BSRIA, PRA
Copper	NEL	BNFMTC
Textiles	NEL	Shirley Institute
Paper & Board	NEL	PIRA
Mechanical Engineering	NEL	PERA
In Progress		
Malting	ETSU	MAGB
Iron & Steel	ETSU	British Steel Corporation

* Figures relate to the Number within the Energy Audit series of
  reports.

Abbreviations:
BGIRA	=	British Glass Industry Research Association
BNFMTC	=	British Non-Ferrous Metals Technology Centre
BSRIA	=	Building Services Research & Information Association
ERA	=	Electrical Research Association
ETSU	=	Energy Technology Support Unit
MAGB	=	Maltsters' Association of Great Britain
NIFES	=	National Industrial Fuel Efficiency Service
NEL	=	National Engineering Laboratory (Dept of Industry)
PRA	=	Paint Research Association
PIRA	=	Paper Industry Research Association
PERA	=	Production Engineering Research Association

Table 2.   UK Government Energy Audit Studies.

Source: ETSU, July 1980

variety of reasons, they could not be attained. Nevertheless, the results have been very encouraging although it has all taken much longer than was originally expected.

I said earlier how ETSU had analysed the Census of Production to arrive at a selection of the top thirty energy using industries where there would seem to be an a priori case for energy conservation, listed in Table 1. This was too many sectors to study so a selection was made of those sectors where the energy intensiveness by value added was greater than 0.1. To this list was added the dairy industry because of its rapid growth in energy consumption, also petroleum refining because its product was of such critical importance. Some of the sectors were broadened out and boundaries were moved a little bit so as to arrive at manageable entities for investigation. The result of all this was the programme indicated in Table 2. Originally, it excluded Iron and Steel, the biggest energy consumer of all, but that has now been added, as you can see.

This programme, with the inclusion of Iron and Steel, is leading to an in-depth understanding of the way energy is used in 70% of the UK's industrial energy consumption. This understanding includes an appreciation of the technological possibilities for improved energy use and, in many cases, the types of research and development or demonstration that are required to exploit these possibilities.

Of the twenty or so Energy Audit Studies undertaken ten reports have now been published and these are being followed up through a variety of means. Indeed, the work of follow-up usually gets underway long before the report is published. Publishing a report requires all sorts of details to be finalised and generally agreed, whereas when one has discovered a promising opportunity one should get on with it straight away. The list of reports published is indicated in Table 2, and I shall return shortly to the broad conclusions coming out of the whole exercise.

Before that, however, I would like to illustrate the findings of this scheme using just one of the reports. Since I quoted Sir Oliver Lyle's remarks about a brewery in the 1930's and since beer production is a more interesting topic for discussion at a summer school than the production of coke, say, I have picked out our report on the Brewing Industry, and I would like to use this to illustrate the general points that I have been making. The main methods by which this study was carried out were as follows:

- analysis of international statistics and literature

- an energy survey of the Brewing Industry conducted by the Brewers Society of Great Britain

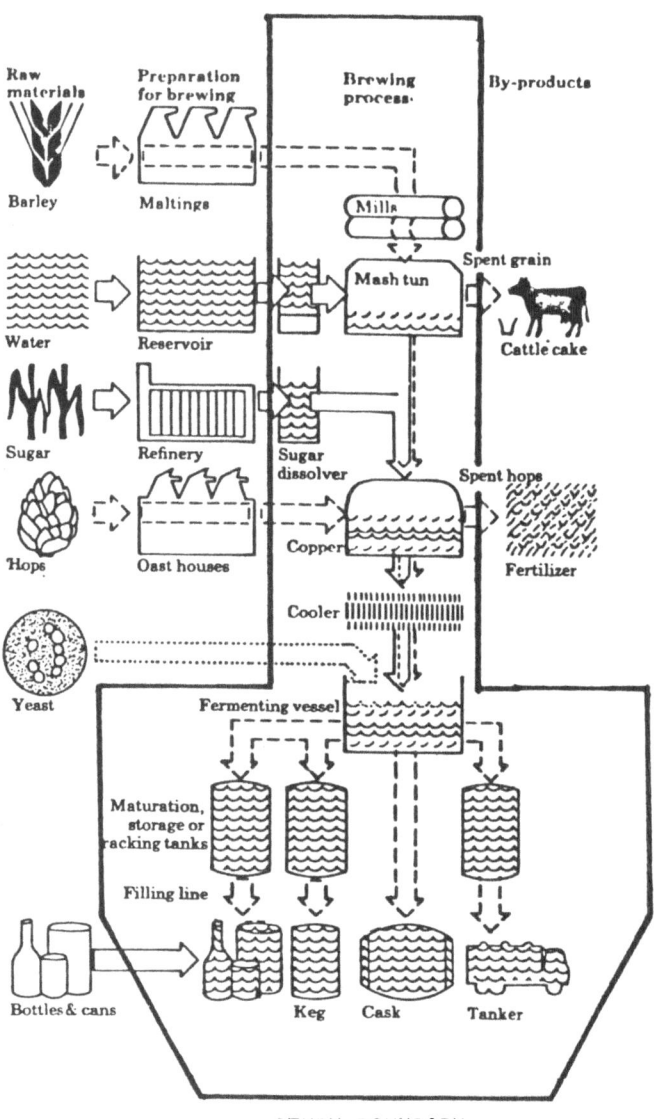

STUDY BOUNDARY

Fig. 10.   The UK Brewing Chain and Study Boundary.
           Source: UK Energy Audit Report No. 8.

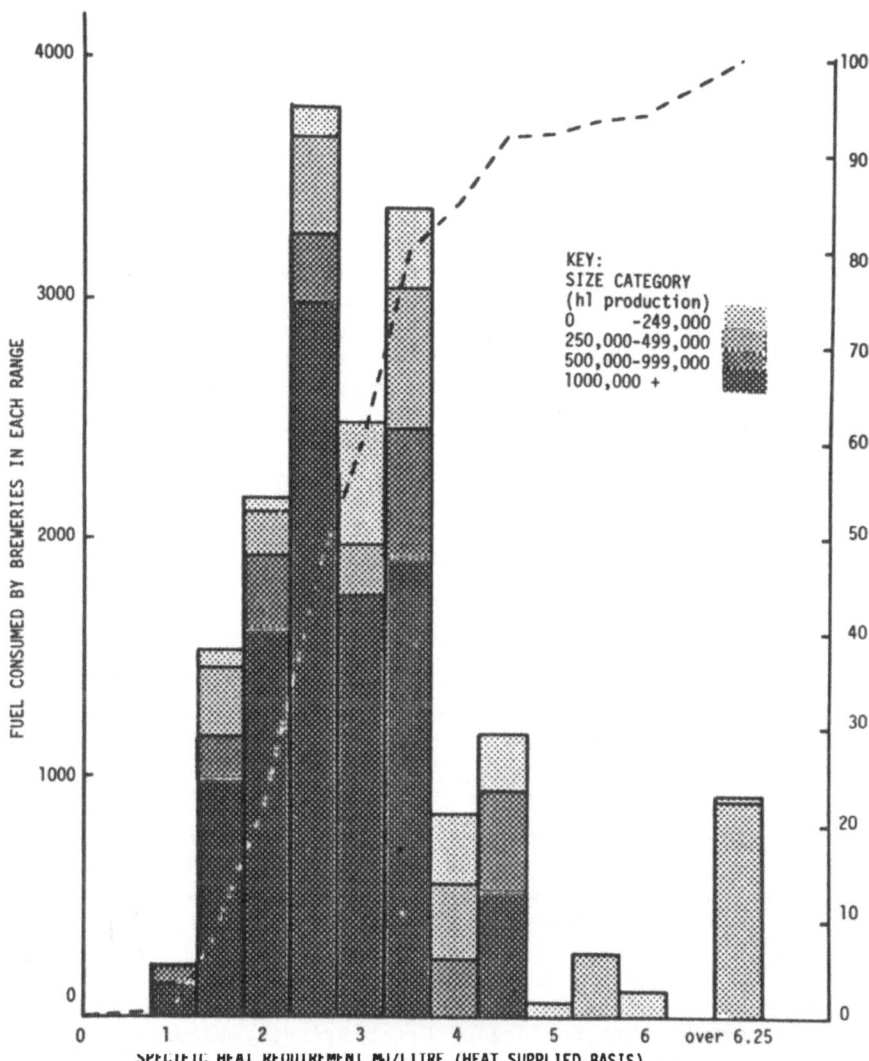

KEY:
SIZE CATEGORY
(hl production)
0   -249,000
250,000-499,000
500,000-999,000
1000,000 +

FUEL CONSUMED BY BREWERIES IN EACH RANGE

SPECIFIC HEAT REQUIREMENT MJ/LITRE (HEAT SUPPLIED BASIS)

Fig. 11.   Distribution of Specific Energy
           Consumption by Breweries According to Size.

           Source: UK Energy Audit Report No. 8.

- independent energy audits of production processes by consultants

- technological discussions with brewers and equipment suppliers

- conclusions and recommendations discussed and agreed with the industry.

I shall try to illustrate each of these very briefly.

First of all, we had to be clear about the system boundary within which we were making the energy analysis. Fig. 10 shows the various steps in beer production and you can see the boundary around the part which ETSU was investigating, i.e. the brewing process itself. Using national energy statistics one can quantify roughly the other energy inputs to beer, as follows:-

Barley production	28%
Malting	18%
Brewing	31%
Packaging	16%
Transportation	7%
	100%

Thus, only 31% of the energy sequestered in beer is expended in the brewing process, i.e. the PER is 31% of the GER. National statistics also allow us to compare the energy consumption in different countries. However, this is about as far as one can go using national energy statistics.

We were therefore very fortunate to benefit from an energy survey by the Brewers Society of Great Britain which produced much more detailed information such as that illustrated in Fig. 11, which shows the distribution in specific fuel consumption, in "heat supplied" terms, among breweries. The results have been analysed also in relation to the production capacity of the brewery. So one can see here the variations in energy consumption per unit of output, which are quite wide, and which suggest a higher energy efficiency for the larger breweries, as one would expect. This survey also yielded information about the different types of beer and different types of packaging of the beer. The data on this are summarised in Table 3, which is now in primary fuel terms. This

TABLE 3

Average Specific Energy Consumption of Breweries by Brewery Size
and Most Common Combinations of Product and Package Types

Size category (hectolitres/year)	Class by product or package (number in group)	Specific energy consumption (MJ/litre)		
		Fuels	Electricity	Total
All	Chilled filtered pasteurised only (4)	2.93	0.33	3.26
	Cask-conditioned only (10)	2.65	0.14	2.79
	Lager only (3)	1.13	0.31	1.44
	Chilled filtered + cask-conditioned (8)	2.17	0.17	2.34
	Chilled filtered pasteurised + lager (7)	2.25	0.44	2.69
	Chilled filtered pasteurised + cask-conditioned (18)	2.89	0.24	3.13
	Three types, not chilled filtered (13)	3.03	0.38	3.41
	Three types, not lager (11)	4.66	0.24	4.90
	All four types	2.07	0.34	2.41
	Bulk only (43)	2.62	0.30	2.92
	Bulk + cans (6)	1.78	0.36	2.14
	Bottles + bulk + cans (7)	2.66	0.47	3.13
0 - 249,000	Cask-conditioned only (7)	2.70	0.11	2.81
	Cask-conditioned + chilled filtered pasteurised (11)	3.09	0.33	3.42
	Bulk only	2.91	0.15	3.06
	Bulk + bottles (20)	3.36	0.34	3.70
250,000 - 499,000	Bulk only (11)	3.06	0.22	3.28
	Bulk + bottles (8)	4.42	0.33	3.75
500,000 - 999,000	Bulk + bottles (5)	3.41	0.46	3.87
1,000,000 +	Bulk only (13)	2.61	0.33	2.94

Source: UK Energy Audit Report No. 8.

tells us, for example, that lager beer requires less energy than
other types of beer and that canned or bottled beer requires more
energy than beer which is supplied in bulk. As a result of this
survey we now have a great deal of detailed information about the
pattern of energy use in the industry.

However, we also need to look at the details of the production
process and that was done through some energy audits of operating
breweries. Fig. 12 shows the results obtained from a modern
brewery which was surveyed in the past couple of years. This can
be compared with Fig. 1, from Sir Oliver Lyle's book. These two
Sankey diagrams are normalised so that the amount of heat flowing
through the "copper" is the same. The copper is the essential part
of the brewing process. They show that the modern brewery is using
very much less energy than was the 1930's brewery. This has been
achieved largely through good energy management, as indicated by
the much reduced amount of heat going into "space heating and
unaccounted losses". In addition, however, the modern brewery is
recycling a high proportion of the heat going through the copper,
and the boiler efficiency is much higher than in the earlier case.
The brewery exhibited here is up-to-date and well run, with the
result that it is achieving a very low specific energy consumption
per unit of output. Nevertheless, there is still scope for further
energy saving.

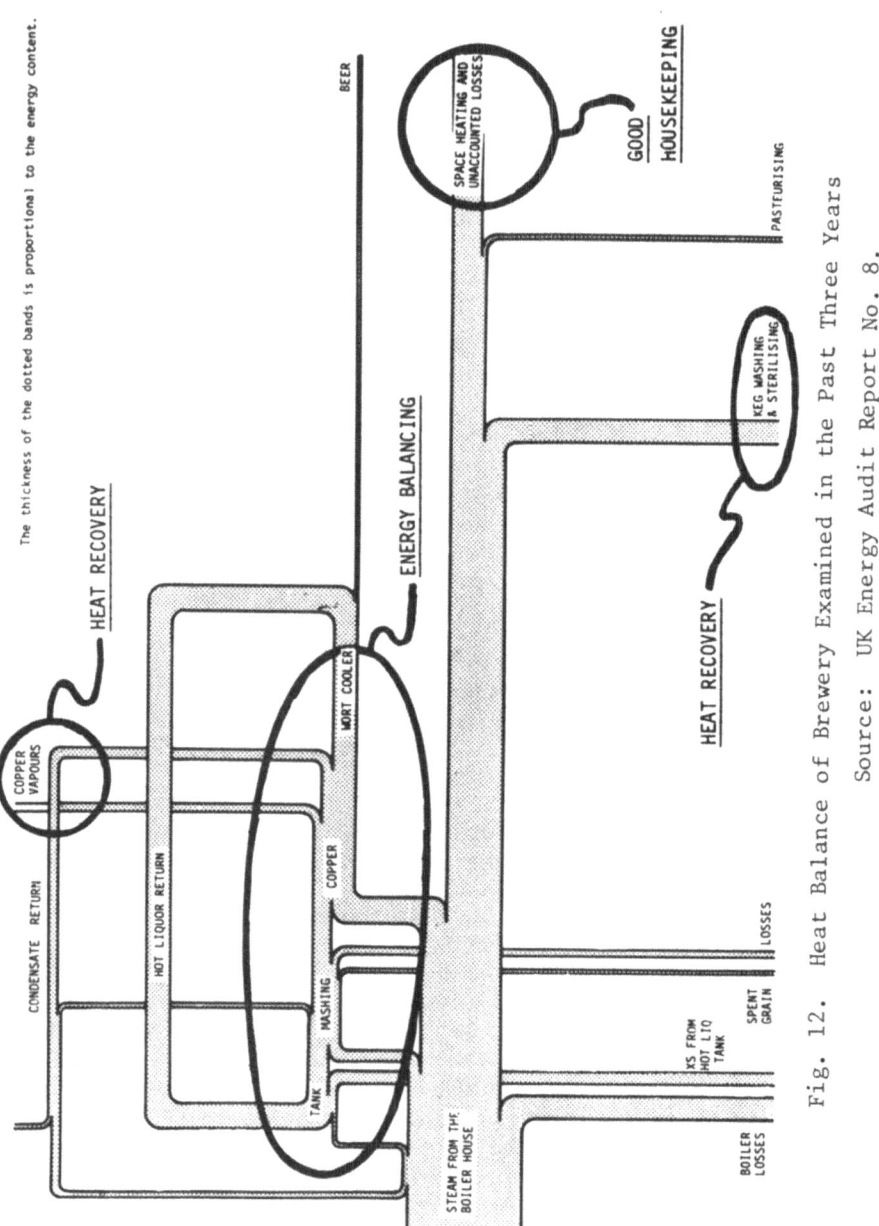

Fig. 12.   Heat Balance of Brewery Examined in the Past Three Years

Source: UK Energy Audit Report No. 8.

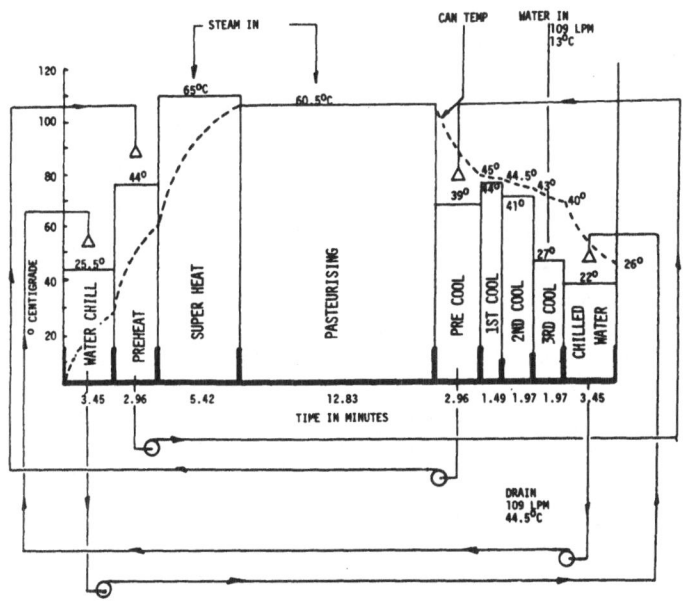

Fig. 18.   Arrangement of a Typical Can Pasteuriser.

Source: UK Energy Audit Report No. 8.

    To investigate the possibilities for further conservation we
had to conduct fairly detailed technological discussions with the
industry about the various possibilities. We had to understand the
various processes, such as the can pasteurising process which is
illustrated in Fig. 13, and we had to explore the realistic possibilities of change with plant engineers and development managers.
As a result of this we identified several worthwhile possibilities
which have been superimposed on Fig. 12. In this figure the
so-called "copper vapours" are being recovered but that is not done
in all breweries, so the technology needs to be more widely
deployed; we also have to look at the energy streams coming out of
the diagram to see if these can be cut down; by this means we identified an important area for heat recovery in keg washing and
sterilising; we now have a demonstration project covering this,
with a potential national energy saving of 50,000 tonnes of coal
equivalent (tce) per annum; despite the substantial reductions
already achieved there is still scope for further saving through
good housekeeping, at the right hand side, and finally we believe
that considerable saving could be achieved through the use of a
computer based model to optimise the overall energy balance in the
brewhouse, i.e. the complicated part of the diagram around the
copper. When one adds up the potential identified for these
various technological measures it amounts to 33% of the total

energy used by the industry in 1976, quite a significant and
worthwhile target to be pursued.

Before going on to look at the aggregated results emerging
from the Energy Audit Studies I have an important message about the
way this sort of exercise is tackled. Our Scheme is a government
scheme, managed by technical staff in government establishments,
and there is a danger that industry sees these people as inter-
fering government 'experts' who don't understand industrial
reality, are likely to come to the wrong conclusions and could
stimulate undesirable government intervention in industrial
affairs. It is therefore most important that the energy auditors
obtain the collaboration of the relevant trade associations and
important or representative companies in the industry. They must
carry out the exercise in a way which is helpful to the industry as
well as to government and their reports should be agreed by both
sides. Obtaining this agreement may take time, sometimes a very
long time, but it is worth it because a spirit of understanding is
then established. I cannot emphasise this too much.

Where we have not given due weight to industry's views and
concerns we have got into difficulties and reached wrong conclusions. Where we have established a close working relationship with
the industry the exercise has proved very fruitful. In the case of
our Brewing Audit we achieved a good collaboration. The trade
association helped us to do it, organised an energy seminar to
launch it and sent copies to all their members. We hope that the
collaboration will continue as we get results from government
supported RD&D projects so that the benefits can be realised by as
many companies as possible.

I can summarise the lessons learnt from this scheme, as
follows:-

(i)   The achievement of the objectives, as set out above,
         is a major national undertaking.

(ii)   It is absolutely necessary to consult with industry
         throughout and preferably to involve companies and
         trade organisations in an active way.

(iii)   Even with such consultation the task is doubly diff-
         icult in some sectors because of commercial con-
         fidentiality

(iv)   Energy conservation is not a thing apart; it is
         integral with production technology, quality control,
         etc. Therefore recommendations cannot be made
         without an appreciation of all the technological
         implications.

(v)   Even then, there are commercial financial, managerial
           and environmental constraints which have to be
           understood and taken into account.

(vi)   Even with the full technical and commercial cooper-
           ation of the industry it is difficult and time con-
           suming to write reports which will have any impact.

(vii)   Even with super reports which identify cost effective
           measures it is difficult to make anything happen
           because energy conservation does not rank high in the
           scheme of things.

(viii)   This is especially true in a time of recession.

All that may seem a bit depressing, but when you stop to think
about it it isn't too surprising, nor should we get depressed about
it. Our whole way of life has been built on the availability of
cheap energy. Now that the situation is changing, and let's face
it - it hasn't changed all that much yet, quite far reaching
changes may have come about. These won't happen overnight and they
won't happen without some kind of struggle.

The Technical Potential

        I would now like to touch briefly on the national findings in
the UK of these two schemes, the IETS and the Energy Audit Studies,
coupled with the results of various other studies both national and
international. We have aggregated the findings of all these
studies to arrive at a global estimate of what we call the "technological potential" for energy saving in industry. What we mean by
this is the potential energy saving which we can identify now as
technically feasible using technology that is more or less known
today and is likely to be economic. Now, it may turn out that some
of these possibilities are not actually feasible in practice, or
they may not prove to be economic, at least not at the present
time. On the other hand, we are not assuming any technological
breakthroughs; and it is almost certain that in some areas people
will come up with new ideas for energy saving that we are just not
aware of at the present time. Therefore, it is quite likely that
in the long run our conclusions will prove to be an under-estimate.
In the short term however, they are probably an over-estimate
because there are various barriers to the achievement of the potential, and I intend to discuss these barriers later.

        Table 4, taken from Energy Paper 32, published by the UK
Department of Energy[4] is a tabulation of the findings of our
investigations for the whole of UK industry. The numbers are in
millions of tonnes of coal equivalent. And where we did not have
good results broad guesses were made. So it is all a little bit

SIC Order MLH	Industry/activity	Process energy	Potential saving	% saving	Category of saving			
					Process change	Process improvement		
						Op/Ctr	Ins/HR	Imp Ef
III								
212	Bread and flour confectionery	NA						
215	Milk and milk products	1.0	0.14	14	0.05	0.05	0.04	
231	Brewing and malting	NA						
IV								
261	Coke ovens (provisional)	2.6	1.4	54	1.25		0.15	
262	Mineral oil refining	11.5	2.0	17	0.5	1.0	0.5	
V								
274	Paint manufacture	0.6						
278	Fertilisers	4.6*	1.8	39	1.6	0.1	0.1	
	Chlorine-alkali	1.68	0.25	13	0.25			
	Ethylene/propylene	NA						
VI								
311	Iron and steel making	21	8.2	39	48	1.24	2.16	
311	Iron and steel finishing	9	2.0	22	0.7	0.3	1.0	
312	Steel tubes	NA						
313	Iron castings	5.3	0.6	11	0.4	0.1	0.1	
321	Aluminium (smelting)	4.9(1)	3.0	61	2.1			0.9
321	Aluminium (other)	1.29						
322	Copper	1.40	1.0	29	0.5	0.2	0.25	
323	Other base metals	0.83						
VII	Mechanical engineering	NA						
IX	Electrical engineering	3.7	0.43	12		0.13	0.13	0.17
XIII	Textiles	NA						
XVI								
461	Bricks	1.16	0.67	58	0.4	0.07	0.2	
461	Refractories	0.63	0.18	28	0.05	0.03	0.1	
462	Pottery	0.58	0.06	10			0.6	
463	Glass	2.0	0.29	15	0.02	0.02	0.25	
464	Cement	4.2	1.5	36	0.8	0.3	0.4	
XVIII								
481	Paper and board (provisional)	4.8	1.4	30	0.5		0.4	0.5
	All above industries	82.43(1)	24.92	30	13.92	3.59	5.84	1.57
	Other industries	49.57	8.0	16			8.0	
	Total	132(1)	32.92	25	13.92		19.0	
	Machinery drives	(18.5)	1.0	5	0.75			
	Paint finishing	(1.5)	0.75	50	0.5	0.1	0.15	
	Waste heat utilisation		0.5		0.5			
	Waste as fuel		3.0		3.0			
	Total		5.25		4.75		0.5	
	GRAND TOTAL (provisional)	132(1)	38	29	18.67		19.5	

-   Covers less than the SIC MLH listed
*   Includes some feedstock
( ) Not additive - see text for explanation
NA Unavailable as audit, but covered through Thrift scheme (see Other Industries)
(1) Includes energy content of imported aluminium

Table 4.   UK Industrial Energy Conservation Potential. (MTCE Primary Fuel Equivalent).

Source: Ref. 4.

rough and ready, but it is something to be getting on with. You can see that the total energy consumed is 132 mtce; the potential savings identified amount to 38 mtce which is equivalent to almost 30% of primary energy consumption. Moreover, these potential savings are broken down into two categories, process change and process improvement. Process improvement refers to such things as the optimisation and control of processes, better insulation and more extensive use of heat recovery. You can see that approximately half the potential is associated with process improvement. This means that one ought to be able to cut down energy consumption by as much as 15% of total energy use by such means as adding heat recovery equipment, better instrumentation control, improved insulation, etc., as already illustrated in the results of the two studies which I have quoted.

Furthermore, such measures are not peculiar to particular industries. Heat recovery and control instrumentation are generic technologies which can be applied across the board. Therefore, in addition to those studies of particular industry sectors we have also carried out technology assessments as follows:-

Heat pumps
Mechanical vapour recompression
Organic Rankine cycle turbines
Waste as a fuel
Fluidised bed combustion
Microprocessor controls
Drying and evaporation
HT process insulation
Heat pipe markets
Ceramic recuperators
Centralised hydraulic power.

These are studies which evaluate the available technology, assess the markets and the development of these markets, and identify ways in which the technology can be improved or the markets expanded. Table 5 indicates the main generic technologies which we think need to be promoted in industry and, associated with each is an estimate of the energy savings which we have derived from our technology studies. The results are broadly consistent with those from the Energy Audit Studies, although the overall potential would appear to be somewhat greater.

These various data are peculiar to the UK and are based on fieldwork and consultations with companies in the UK. How do our conclusions compare with the findings of other countries? Table 6 is taken from a published report by Shell[5] which was concerned with Western Europe. They too find this notion of technological potential useful and if you look at the figures for the industrial sector you will see that they are very much in line with our

Priority	Technology	Requirement	UK Potential
*****	Waste heat recovery and utilisation	RD&D	6-10 mtce
****	Instrumentation and control	RD&D	3-4 mtce
***	Waste as a fuel	D&D	3-6 mtce
**	Heat pumps	RD&D	1 mtce
**	Drying and evaporation	RD&D	0-5 mtce
*	Industrial CHP	D&D	1 mtce
	Machine drives	D	4 mtce
		say	18-28 mtce

Table 5.   Generic Technologies Available Today and
their Potential.

Source: ETSU.

	Technical potential	1976	1985	2000
Transport				
Cars	20 — 35	3 — 5	5 — 20	15 — 25
Trucks	10 — 15	0 — 2	2 — 5	5 — 10
Ships	30 — 40	4 — 6	5 — 10	10 — 25
Aircraft	20 — 30	5 — 7	5 — 20	10 — 25
Industry				
Iron & Steel	25 — 35	0	10 — 15	15 — 30
Other furnace	25 — 35	3 — 5	10 — 20	15 — 30
Chemicals (fuel)	15 — 25	0 — 2	5 — 15	15 — 20
Other	20 — 35	4 — 8	10 — 15	15 — 25
Domestic				
Residential	40 — 60	3 — 10	10 — 20	20 — 40
Comm/Public	40 — 50	3 — 6	10 — 15	15 — 35

Table 6.   Energy Savings Estimates - Western Europe -
% reduction from 1973 Practice.

Source: Ref. 5.

conclusions, 25-35% or thereabouts. A similar study[6] by the Ford Foundation in the United States concluded that with existing technology it should be possible to reduce the specific fuel consumption in several industries - iron and steel, petroleum refining, paper aluminium, copper and cement industries - by one-third, again in line with our findings. Thus, I think there is agreement between investigators in several countries, approaching the problem in slightly different ways, that broadly speaking it is technologically possible to achieve around 30% saving in specific energy consumption over some future period.

If we look again at the Shell table we can see that they expect the bulk of this technical potential to be achieved by the year 2000, at least in industry. I am a little less sanguine than that. Certainly in the UK, industry has many other problems to worry about. The achievement of 30% energy saving, equivalent to 40 mtce per annum, calls for an investment of perhaps 20 billion dollars. Even over a 10-20 year period that is a lot of invest-ment, and even if the finance is available there are additional requirements in terms of management time and technical information. Therefore, I think that even with a fair wind it will take us well into the next century to achieve the sort of potential we have identified. The easiest part is that concerned with process improvement. When one turns to process change there may be complications; the product may be unacceptable, there may be non-energy aspects that create difficulties, etc. I shall return to this shortly. The important thing to note is that the realisation of the identified potential, calls for resources, the solution of problems, the motivation of people and all sorts of other things which may be difficult to achieve. So we should be careful not to get too carried away with these impressive figures for potential. Before turning to the barriers and how we can best overcome them I should like to digress now into a brief general discussion on the problems of technological innovation.

TECHNOLOGICAL INNOVATION

The General Problem of Innovation

People who habitually talk in English tend to use the word "innovation" as synonymous with "invention". However, the subject of innovation encompasses a much wider context altogether than is implied by the term invention. This is rather important, because if we misunderstand the context we will actually get our whole approach wrong. Therefore, I would like to begin by quoting you a definition of the term innovation from a report produced by the British Government in 1968[7], the so-called "Zuckerman" report. It defined innovation as, "the technical, industrial and commercial steps which lead to the marketing of new manufactured products and to the commercial use of new technical processes and equipment".

Thus, innovation is not something that is carried out solely in a
laboratory; it involves production, marketing and sales and even
the after-sales service and training that is so essential to many
new-fangled things. That particular report concluded that
successful innovation requires:

- a direct linkage of R&D with production and marketing into
  a single interacting operation

- planned programmes of innovation related to market
  opportunities

- effective technological management

- short lead-times

- a balanced use of scientific and technological resources
  over all stages of the innovative process.

The key requirements is the integration of research, development,
production, marketing and sales.

That report was a single effort in a single country at a
particular point in time. By itself it would be of little
world-wide significance except that the problem of innovation is
widely recognised and has been the subject of extensive study and
of a vast literature over the entire period since the Second World
War. I have attended numerous conferences where people from
private industry have grappled with this problem and have exchanged
views on ways and means of promoting and controlling innovation; it
is also a matter of concern to research organisations whether they
be universities or commercial enterprises; and it is of great
concern to national governments. I am sure that most of you must
have read at one time or another articles in your newspapers about
the problems of innovation, the amount of effort devoted in your
country to R&D and the role of government in supporting industry to
develop and promote new products.

In the UK, for example, many leading companies have set up
corporate laboratories or new venture groups to explore new technical ideas and how to exploit them; merchant banks have toyed with
innovation as a means of making money; the research associations
and private institutes have experimented with subsidiary companies
to exploit new ideas; and the British government over a period of
30-odd years has set up various institutional structures to promote
innovation; these range from the National Research and Development
Corporation which was set up shortly after the Second World War to
encourage the exploitation of ideas by private inventors to the
Rothschild "customer contractor principles" for guiding government
sponsored applied R&D[8]. The governments of other industrialised

"Show us the mousetrap."

Fig. 14.   The Mousetrap Story.   Source: Ref. 9.

countries have been no less concerned with this problem. Why is it
such a problem?

        I think that the essential difficulty of innovation can be
illustrated by a story, a true story about an invention that
failed. This story comes from the USA. The American philosopher
Ralph Waldo Emerson is credited with the statement that, "If a man
write a better book, preach a better sermon, or make a better
mousetrap than his neighbour, though he build his house in the
woods, the world will beat a path to his doorstep." as illustrated
in Fig. 14, a cartoon from a July 1980 issue of "Punch"[9].
Emerson was writing in the last century. In the present century an
American entrepeneur came across it and thought he would try to
develop the best possible mousetrap and put it on the market. He
therefore carried out research into the behaviour of mice,
evaluated various designs of mousetrap, and after extensive R&D
arrived at a design for a very superior mousetrap. As I understand
it, this mousetrap consisted of a shiny little black box into which
the bait could be easily placed and there was no danger of the
mousetrap catching the owner's fingers. It was very effective at
attracting mice, it killed them in a neat way and once the mouse
had been caught and killed the owner could simply take the box,
hold it over a bucket, press a button and the dead mouse would drop
out. It was a bit more expensive than the conventional mousetrap,
7 cents as opposed to 3 cents, but it could be used many times, and
it was very effective at catching mice.

This mousetrap was a commercial failure and was withdrawn from the market after a fairly short period. The reasons for its failure were as follows:

- in the typical US household it is the wife who is concerned about the mouse and who pesters her husband to lay a trap

- the husband lays the trap but it is usually the wife, who is there in the house during the day, who finds the dead mouse in the trap and has to deal with it

- the wife doesn't like extracting the mangled remains of a mouse from a trap and so she wraps the whole thing up in tissue paper, throughs it into the dustbin and buys some more mousetraps

- this means that they go through quite a lot of mousetraps but since they are not very expensive no-one is worried

- although the new mousetrap wasn't very expensive it did look rather expensive, so the wife felt guilty if she threw it away

- in addition it was clinically repugnant to the housewife with its push button ejection of the dead mouse.

- so people didn't like the new mousetrap and they didn't buy it.

The moral of the story is that a well-conceived technological approach will fail if it doesn't take account of the relative benefits and costs as perceived by the user and also of the user's qualitative needs and sensitivities. So Emerson was proved completely wrong. The world won't beat a path to the doorstep of an inventor with an exciting new technological device.

There are numerous other more serious cases that have been extensively documented in the literature some of which, I am sure, must be familiar to you. Two examples of major innovations which consumed enormous financial resources, one of them a failure and one of them a success, are Corfam, an artificial leather developed by Dupont, and the Xerox process for reproducing text.

The trouble with leather substitutes is that they don't breath and they don't look like leather. So Dupont developed at great cost a man-made material with properties closely resembling those of leather. Not only that, they researched the markets very carefully and they carried out thorough user trials with 20,000 pairs of shoes. Over the four years 1964-68 they actually built up sales to 40 million per annum. But it wasn't enough, so they pulled out of the market with a loss of $250M. There were actually several reasons for the failure but I would like to mention only two of them. The first reason concerns the nature of the market. The US market is dominated by women's shoes and it turns out that where shoes are concerned "la donna e mobile". Fashions change rapidly and so the long lasting quality of Corfam was irrelevant; in addition, shoes are often cut-away so there was little advantage in "breathability" and an unanticipated competitor, in the form of PVC, entered the market. PVC is shiny and cheap and proved very attractive for women's shoes compared with the more costly Corfam. The second reason was the strong response of the leather industry to improve its product and reduce costs. This response, coupled with the reluctance of shoe manufacturers to change their production methods (in which they had made large investments) and the sudden availability of cheap leather from Eastern Europe strengthened the position of the traditional competitor - leather. This response by the leather industry could perhaps be interpreted in terms of the "sailing ship effect", a term coined by S. C. Gilfillan in 1935 who pointed out that all the greatest improvements in the design of sailing ships came after the invention of the steamship. It would seem that old products don't die without a struggle and sometimes it is the new product that dies.

In the case of Xerox, which was invented and patented by a private individual before the War, the technological development was rather difficult and the initial performance was very poor; not surprisingly, therefore, when the process was offered by Battelle for licensing after the War none of the major enterprises who were approached to commercialise the invention on a large scale were interested in taking it up. It was a more modest enterprise, the Haloid Corporation of New York, who took it up and the breakthrough came with their realisation that while customers would not be prepared to pay the capital costs of a machine to operate this rather dodgy new process they might accept it through a leasing arrangement whereby the financial outlay for the user was minimised and the benefits and attractions maximised. That way the user became gradually hooked on the new device. Moreover the supplier could make his profit not so much from the machine itself as from the vast amounts of paper that each machine consumed. This marketing strategy, as we all know, was highly successful even although the early machines were not very impressive. These two tales further illustrate the theme that technical excellence is not enough, may even not be necessary; you have to get your marketing right.

These are just three examples. In fact, if one collects statistics on new ideas put forward by scientists, private inventors and others then one finds an astronomical number of 'inventions', some of which have been pursued to a considerable degree. But only a tiny fraction of those ideas have ever achieved significant profit in the marketplace. In its 30 years of

Fig. 15a.

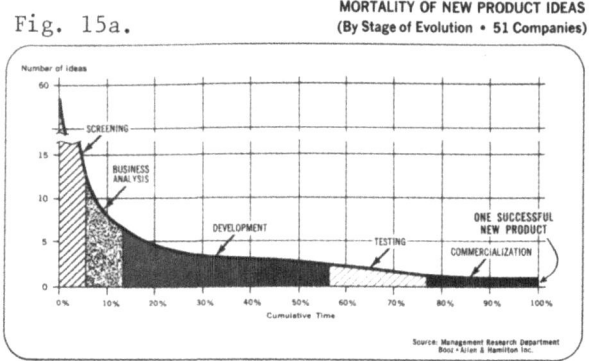

Fig. 15b.

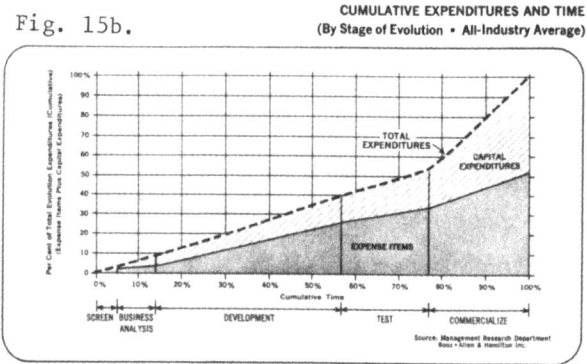

Fig. 15c.

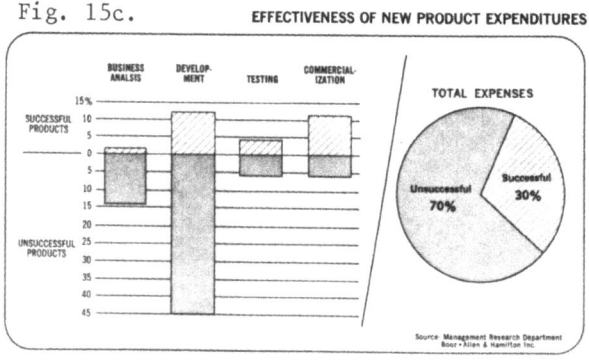

Fig. 15. Source: Ref. 10.

existence the UK National Research and Development Corporation has
received over 40,000 inventions. Of these less than 7,000 were
accepted after evaluation. Of these only 350 are currently
bringing in revenue, and only one has been a world beater, the drug
cephelasporin, which originated from some researches by a doctor
here in Sicily. The others have 'died' at one stage or another.
NRDC's experience may not be typical; nevertheless this notion of
"product mortality" is further illustrated in Fig. 15a from a study
of 51 companies conducted by the American consultants, Booz, Allen
and Hamilton[10]. Out of 60 initial product ideas only one was
commercially successful.

One of the reasons for the high failure rate is that as one
proceeds from a bare idea through to a marketed product the costs
escalate rapidly. A rough rule of thumb which I use is that if it
takes one dollar to research an idea, say to the stage of a working
laboratory prototype, then it might take ten dollars to develop
this idea to a properly value engineered and reliable product, and
possibly one hundred dollars to put this product into production.
Fig. 15b also from Booz, Allen and Hamilton, shows the actual
expenditure by American industry on the various stages indicated in
Fig. 15a. Therefore, a company which is in the business of
technology and which has research facilities will be generating
within itself a great many new ideas which need to be investigated
and pursued; on the other hand the pursuit of those ideas beyond a
minimum paper exercise will entail enormous resources. Fig. 15c
shows how the bulk of expenditure is on unsuccessful products.
Obviously, the company has to achieve some balance between the
dampening down of creativity and going out of business through
excessive expenditure on RD&D. Because of this a great deal of
effort has been expended on developing ways of:

- generating creative ideas and concepts;

- evaluating the merits of these ideas and selecting potential winners;

- devising methods of feedback so that promising ideas are
  selected out and pursued and poor ideas are eliminated at
  the earliest possible stage.

Private industry has devoted a lot of thought to this and so has
government. This is something that you might like to pursue
separately.

Some Pointers for Energy Conservation

For our purposes I would like to come now to one or two
general lessons that have emerged from all this activity. Although
they all come from some years back I think they are of relevance

to our subject and I have actually made use of them in my own
approach to the promotion of energy conservation technology in
industry.

The first elementary but important notion was perceived by a
former student and professor of my own university, Adam Smith, 200
years ago, when he said that, "Consumption is the sole end and
purpose of production, and the interests of the product ought to be
attended to only as far as may be necessary for promoting the
interest of the consumer. This maxim is so perfectly self-evident
that it would be absurd to attempt to prove it." The same notion
has been expressed in more colloquial terms through such phrases as
"the customer is always right" and "only the wearer knows where the
shoe pinches;" politically you may disagree with those statements,
but if you wish to innovate then you have to pay close attention to
what I would call the "use system" in which the technology is to be
deployed. The Better Mousetrap Company failed with their product
because they didn't fully understand the use system of the American
household. In the same way, if you design an instrument or a heat
exchanger for use in industry which cannot be accommodated easily
within the use system of the factory, i.e. if it isn't capable of
being operated and maintained by the people who are there, if it
disrupts the production process, etc. etc. then it will simply not
be bought. Therefore, in promoting energy conservation technology
and particularly in promoting retrofit technologies for process
improvement, it is most important that in our researches we should
acquire an understanding of the use system and develop the product
accordingly. All this may seem very obvious but it is all too
frequently forgotten, and it is one reason why demonstrations are
so important.

My next theme is a strategic one about how a company should
tackle the expansion of its business through innovation. In a book
which was published many years ago[11] the American consultant
Igor Ansoff showed a simple little matrix which illustrates an
important concept. It is shown on Fig. 16. Ansoff was concerned
with how a company can achieve growth. The simplest way is to
increase its penetration of its existing markets with its existing
products, as indicated in the top left hand box. The most
difficult way is to diversify, i.e. to develop new products for new
markets, as shown at the bottom right. My personal view is that
the most difficult way of going about diversification is through
innovation. In my earlier work at Harwell I had contact with a lot
of companies who were considering new ventures with novel technical
products. It proved to be a very hard way of making money and I
can scarcely recall any of those ventures that was outstandingly
successful. The more normal way for a company to achieve growth is
that indicated in the top right hand box, viz. the development of
new products for existing markets. The markets and the sales
organisation which a company has are one of its major assets and if

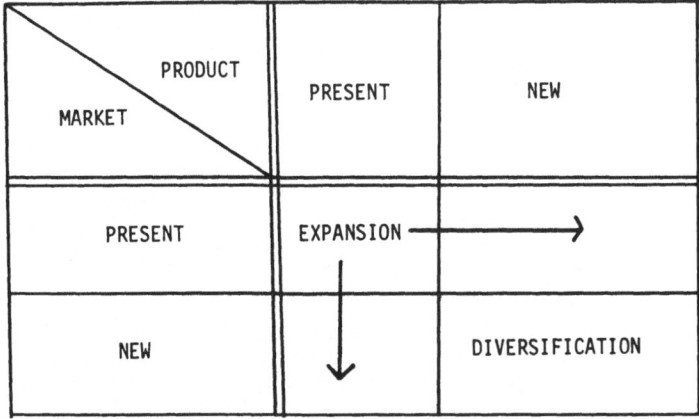

Fig. 16.   Ansoff's Corporate Strategy Matrix.

Source: Ref. 11.

it can bring reliable new products on stream for those markets it
stands a good chance of being successful. The alternative approach
of tackling new markets with existing products is not so common
because although it may seem easier at first sight, there are a
great many hidden pitfalls. The Stanford Research Institute has
documented many amusing cases of how American companies have
totally failed to understand European markets and fared badly as a
result. This simply emphasises once again that the marketing
aspect is often more difficult than the technical aspect. The
general message is that one must understand one's existing
strengths and build on them and not attempt to tackle too many new
things at one time.

This sort of consideration is very relevant to the development
and exploitation of heat pumps, for example. Various types of
company are interested in getting into the heat pump field:

-   companies in the central heating business who supply
    boilers and radiators to the domestic market but have no
    experience of compressors;

-   companies in the refrigeration business that have ex-
    perience in compressors but not of the central heating
    business;

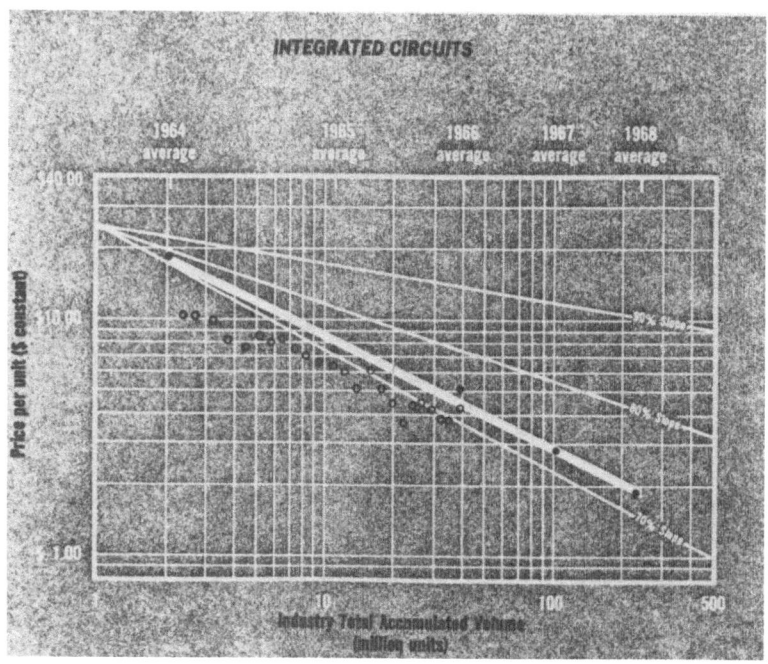

Fig. 17.   Experience Curve - Integrated Circuits.

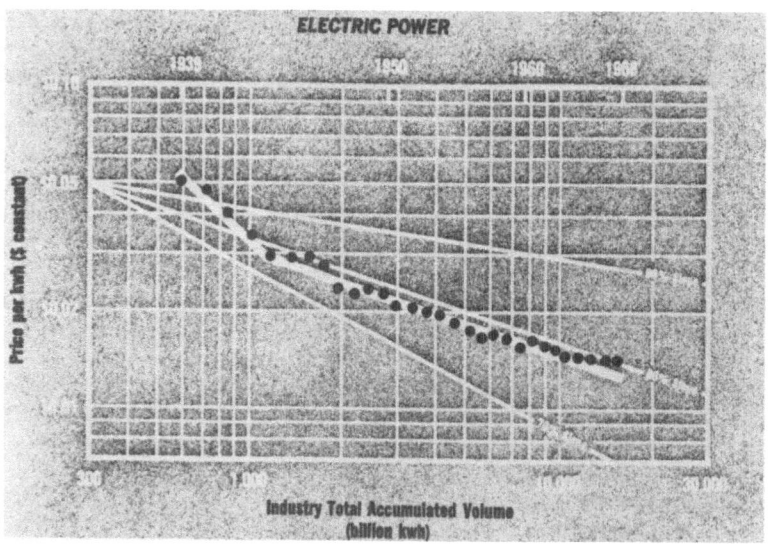

Fig. 18.   Experience Curve - Electric Power.

Figs. 17 & 18. Illustrations of the Experience Curve.

Source: Ref. 12.

- companies in the engine business who can supply prime movers but have little experience of central heating or compressors;

- other companies with good financial and technical resources who are tempted to diversify.

One thing is pretty clear. They won't all succeed, so where would you put your money? These are the kind of questions that have to be considered and, in particular, where national governments or financial institutions are supporting such developments they need to have some kind of view as to the companies that stand the best chance of innovating successfully.

My next theme is one which was developed some years ago by the Boston Consulting Group in the USA, the so-called experience curve theory. The basic idea is that as a company or an industry accumulates experience of a product the costs of the product come down. Put more precisely, "Costs appear to go down on value added at about 20-30% every time total product experience doubles for the industry as a whole as well as for individual producers"[12]. The measure of total product experience is the number of units sold, and note that this is the integrated number sold since the product started not the number sold in any one year. When one looks into this effect one finds out that there are good reasons for it, such as economies of scale, but it is not my purpose to go into these reasons here today. Essentially the effect is due to more efficient use of the factors of production. The area where this theory has been demonstrated most dramatically is in semiconductor technology, where we all know of the falling cost of calculators and other products. In fact, the Texas Instruments Corporation based its entire corporate strategy on the experience curve theory and by this means was able to dominate world markets for a time by underselling the competition. Figs. 17 and 18 give just a couple of examples of products whose costs have come down in accordance with the experience curve theory. One of them is very relevant to our theme, it is electric power supply in the USA, and you can see that over a period of 30 years the price per kilowatt hour has fallen consistently in line with the theory. The other shows the much more rapid fall in the price of integrated circuits over the 4-year period 1964-68.

Much the same considerations should apply to the cost of making heat exchangers, boilers, instruments and controls etc., and also to the general efficiency with which energy, one of the factors of production, is used in industrial processes. This has been investigated in the UK by the Systems Analysis Research Unit of the Department of the Environment. They have found that energy consumption per unit of output has declined in a more or less systematic way for several industries over a 10-15 year period, as

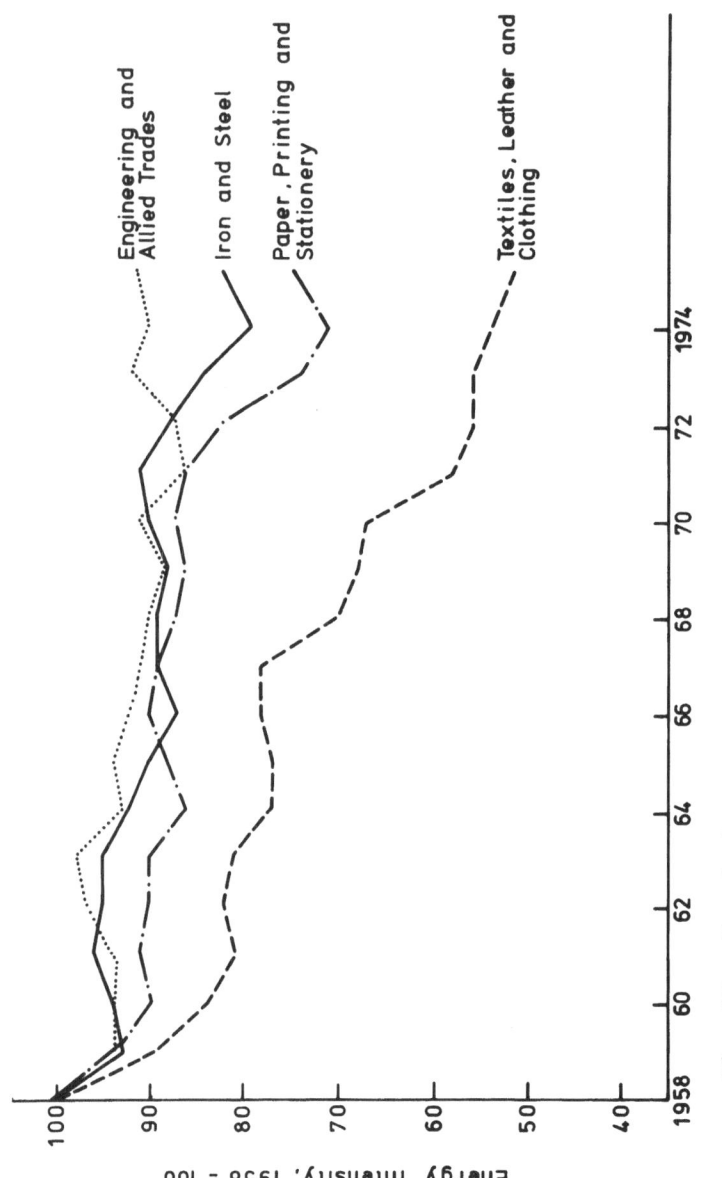

Fig. 19.   Trends in Energy Per Unit of Output (Value Added).

Source: Systems Analysis Research Unit, UK Dept. of Environment.

illustrated in Fig. 19.   They put this down to two effects,
capital/energy substitution, i.e. factor substitution, and the
experience curve theory. We can therefore expect similar trends in
the future.   For example, electricity supply utilities use process
control computers for boiler management; through the combined
effects of fuel price increases and the falling cost of
microprocessors the same concept, in miniaturised form, might be
applied quite soon to the tens of thousands of industrial boilers
with capacities one hundred times less than that of a major
generating station, and perhaps even eventually to millions of
domestic boilers which are down in size again by another factor of
a thousand.   These two effects, rising real energy prices plus
increased experience in the technologies which are today applied
only on a large scale will lead to those same technologies
increasingly being applied in the future on a small scale and this
too is something that we must take into account in formulating our
strategies.

My fourth theme, and this is really a combination of themes,
is to tell you about the results of an investigation into the
factors affecting success in innovation. There have been many such
studies and many papers published. The one I am going to tell you
about was carried out by the Science Policy Research Unit at the
University of Sussex in England. It was called Project
SAPPHO,[13] and it was an investigation into the factors
affectin success and failure in carefully selected pairs of
innovations. Each pair consisted of two very similar products one
of which had been successful and the other of which hadn't. There
were 29 such pairs, all of them in the category of scientific
instruments or chemicals. Most of them were from the UK but
innovations from seven other countries were included. A wide range
of hypotheses, and in particular various hypotheses which had been
claimed in the published literature at the time, were tested out by
statistical analysis on a computer. It was a big exercise but like
most well conceived exercises the results can be expressed very
simply. There were five main conclusions, as follows:

  (i)   successful innovators were seen to have a much better
          understanding of user needs

  (ii)   successful innovators pay much more attention to
          marketing

  (iii)   successful innovators perform their development more
          efficiently than failures but not necessarily more
          quickly

  (iv)   successful innovators make more effective use of
          outside technology and scientific advice

    (v)   the responsible individuals in the successful attempts
           are usually more senior and of greater authority than
           their counterparts who fail.

The first three of these conclusions tie in with what I have
already said. The last two are additional findings which we need to
take note of. In particular, the last conclusion won't surprise
the cynical amongst us but it is often very difficult to satisfy
that condition in a particular organisation.

My job is concerned with the provision of financial assistance
by government to companies and research organisations for research,
development and demonstration in the field of energy conservation.
For example, the UK Department of Energy is assisting companies who
wish to develop and market heat pumps. The kind of things I have
been talking about here are all very relevant to the odds which we
might put on particular proposals which come before us; similarly
if I were working in a company with ideas coming up to me from the
laboratory much the same considerations would apply; and if I were
in a university seeking companies to exploit particular ideas that I
had developed, again I would make use of these concepts. In short,
anybody who is engaged in the business of innovation needs to understand these lessons. Happily, many of the senior and experienced
people that I meet do understand them and make them part of their
judgement. However others do not and even those who do often find it
difficult in practice to make decisions along the lines indicated.

BARRIERS TO ENERGY CONSERVATION TECHNOLOGY

I have talked about the very large potential for energy conservation technology in industry and I have described the general
nature and difficulties of innovation. There is a further matter
that I need to consider before coming to discuss strategies for
achieving the potential. In addition to the general difficulties
of innovation there are particular barriers to energy conservation
technology. Figs. 8 and 9 illustrated the results from the Industrial Energy Thrift Scheme in the UK, and I mentioned that the
Research Associations involved in the work have carried out
follow-up visits to see what has happened as a result of their
recommendations. Quite a lot has happened but they also found that
many recommendations had not been a implemented. They therefore
investigated the reasons for this and Fig. 20 illustrates some of the
obstacles to improved energy utilisation which they identified. I
would argue that all of these obstacles can be categorised under
three main headings, technical, economic and attidudinal. I would now
like to consider these three in turn.

Technical Barriers

The American economist J. K. Galbraith has pointed out[14]

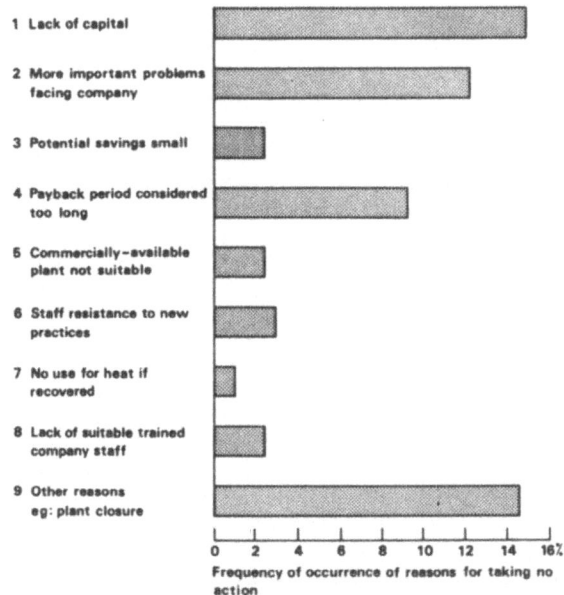

Fig. 20.   Barriers to Improved Energy Utilization -
Analysis of IETS Follow-up Enquiries.

Source: Ref. 22.

that no war plane was used in the Second World War which had not
been essentially designed before the War started.   Thus, even under
the extreme pressures of war, when technological resources can be
dedicated in a way which cannot be achieved in peace-time, it
wasn't possible to develop and deploy such new technological
weapons in the space of six years.   This illustrates one of the
main technical barriers - lead times.   In energy supply it takes 10
years or more to plan and build a new power station or to open up a
new coal mine and it takes a similar time for an industry to change
over to a new production process.   On matters such as heat recovery
the timescales can be much shorter but where new technologies are
involved they must still go through a long process of research,
development and demonstration before they can contribute to conservation on a commercial scale.   Hence we need to reduce lead times.

Assuming that an industrialist wishes to achieve lower energy
efficiency and is prepared to spend money on it, there are two
things he can do.   He can either replace existing plant (i.e.
equipment, processes, buildings, vehicles, etc.) with new plant
which has a lower energy requirement, or he can attempt to modify
the existing plant so as to reduce its need for energy.   Much
industrial equipment has a very long life, especially such things
as buildings and boiler plant.   There are boilers operating in UK

industry which were installed nearly 100 years ago. Some process
plant also has to last a very long time; cement kilns, for example,
are expected to survive 30 years or so, and glass furnaces operate
continuously for up to 12 years. Hence the replacement of existing
plant may not be timely. That means that the industrialist will
tend to look in the first instance at plant modification, i.e. such
things as heat recovery, insulation, better control of the plant,
etc., the sort of things which I have identified already as process
improvement. Even there, however, there will still be delays
because of the time required to:

- modify, add or replace capital equipment (e.g. heat
  exchangers)

- demonstrate the technical and cost effectiveness of
  technologies which may previously have been uneconomic

- develop appropriate new technologies or adapt existing
  technologies to new applications.

The industrialist will be reluctant to put in heat recovery equipment, say, unless he has complete guarantees from the manufacturer
and, preferably, customer references that the equipment will
actually work in his situation. If he doesn't have these he won't
take the risk.

Stated in the abstract this may seem a little exaggerated.
After all, will the industrialist not welcome the opportunity to
be first in the field with something that is going to save him
money and possibly give him an advantage over his competitors?
The answer is usually no, and this becomes very clear when you go
around and talk to companies about putting in energy conservation
equipment and hear the stories of things that have gone wrong,
the problems that have been caused and the huge costs that have
resulted. I could give many examples and they are all consistent
with Project SAPPHO which said that, "The successful innovators
eliminate technical defects from the product or process before
they launch it. They usually employ a larger development team on
the project, and spend more money on it." The plain fact of the
matter is that energy price changes have been occurring at a rate
which is faster than the usual rate for technological innovation,
particularly in the engineering type of development which is so
relevant to energy use in industry.

Hence, if a company, industry, or country, wishes to bring
energy conservation technology on stream they must accelerate the
rate of development by increasing the technical resources deployed
on RD&D and obtaining or making available more technical information. The prime requirement of any government intervention in
this respect must be to achieve a shortening of lead times.

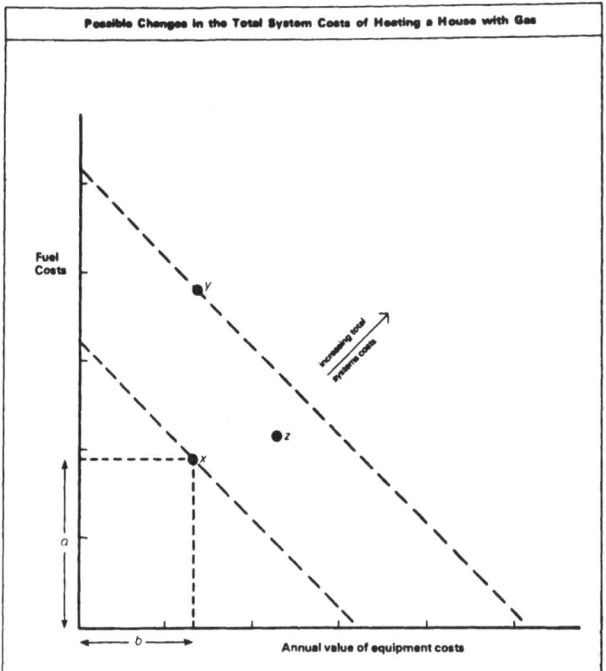

Fig. 21.   Schematic Representation of System Costs.
Source Ref. 24.

Economic Barriers

Faced with a rise in energy prices plus the aforesaid technical difficulties the simplest thing for a manufacturer to do is
to pass on the increased cost of production by putting up the price
of his product. If this leads to a reduction in demand it will
still give him rather more time to tackle the technological problems. In the meantime, of course, it is the customer who bears the
brunt of the energy price rise. If prices continue to rise then it
becomes increasingly difficult for the manufacturer to do this and
so he has to begin to think about his production costs.

However, unless he really believes that energy prices will
continue to rise he may still be tempted to minimise the first cost
of new plant by excluding energy conservation features. As an
example of this, boilers are today available with instrumentation
which will ensure their achieving maximum combustion efficiency.
In the UK, however, many industrial purchasers of boilers chose not
to have this instrumentation because it adds to the first cost of
the boiler. This may be a false economy if fuel prices continue to
rise; on the other hand it may be a good economy if they don't.
This is illustrated by Fig. 21 which plots the annuitised value of

equipment along one axis and the annual cost of fuel up the other
axis. If fuel prices increase from say "x" to "y", which represents more than a doubling, then it may pay overall to develop and
invest in a technology with capital costs corresponding to "z".
Although the first cost of this is much higher than the technology
represented by "x" the fuel costs are much lower and so the overall
system cost is lower. Hence, from our R&D standpoint we need to
assess technologies in terms of their life-time system costs, even
although the user or purchaser of the technology may not look at it
this way. One difficulty is that one doesn't know for certain what
the costs of fuel will be in the future; all one knows is what the
equipment costs now; hence one needs to have a view about the
future to make decisions of this sort.

Next, if we compare energy conservation with energy supply it
is pretty clear that supply is highly concentrated, lends itself to
the economies of scale, involves high technology, can support substantial programmes of RD&D and in many cases has the backing of
government. By contrast energy conservation is highly disaggregated; even in the industrial sector we are concerned with
hundreds of thousands of users; investment decisions have to be
made by every user, usually without support from government or
anyone else, and what the individual user perceives as "economic"
in conservation may, as in the case of boilers, actually exclude
some measures or investments which could be judged as cost effective in terms of national resources. Another factor is that the
person making the investment decision may not be the same person as
benefits from the energy saving. This problem is particularly
acute in the domestic sector but it applies also in industry to a
lesser extent.

Finally, the benefits of energy conservation investment – the
savings in energy achieved, will not usually be visible within a
company. Although in theory such savings should reveal themselves
in terms of profit, there will be no obvious connection between any
profit achieved and the investment made in boiler control, heat
recovery, etc. For one thing, in a period of rising energy prices,
unless one does some careful inflation accounting, the fuel bill
will have gone up in any case and casual analysis will not show any
cash saving. Shell, in one of their reports,[15] have recognised this and speak of "the very real psychological barrier inherent in the need to spread a capital investment over a long period
of time, divided by an invisible commodity – the energy saved."

The result of all these considerations is that despite an
obvious theoretical benefit from investments or developments in
conservation technology during a period of rising energy prices, it
is often difficult to assess what the benefit might be in advance
and even where an investment has been made it may be difficult to
see in retrospect the financial impact of the benefit. It is not

therefore surprising that energy managers in industry often have
great difficulty in persuading their management to invest in
equipment which they think will benefit the company, and it
suggests that we have to give particular attention to economic
assessment, cost reduction and value engineering in our development
programmes.

Attitudinal Barriers

Perhaps the most deep-seated barrier to energy conservation in
industry is something to do with attitudes. I don't wish to
suggest that there is anything reprehensible about those attitudes;
far from it, I am very sympathetic to them for they are similar to
my own personal attitudes when it comes to energy conservation
investments in my house. Nevertheless, if we are seriously bent on
developing and promoting energy conservation technology in industry
we must understand those attitudes and develop our technology
accordingly.

The most important fact to understand is that most companies
are not in the business of saving energy; they are in the business
of making and selling goods or supplying services of one kind or
another. Furthermore, they are usually concerned with making and
selling more goods, either because they want to expand or because
they are struggling to survive. Therefore, the saving of energy,
like non-destructive testing or environmental control, is a
necessary evil so far as the business of production is concerned.
It may be a "good thing" for society but it costs money, it takes
up management time, and it may actually hinder the output of goods.
Therefore it is not something that a company wishes to take up
unless it absolutely has to. This attitude is likely to be most
acute when it comes to new technology. As a senior manager in a
leading UK chemicals company said to me, "In the field of chemicals
we are in the risk business; when it comes to raising steam we are
not." Thus, even if company management is persuaded that it must
invest in heat recovery it will be reluctant to invest in equipment
that is not fully proven and guaranteed, even when there is some
attractive new option available which, on paper, ought to give them
a much better return on investment. The risks are too great, for
if the company puts in the new equipment and it doesn't work their
production may be brought to a halt, valuable technical resources
may be tied up in putting it right, and generally speaking the
business of the company will be disrupted. That is the last thing
they want.

The result of this is that companies expect a higher return on
investment from energy conservation than they do from other investments. This is illustrated in Fig. 22, taken from the Harvard
Business Review[16], and showing the results of an investigation
in the USA. The figure compares the expected, after-tax ROI for

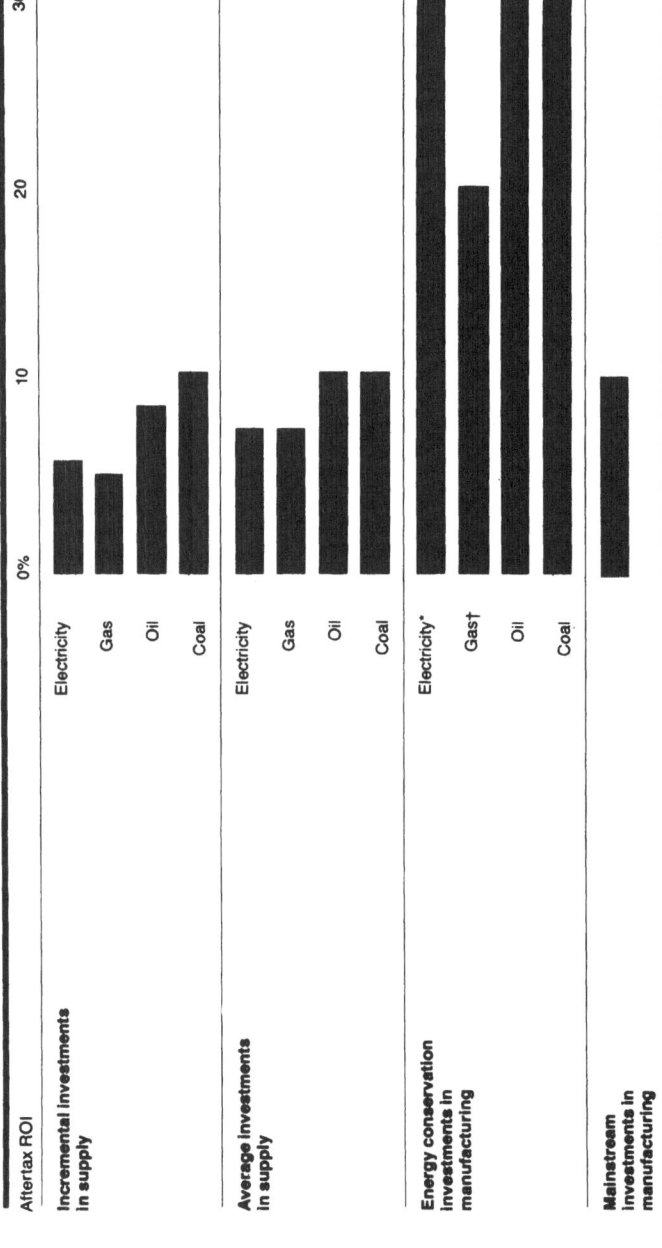

Fig. 22. Estimated Prevailing Returns from Investment in Energy Supply and
Manufacturing in the USA.   Source: Ref. 16.

investments in energy supply, mainstream production and energy
conservation. Conservation requires an ROI much greater than any
of the others, or put another way, conservation equipment has to
achieve a very short payback time, typically less than two years.

These various barriers which I have briefly described will be
most serious in firms or in industries where energy does not constitute a high proportion of production costs. Such firms are
unlikely to have technical staff who are conversant with heat recovery technology, boiler control, energy auditing, etc. Hence they
will require technical education and information and when they do
decide to purchase equipment they will be looking for a turnkey
contract which minimises their technical and managerial involvement. By the same token, since fuel costs are not a serious factor
in the business they will be less inclined to consider such sophis-
tications as life-time system costs; so they will tend to go for
the lowest first cost and hope for the best. Finally, the firm's
management is unlikely to be conversant with energy matters; they
will therefore tend to keep old boilers running for as long as
possible, set very short payback times for conservation investments
and generally eschew innovative approaches to conservation. Therefore, on top of all the difficulties which I have discussed already
concerning innovation in general, we are faced with additional
difficulties in the case of conservation innovation and those additional difficulties are likely to be most severe in the area of
process improvement which I have identified already as the area of
greatest short term potential. Faced with these various difficulties I would now like to consider strategies for promoting
energy conservation technology in industry.

STRATEGIES FOR ENERGY CONSERVATION TECHNOLOGY

I come now to the consideration of strategies for promoting
energy conservation technology in industry. These strategies are
relevant to any organisation faced with investment in conservation
- a big energy user or an energy conservation equipment supplier or
a government. My own thinking and most of my examples tend to be
government oriented, but I would emphasise that the same basic
questions have also to be faced by commercial enterprises and
international organisations, although their answers might have a
different emphasis. And speaking of international organisations, I
would also emphasize that one cannot mount a comprehensive programme on energy conservation technology without taking advantage
of developments from different countries. The UK has pioneered in
fluidised bed heat recovery; Japan has pioneered in organic
Rankine cycle systems for generating electricity from low grade
heat; Germany and the USA are collaborating on high temperature
materials for gas turbines etc., and the USA and Germany are probably leading in heat pumps. Both the private entrepreneur and the
government apparatchik need to range wide in their enquiries so as

to take maximum advantage of developments elsewhere.

Energy Flows

   The first thing to bear in mind is that energy flows down a
chain from the supply end to its final dissipation or sequestration
in durable hardware and it is possible to conserve energy at any
stage along the chain. We can list the main ways in which energy
can be conserved in supply, in its use and also in the use systems
of the final product within which the energy is sequestered, as
follows:

   (a)   Conservation in Energy Supply:

-   substitution of less energy intensive supplies:

    .   renewable sources for depletables
    .   lower grade for higher grade fuels

-   improved delivery of useful energy:

    .   combined heat and power
    .   peak smoothing
    .   improved conversion efficiency
    .   heat pumps

-   reuse of materials:

    .   waste as a fuel

   (b)   Conservation in Energy Utilisation:

-   substitution of less energy intensive processes:

-   improved process efficiency

    .   improved operation
    .   improved control
    .   improved insulation
    .   improved energy recovery

-   reuse of materials

   (c)   Conservation in Product Utilisation:

-   substitution of less energy intensive products:

    •   by choice of system

    •   by choice of design

    •   by choice of material

  –   enhanced product life

  –   reuse of products

When considering new technology one has to bear in mind that developments aimed at one point in the chain may have repercussions on developments at another point. Such considerations apply particularly to the domestic sector where, for example, the benefits of heat pumps, which are expensive, may not be sufficiently realised due to the wider take-up of insulation, which is cheap, or to the adoption of CHP/district heating. Therefore, if one is planning a new development one has to assess parallel developments upstream and downstream. This means that one may have to look at developments aimed at a totally different market.

Some of the items listed above may seem a little far-fetched, such things as 'improved design' and 'enhanced product life'. I would therefore like to digress briefly to give an example which incorporates several of the items listed here, which has proceeded imperceptibly over 2-3 decades and can be said to have resulted in quite significant energy conservation. This examples concerns milk bottles. In the UK 93% of all households have their milk delivered to their doorstep, most of it in bottles. Fig. 23a shows the energy sequestered different possible containers for milk. This shows that the 12 ounce glass bottle, which weighs 340 grammes and which has been standard for many years, is the most energy intensive container of all. But the glass bottle can be re-used, and over the past 30 years or so the glass makers have carried out extensive research along two main lines:

  –   to develop improved surface coatings for bottles so that they don't get scratched so easily and therefore can be used more often

  –   to develop a better understanding of the distribution of stresses in glass so as to achieve lighter but more resiliant designs of bottle.

A great deal of money has been spent successfully on such research with the result that the strength of bottles has been greatly increased, enabling the companies to reduce the weight of glass used; in parallel with this the product life has been extended, i.e. the number of trips which each bottle can make has been increased, and the recyling of glass has also been introduced.

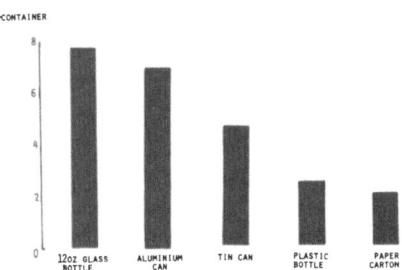

Fig. 23a. The Energy Sequestered in Different Containers.

Fig. 23b. Progress in Reducing the Weight of Milk Bottles.

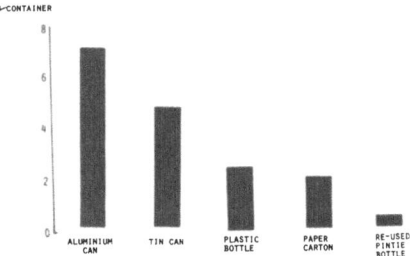

Fig. 23c. Energy Per Pint of Milk with re-used "Pintie".

Fig. 23   Source: Ref. 17.

Fig. 23b shows four bottles. The one at the left is a 500 g
version from the 1940's, then a 440 g version from around 1960,
then a 340 g version which is still in use, and the differently
shaped one at the end is the new 240 gramme "Pintie" bottle which
is just being introduced and promoted throughout the country. So
research into milk bottles has led to an improved design of bottle,
which uses less than 50% of the material previously used, and this
improved design has an enhanced life, so much so that bottles can
now do up to 40 trips each, and even then much of the glass is
recycled into new bottles. Therefore, the amount of energy
required to deliver a pint of milk in a multi-trip "Pintie" bottle
is much less than the energy sequestered in the bottle itself.
Fig. 23c shows the actual energy required per pint of milk
delivered. We can now see that the glass bottle achieves the
lowest energy requirement. I have estimated that by the time the
"Pintie" bottle is used throughout the UK a national energy saving
of over 200,000 tonnes of coal equivalent per annum will have been
achieved. That's quite a significant saving. Furthermore, the
experts believe that they can get the bottle weight down to 100
grammes, so there may be further energy savings over the next 10-20
years.

Finally, there is a further effect downstream in the transport
of the milk bottles. The new 'Pintie' design allows more milk
bottles to be packed in a lorry. This means that less petrol is
used in milk delivery. The effect is small but the point is that
it is a beneficial effect rather than a deleterious one, so that
all the energy effects of these various bottle developments are
additive in terms of energy saving, rather than cancelling one
another.

This illustrates how subtle changes in the way products are
made and used can achieve quite substantial savings and in the long
run we may see major energy savings come about through such developments. However, as in the case of the milk bottle, such changes
are likely to take decades and because of their complexity they
should in any case be approached with caution. In the meantime,
there are many more immediate ways of conserving energy, and so far
as industrial energy use is concerned, the most fruitful area for
RD&D in the short-term is the area of energy utilisation, i.e. the
actual process use of energy in industry - or Item (b) above. Most
of the 30% technological potential which was discussed earlier is
concerned with this part of the chain.

Priorities for Improved Energy Utilisation

What everyone would like to see in this area is a rapid
advance in the state of the art with regard to efficient energy
use. The basic options for achieving this are as follows:

(a)   The Reapplication of Existing Technology:

-       by its more widespread use in existing application
        areas

-       through the transfer of proven technology into new
        application areas

(b)   The Improvement of Extension of Existing Technology:

-       improved performance or reapplication on a changed
        scale in existing application areas

-       improvement or adaptation or change of scale to
        enable transfer to new application areas

(c)   The Development of New Technology:

-       for the replacement of existing processes

-       for introduction into applicaton areas that have
        not hitherto used any comparable technology.

Because energy conservation is inherent in the economic use of
energy and, as I emphasized at the very beginning, has been
practised for the past hundred years or so, technologies which are
already in use must outweigh in immediate importance those still to
be invented or applied. Therefore the re-application and
improvement of existing technology in new contexts is of paramount
importance in advancing the state of the art. To some people this
may not seem so exciting as, say, nuclear fusion research but it
still requires considerable technological effort - research,
development, design and demonstration, and it may even involve
quite basic research to overcome problems such as the fouling of
heat exchangers. I shall illustrate all of these points later.

Next, because, as I have emphasised already, energy conservation is highly disaggregated, much of the required RD&D has to do
with making the same basic technology work effectively and reliably
in different environments. This is the problem often referred to
as "technology transfer". In energy conservation we need a lot of
technology transfer. The transfer of technology from one sector to
another may be achieved through well conceived and promoted demonstrations of commercially available equipment, but in some cases it
will necessitate R&D to improve the design or performance so as to
enable the equipment to operate at higher temperatures or with
higher dust loadings or with more corrosive gases, for example.
Also, because of the experience curve phenomenon which we have
discussed already it should become possible to apply successful
technologies down market on a reduced scale where the potential may

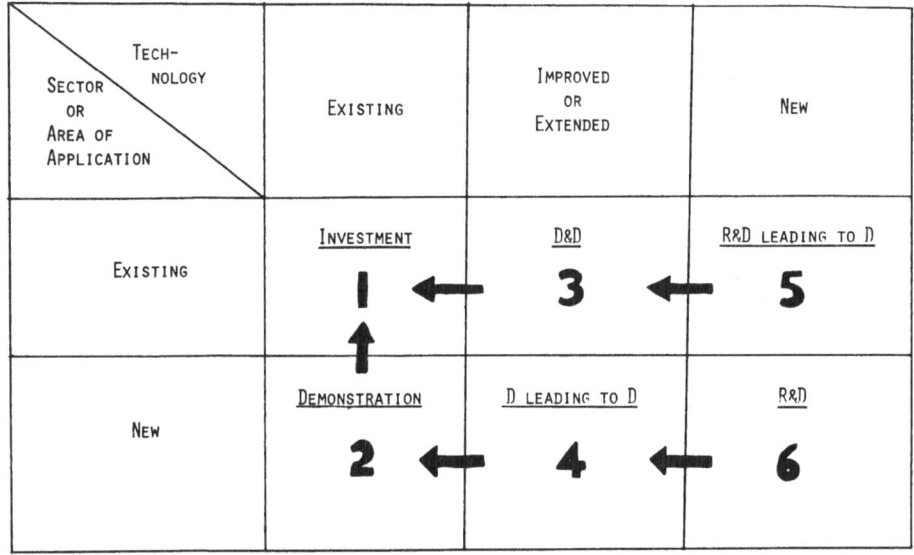

Fig. 24.   Technology-Application Matrix.

Source: Ref. 4.

be very great; however, this may necessitate entering markets
where the technology hasn't been used before; so once again
development work and demonstrations and vigorous marketing will be
required.

This emphasis on making the most of what already exists, and
not doing too many new things at the one time, is similar to the
emphasis placed by Ansoff in a company's introduction of new
products. Following Ansoff, therefore, I have devised a matrix to
put these ideas in a simple way. It is shown in Fig. 24 which is
taken from Energy Paper 32. Along the top I have indicated the
status of the technology and down the side the area of application.
The large numbers in each box indicate the priority which I consider should be given to that box. Thus, the increased sale of
existing technology into existing applications should have top
priority. After that demonstrations of existing technology in new
applications and improvements to existing technology should be
given high priority, and so on until we come at the bottom
right-hand corner to new technologies for new applications. This
is the highest risk area and is similar to Ansoff's diversification
box. While we should not dismiss this possibility it should not
take high priority in our allocation of resources. It's like a

horse at 100-to-1; it might just win, but only a crazy man would
put a lot of his money on such horses.

If a government wishes to promote energy conservation technology in industry then Fig. 24 suggests the kind of priorities it
might adopt for the allocation of finances and effort. The means
which a government has at its disposal range from fiscal and statu-
tory measures to financial support for research and development.
If my arguments were to be taken at face value then the government
would give high priority to financial incentives for investment,
because that corresponds to box 1 in Fig. 24, followed closely by
demonstrations and associated information, publicity and training.
This last is a very important factor in technology transfer because
if companies don't know what's available or how to use it then they
will be disinclined to invest. The government would also provide
support for R&D but, on the above arguments, the amount of money
made available for this would be less than for investment and/or
demonstrations. Alternatively, it could be argued that governments
should allocate their financial resources in inverse proportion to
the priorities I have suggested. The argument for this is that the
high priority items should be pursued by the companies themselves
in any case and it is government's role to back the more risky and
long term things which are less likely to be undertaken on a
straight commercial basis. The choice between these two alternative government approaches, or any compromise approaches, is
likely to be made on the basis of political considerations and the
particular concerns of individual countries.

As it happens, many countries do provide grants or taxation
incentives for capital investment (i.e. deployment) and also have
demonstration programmes and R&D programmes. The IEA has investigated the approaches of member states and these have all been
analysed and tabulated in a book[18]. In addition to national
activities there are international programmes, such as those
organised by the EEC, and the collaborative arrangements carried
out under the IEA. In both cases the emphasis is very much as I
have given here. Thus, the rationale which I have articulated is
being broadly acted upon by various nations and international
bodies although the detailed reasoning behind their schemes may
differ from what I have said and there may also be political
factors.

Assuming that a programme of RD&D is to be established for the
promotion of improved energy utilisation, what sort of things
should be encouraged? My own view is that one should give high
priority to the generic technologies which I cited already, heat
recovery, instrumentation and control, etc – see Table 5. I have
already emphasized the importance of demonstrations for such technologies, so let us consider what is involved in a demonstration
programme.

Demonstration Projects

In my presentation I have laid great stress on:

- careful and detailed energy auditing and reporting

- ensuring that the user can live with the technology

- proper attention to the marketing function in innovation

- promotion of technology transfer

A demonstration programme is about all these things and in Europe, for example, Germany, France and the UK all have major programmes, as well as the EEC. Let us look at what is involved.

First of all, what do we actually mean by a demonstration? A demonstration is a full scale trial of a product or a process which will establish its technical performance and its economic viability under working conditions. The objective of the demonstration, if it's successful, is to stimulate numerous other installations of the same sort, what we in the UK refer to as "replication" and the EEC describe as "diffusion". Basically, therefore, the demonstration is a sales promotion exercise although it is usually a technology which is being 'sold' rather than a proprietory product. Nevertheless, the equipment supplier has a common interest with the government, if the demonstration is supported by the government. He wants to sell more equipment and the government wants to see more firms save energy through the use of such equipment.

The organisation which stands to loose most if the demonstration isn't successful is the equipment user or "host" company. For that reason government aid tends to be directed to the host company; by the same token, if it is the equipment supplier that is promoting the demonstration he will usually supply the equipment free. In the UK, for example there are various demonstrations of fluidised bed combustion boilers which are being promoted by the National Coal Board and in those cases the NCB is underwriting the demonstration in some way. In other words, the user has to have an incentive to participate in the demonstration since he could lose out through disrupted production, excessive costs or failure to achieve the predicted energy savings, and it is the promoter of the technology who has to provide the incentive.

The equipment supplier will be disposed to show that his equipment saves energy, is easy to use and reliable. The user, on the other hand, may have some incentive for disagreeing with this and exaggerating the problems; he may wish to play down the savings achieved so as to maintain a commercial advantage over competitors. It is therefore most important that the technical performance,

reliability, economics and energetics should be established by a
competent and independent organisation. We usually refer to this
as the monitoring organisation. It is their job to carry out an
energy audit, preferably before and after the new equipment is
installed, and to assess generally the pros and cons of the
equipment. Quite clearly, therefore, a successful demonstration
requires collaboration between various parties, as follows:

-   the energy user or "host" (who may be a reluctant
    participant.

-   the equipment supplier (who stands to gain in the market
    place from a successful project).

-   an independent assessor (to monitor the results and
    ensure objectivity).

-   the government (whose interest is in the rational savings
    which could accrue from widespread diffusion of the
    technology).

-   (sometimes) an energy supplier (whose interest is in
    promoting his product through advanced utilisation
    equipment)

Unless all the parties can agree and co-operate in a reasonably
amicable way the demonstration could be ineffective.

Assuming that the equipment works and that the monitoring
organisation carries out its task in a competent way the most
important part of the exercise still remains, that is the promotion
of the results of the demonstration. Project SAPPHO concluded that
successful innovators were more effective in their marketing. So,
in addition to doing their technical development more thoroughly
and their market research more percipiently, they still had to
promote their product in a more vigorous and effective way than
their competitors. Since the whole purpose of the demonstration is
to achieve widespread replication it will have been pointless
unless a really effective marketing campaign can be carried out to
stimulate the necessary investment. Since subsequent investors may
have no financial aid to take up the technology they have to be
convinced entirely by its technical and economic performance; in
other words, they have to be sold on the benefits of the technology, and the board of the company has to be convinced to make
the necessary investment in its entirety.

The magnitude of the marketing task can be best appreciated in
terms of numbers. Suppose a country has an industrial demonstration programme with a budget of $25M over several years.
Suppose that this money is made available in the form of grants of

up to 25% of capital cost, and that the national objective is to
save $10 worth of energy per annum for every $1 spent on the programme, quite a reasonable aim. The first obstacle is to persuade
industry to invest $75M in risky projects of the heat recovery type
(i.e. industry puts up $3 for each $1 of government aid). These
projects will themselves achieve only a small fraction of the total
objective; indeed some of them will be failures and will save no
energy. Hence the national impact has to be achieved by persuading
many other companies to invest in the successful technologies
without any financial aid. If the programme aim is $250M of energy
per annum this can be related to the total capital investment
through the payback time. For heat recovery payback times of 2
years or less are sought, but there will be some technologies where
longer payback times are accepted and also some where a long energy
payback time is offset by improved productivity or savings of
labour, water, space etc. Hence, an overall payback time with
respect to energy of 4 years might be reasonable   That means that
the $25M programme has to stimulate $1000M of capital investment.
That is a very substantial and difficult undertaking indeed.

Research and Development

Turning now to R & D we are faced with an enormous range of
technological possibilities. Energy is used in every industrial
activity; hence there is potential everywhere for conservation;
and lively minds can formulate ideas for R & D into almost anything. For example, the EEC received an R & D proposal this year
for some quite interesting research into more energy efficient ways
of cooking a rare species of shell fish. The proposed programme
looked as though it might indeed lead to energy savings but even if
technical success were achieved and all the shell fish of this
species were cooked by the new means each tonne of coal saved would
have cost at least $60 in R&D alone. We therefore concluded at the
Advisory Committee for Programme Management that this proposal did
not represent good value for money to the Community. Companies and
governments have limited funds for energy conservation R&D and so
we need some priorities or guidelines.

My first guideline is that the potential benefits in terms of
energy saved must be large in relation to the cost of the R&D.
Given all the difficulties discussed earlier, there is a strong
likelihood that any given piece of R&D won't bear fruit; and the
further back it is along the innovation chain then the higher the
probability is that it won't lead anywhere. In the UK
demonstration programme the Department of Energy is aiming overall
for at least $5 per annum of national energy saving for each $1 of
public money spent on the programme. In the case of national R&D
one has to think in terms of $30-100 per annum saving for each $1
spent on R&D, and even higher ratios for speculative, preliminary
research.

Next, I believe that we need to distinguish R&D which is aimed primarily at energy saving from R&D which may contribute to it only in an incidental way. My earlier story about milk bottles illustrates what I mean by 'incidental'. Althouth the R&D carried out by the glass industry has led to significant energy savings that is not why they did it. They did it to defend their markets against competitors and they succeeded partly because of the resilient 'use system' by which milk is distributed in the UK. I am sure that if their research had led to increased energy use they would still have pursued it provided the economics worked out. Any organisation which funds R&D does so for a purpose. The R&D will be pursued vigorously or stopped depending on how well it appears to be achieving its primary purposes. Where energy saving is not the primary purpose we must be very cautious about supporting R&D out of conservation 'funds', both at company level and at national level, because the final decisions will be taken on the basis of other factors than energy, even when the energy benefit is substantial. Thus, in addition to a high energy saving/cost ratio energy saving should rank high, if not highest, in the technical requirements for success.

My third idea concerns the business relevance of the R&D and its results. I stressed earlier that companies are in the business of making and selling products. Anything that is central to the production process and the resulting products therefore commands strong attention whereas the efficient operation of the boilers and new methods of waste heat recovery probably don't. Energy using companies are therefore likely to devote their own R&D efforts to mainstream activities which will improve their productivity, product quality etc., and not to energy saving technology which is peripheral to the production process. Of course the development and production of boilers, heat exchangers etc. is the mainstream activity of the firms making boilers, heat exchangers, heat pumps etc., i.e. the process plant industry. But that industry is highly disaggregated, many of the companies are quite small and by comparison with the producers of steel, cement, non-ferrous metals, chemicals etc. etc., their R&D facilities are rather limited. Yet, as I have already shown, this is the area of greatest short term potential. Hence, I would argue that any government support of conservation R&D should be focussed strongly on this area and, following the Zuckerman principles, a good proportion of that R&D should be closely related to the generic technologies covered by demonstration projects. In other words, R&D support should be aimed at the development of a strong "energy conservation supply" industry.

I can sum up these ideas with Fig. 25, another little matrix. The lower right hand box is of little interest to us. It would include such things as health and safety, which may be very worthwhile but have practically nothing to do with energy conservation.

RELEVANCE TO ENERGY CONSERVATION & EQUIPMENT SUPPLY \ RELEVANCE TO PRODUCTIVITY QUALITY MARKETS ETC.	CENTRAL TO END PRODUCT/MARKET OR PRODUCTION PROCESS	INCIDENTAL TO END PRODUCT/MARKET OR PRODUCTION PROCESS
PRIMARY OBJECTIVE OF RD&D	**?**	OUR IMMEDIATE CONCERN
ENERGY SAVING INCIDENTAL	POSSIBLE INTEREST BUT MORE PROBABLY FOR DoI RB's	PROBABLY OF NO INTEREST TO DEN OR DoI

Fig. 25.   Energy Conservation R&D Relevance Matrix.

Source: ETSU.

The lower left hand box would include the R&D to make better milk
bottles; it might save energy but that is not its purpose and so
there is no case for support from the energy budget.   The top right
hand box is where I believe there is a need for much more R&D in
the face of the world situation.   This box covers heat recovery,
process insulation etc;   all the key generic technologies which I
listed earlier.   Whereas government support of demonstration
projects should be aimed at the energy user I believe that R&D
support should be aimed primarily at the equipment suppliers to
enable them to increase their technological effort.   The emphasis
should be on hardware development but market research, system
studies and energy auditing also need to strengthened.   The last
box, marked by a question mark, is where energy saving will come
about in the long term through the introduction of new production
processes and techniques, the substitution of materials and so on.
There could be major developments of this kind but their success
will depend on many more factors than just energy efficiency.   It
is questionable if government should get much involved here because
commercial considerations will dominate and major industrial
changes could result.

I happen to believe, on the evidence available so far, that
despite my earlier emphasis on heat recovery etc., more than half

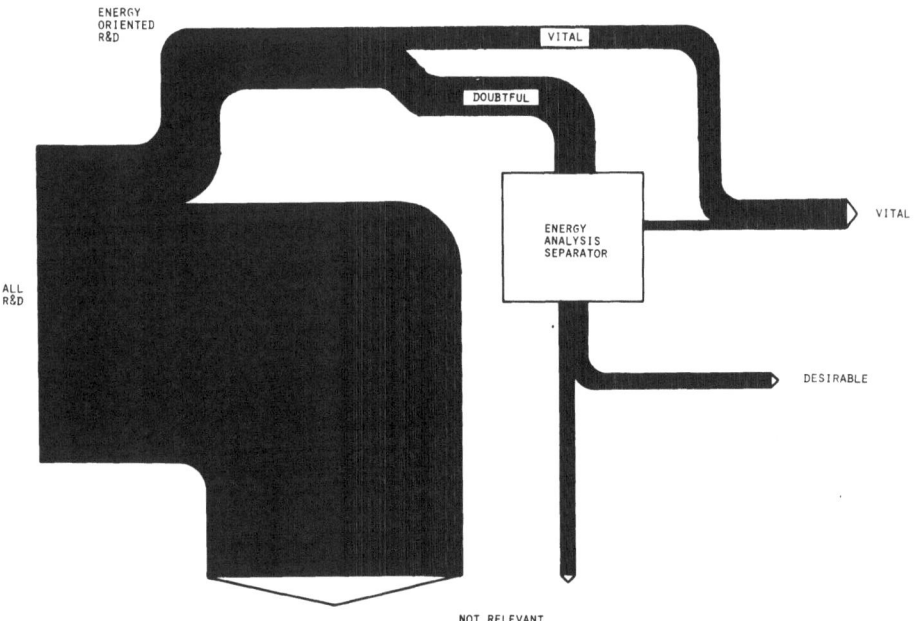

Fig. 26.   Separating the "vital" from the "incidental."
Source: Ref. 17.

the improvements in industrial energy efficiency which will come
about in the long term will be of the 'incidental' kind. They will
be influenced by energy prices but not dominated, and it will take
a very long time (several decades) for the energy price mechanism
operating alongside other prices and pressures to select out the
energy efficient processes, systems etc. Because the effects of
such technological change can be quite complicated I believe there
is a good case for national and international bodies sponsoring
much more study of the energy implications of new materials,
processes, products and systems, what I call "energy analysis
research". Fig. 26 illustrates what I mean, in the form of a
Sankey diagram. Coming in on the left is the stream of new R&D
ideas. Out of these a small proportion will be seen, a priori, to
have a potential energy impact. Some of these will be what I
call "vital", i.e. they will be aimed directly at energy saving,
with a good benefit/cost ratio, e.g. the development of an advanced
heat pump. Good projects of this kind should be supported under
government programmes. The rest of the stream is categorised as
"doubtful", but some of these ideas may in the end turn out to be
very important. So I have put them through an "energy analysis
separator", i.e. their energy impact is investigated at low cost,
on paper or in the laboratory. A very few of these will also turn
out to be "vital", rather more may prove to be "desirable", i.e.

leading to incidental energy savings and the remainder will be
irrelevant or possibly even "undesirable" from an energy point of
view. Such studies will gradually build up a much more detailed
understanding than we have today of the energetics of industry and
society and should contribute gradually to better informed policies
and decisions on energy supply and use.

Conclusions on RDD&D

To wind up this discussion of strategies I thought it would
be worthwhile looking at the conclusions reached in a recent, major
study of energy in the USA, "Energy - The Next Twenty Years"[19],
sponsored by the Food Foundation. Their eighth recommendation,
addressed to the US Government, was, "Vigorously pursue conservation
as an economical energy source", with four sub-headings as follows:

(i)   temporarily subsidize conservation investments

(ii)   use federal facilities as a conservation proving
          ground

(iii)   increase non-hardware conservation research

(iv)   aggressively demonstrate and "market" conservation
          to consumers

Also their third recommendation was, "Use science and technology to
generate and define basic options while relying primarily on the
private sector to develop and deploy technology", with four
sub-headings as follows:

(v)   increase emphasis on non hardware research

(vi)   develop competing processes and generic
          technologies

(vii)   rely on private sector selection and management

(viii)   pursue large scale demonstrations sdelectively and
          with great care.

These conclusions are very similar to what I have been saying.
Although the Ford Foundation's study group considered in general
that "deployment" is best left to the market place they argued item
(i) on the grounds of current market imperfections. I have not
discussed (ii) because it is more relevant to buildings and transport than to industry but I can tell you that it is part of the UK
government's policy. I have covered items (iii) and (v) and would
hold up the Audit Studies and IETS as illustrations of good
non-hardware conservation research. In common with the Ford

Foundation I have placed great stress on (iv) and I would agree
with their caveats (vii) and (viii) insofar as they apply to
conservation. I would, however, place rather more emphasis on (vi)
and include it in the energy conservation recommendation. This is
to accelerate the rate of development of heat pumps, heat recovery,
microprocessor controls etc. which are all part of the "energy
conservation supply" business. This may not be so important in the
USA where there is a lot of venture capital to support innovation
but I believe it relevant to the UK and some other European
countries, as was perceived by Shell in their report "Energy
Efficiency"[15].

Thus I believe that the strategic conclusions that I have
reached are broadly the same as those reached by other investigators such as the Ford Foundation, Shell, the Commission of the
EEC and numerous governments. That doesn't prove I'm right but it
does suggest a consistency of understanding between industrial and
government investigators and perhaps the beginning of a rational
approach to energy conservation technology.

CASE STUDIES - HEAT PIPE HEAT RECOVERY

What I would like to do now is take a few closely related case
studies and describe them to you in detail to illustrate some of
the points that I have been trying to make in these lectures.
These case studies involve the re-application of existing
technology, the improvement of existing technology and the
development of new technology, to use my earlier jargon, and they
also illustrate the part that research organisations can play in
energy conservation innovation for industry.

Earlier on I showed data indicating that the most important
generic technology for energy conservation in industry is waste
heat recovery. This term covers a great variety of situations and
types of equipment. There is a useful reference book which was
published last year and which lists over 300 different types of
equipment[20]. However, quite a high proportion of the waste
heat which is available in industry is in the form of hot gas
coming out of kilns, furnaces, cupolas, ovens, boilers, etc. What
one would like to do is to extract heat from this hot gas so that a
much cooler gas is dissipated to the atmosphere, with the recovered
heat transferred to another gas or liquid or solid for use
elsewhere. The ideal arrangement is one in which heat is extracted
from the exhaust gas and used internally to pre-heat combustion air
to the burners of the kiln, furnace, or whatever. That achieves
heat recovery; it can also improve the combustion efficiency of
the device. That kind of situation calls for a gas-to-gas heat
exchanger. If it is desired to transfer the heat to a liquid then
the requirement would be for a gas-to-liquid heat exchanger. For
simplicity I am going to concentrate here on gas-to-gas exchangers.

There are basically four main types of heat exchanger commonly in use for gas-to-gas heat recovery. First of all there is the plate recuperator which is shown in Fig. 27a. This is simply an arrangement in which the hot exhaust gas flows through a configuration of metal plates with the cold air supply flowing through on the other side of the metal. Heat is therefore transferred through the metal from the hot gas to the cold supply air so that a cooler exhaust emerges and the supply air is warmed up. The second type is the Ljungstrom heat wheel which was patented around 1920 and is shown in Fig. 27b. It is a rotating disc which is permeable to gases; it is usually made up of a wafer-like material or metallic mesh with a large surface area. What happens is that the wheel rotates picking up heat from the hot exhaust gas stream and transferring it to the cold air supply. If the gas flows and the rate of rotation of the heat wheel are properly balanced then very effective heat recovery can be achieved. The third option is the run-around coil, shown in Fig. 27c. This consists of two gas-to-liquid heat exchangers at each end with the liquid flowing through pipes between them. Thus the hot air flows through one heat exchanger transferring its heat through the pipe to the fluid, the fluid then flows to the other heat exchanger where heat is transferred in the reverse direction to heat up the supply air.

Finally, there is the heat pipe heat exchanger, which I particularly want to talk about. This one operates in a slightly more complex way as illustrated in Fig. 27d. There is a fluid inside the pipe which vapourises at the hot end, absorbing its latent heat of vapourisation and condenses at the cool end, giving out its latent heat. The condensed liquid then returns, sometimes via a wick, to the hot end where the cycle begins again. So heat is transferred along the pipe from one end to the other by means of this evaporation/condensation cycle. If now the pipe has one end in a hot exhaust gas and the other end in the cool supply air stream, heat will be transferred into the pipe from the hot gas, along the pipe via the pipe's internal fluid, and then out into the cool gas at the other end.

Demonstrating Existing Heat Pipe Technology

One of the demonstration projects which has been supported by the Department of Energy in the UK concerns the use of heat pipes for heat recovery in the non-ferrous metals industry. The project was proposed by BNF Metals Technology Centre, which is the research association dealing with the non-ferrous metals industry, and it involves demonstrations of heat pipe systems in four different applications. BNF had been contracted in 1975 by the UK Department of Industry to carry out the Energy Audit Studies on the non-ferrous metals industry, which were listed in Table 2, and in the course of that exercise they identified considerable potential for heat pipe heat recovery in the industry. Originally they put

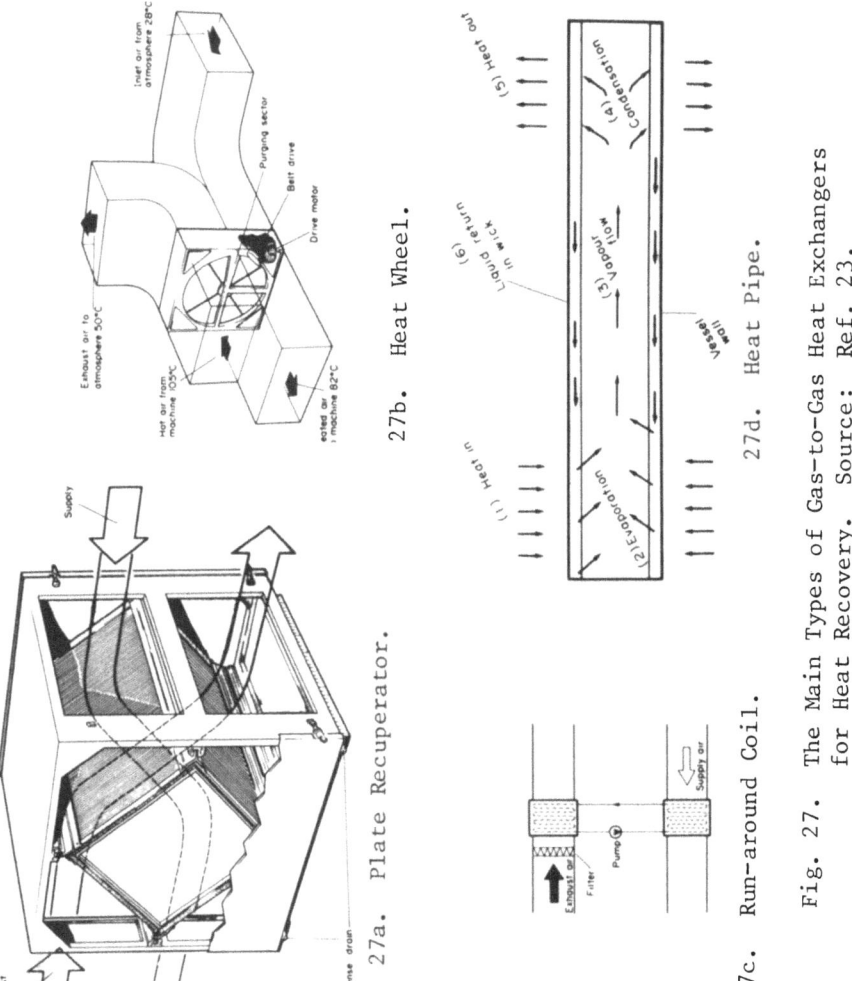

27a. Plate Recuperator.

27b. Heat Wheel.

27c. Run-around Coil.

27d. Heat Pipe.

Fig. 27. The Main Types of Gas-to-Gas Heat Exchangers for Heat Recovery. Source: Ref. 23.

the potential at 350,000 tonnes of coal equivalent, based on installations at 400 different works. Then they picked out three particular sub-sectors of the industry as being the most worthwhile and further investigation suggested that reasonable targets for energy saving through heat pipes in those sectors were as follows:

Copper Sector	0.78M GJ
Aluminium Sector	2.00M GJ
Lead Sector	0.39M GJ
Total	3.17M gigajoules

These target energy savings are equivalent to ~100,000 tce per annum and are associated with four demonstration installations, there being two in the aluminium sub-sector.

For each sub-sector BNF investigated several works by carrying out an energy audit and also making a techno/economic assessment of the scope for heat recovery. In the case of the copper sector, and on the basis of such an investigation, they recommended a demonstration installation with IMI-Yorkshire Imperial, at their Works at Kirkby in Lancashire. The Company agreed to put in demonstration equipment under the terms of our national Demonstration Projects Scheme, and the project is now in progress.

However, before finalising the type of installation, BNF investigated various system configurations, and I should like to explain these briefly. IMI-YI operates six sklenar furnaces at this works so a lot of waste heat was available. It was possible to use some of the recovered heat to pre-heat combustion air; it was also possible to heat water for general use in the Works; and it was possible to use recovered heat for space heating in an adjacent block of offices. Heat pipes could be used for all three requirements, with the heat distribution controlled by a microprocessor system, so it looked a very comprehensive demonstration. Unfortunately the space heating component necessitated transmitting the heat a considerable distance, and that in turn necessitated a rather costly system of ducts. Four different options were therefore investigated and these are summarised in Table 7. This illustrates several points.

System 1, which would have pre-heated the combustion air on one furnace, as well as supplying hot water and providing space-heating, was over-ambitious and too costly. The high cost arose not from the heat pipes or the nature of the technology but from the fact that the source of heat wasn't close enough to the

	SYSTEM NUMBER			
System feature	1	2	3	4
Combustion pre-heat (no. of furnaces)	1	6	1	1
Water heating (heat exchanger)	HPHE	HPHE	HPHE	finned tube
Space heating (heat exchanger)	HPHE	-	-	-
Heat recovery (gigajoules per annum)	53,000	59,000	32,000	32,000
Installed capital cost ($000 - 1980)	520	402	326	253
GJ p.a. saved/$000 cost	101	147	98	126

Table 7.   BNF/IMI-YI Heat Pipe System Options.
Source: ETSU.

requirement for space heating. Therefore the space heating idea
was dropped. System 2, which shows the highest heat recovery of
all over the year, and gives the best heat/cost ratio, would have
required the pre-heating of combustion air on all six furnaces.
This was considered too risky by the Company because the technology
hadn't been fully proven, and it was also considered unnecessary by
us in Government for the purpose of demonstration. Therefore the
project was cut down in scale so as to pre-heat combustion air on
only one furnace. I should add, however, that if the demonstration
is successful then I would expect the Company to invest in
combustion air pre-heating for all six furnaces in due course.
System 3 is the more modest scale project and here you can see that
the ratio of heat recovery to cost is once again rather poor. In
fact it wasn't really attractive enough for the Company to make the
investment, given the uncertainties. The reason for this is that
the air-to-water heat pipe heat exchanger is a rather novel and
costly unit, developed by Industrial Research and Development Ltd.
and not yet in production. In the final system, System 4, this
heat pipe heat exchanger is replaced by a more conventional finned
tube heat exchanger. This brought the cost down considerably, it
also reduced the number of novel components and hence the risk,
and the economics are now quite attractive. Also, the amount of
heat recovered per annum is still pretty good. This is the system
that is being installed, using a commercially available Q-Dot heat
pipe, made in the USA.

I think this illustrates the need for very careful attention to the energy auditing, the costing, the technological performance and the risks, and in this connection I must emphasise that the system cost may be very different from the cost of the basic heat exchanger. In the final system only 5% of the cost is for the actual heat pipe unit itself and only 2% for the air-to-water heat exchanger. All the rest has to do with the structural steel work, ducting, controls, design and commissioning, etc. Thus, although heat recovery looked initially straight-forward and very attractive at this Works where they had six furnaces providing lots of waste heat and plenty of scope within the Works for using it, it took a lengthy investigation and some rather careful evaluation before a satisfactory system configuration was arrived at. Even now it may not work as well as expected. That is why it is a demonstration project. But if it does work, then there are already other companies following the progress of the project, and making preliminary energy audits of their own works so that they will be in a good position to install similar systems rapidly as soon as the results are known. Thus, we have already started the "marketing" of the project and with a little bit of luck we might actually achieve a reasonable proportion of the target savings which are aimed at.

BNF have carried out similar exercises in the aluminium and lead industries and we are now in the process of finalising demonstration installations with companies in those sub-sectors. In one case we may even get a heat wheel and a heat pipe installed at the same works so that we will be able to weigh up the pros and cons of the two systems. The results from all these demonstrations will be of great value to companies in the non-ferrous metals industry and we hope that significant savings will eventually be achieved through the wide-spread adoption of heat pipes and heat wheels as appropriate. Taking the two together the scope amounts to perhaps 500,000 tonnes of coal equivalent per annum in the UK metals industries as a whole.

Developing Improved, Lower Cost Heat Pipe Heat Exchangers

The Harwell Laboratory in England, where I work, made an assessment of the various gas-to-gas heat recovery systems in 1977 and concluded that the prospects for technical improvement were greatest for heat pipe heat exchangers. They obtained funding from the EEC and the UK Department of Energy and started an RD&D programme to develop and demonstrate improved heat pipe heat exchangers, i.e. heat exchangers which would achieve the same heat recovery but in a more compact form and at lower cost, so that they would be economically acceptable in industries with low machine utilisation factors. Towards this end, through sub-contracts with various research associations, they assessed seven potential applications in six different industrial sectors. In each case an

INDUSTRY	MACHINE	KW DUTY OF EXCHANGER	EXHAUST TEMP (^{0}C)	POTENTIAL REPLICATION FACTOR	
				UK	EEC
Laundering	Conditioner	65	78	400	2000
Laundering	Tumble Drier	40	80	1200	5000
Paper	Glazing Hood	140	110	250	1500
Wool	Wool Drier	25	66	100	500

Table 8.   Heat Pipe Markets — UK and EEC.

Source: AERE Harwell.

economic and technical assessment was made of the prospects for the four main types of heat exchangers which I have already described, plate recuperators, heat wheels, run-around coils and commercially available heat pipes. The conclusions from this exercise were that only the heat wheels and the heat pipes could achieve pay-backs of less than two years, with low machine utilisation. It was also concluded that the overall system costs could be brought down if the physical size of the heat exchanger could be reduced thereby reducing ducting costs but at the expense of more powerful fans. The increased pressure drop through the system would be overcome by additional fan power. Four of the applications were assessed to be suitable for heat pipe heat exchangers, as shown in Table 8, and the companies concerned agreed to accept prototype units for demonstration trials in their works.

In parallel with the assessment of these applications the team at Harwell were working on an improved design of heat pipe heat exchanger. They determined that commercially available HPHE's were not fully optimized, i.e. the heat pipes were capable of transferring more heat than the gas side surfaces could supply. First they designed an ovate heat pipe with enhanced internal surfaces which is three times more effective at transferring heat than the conventional plain bore circular section heat pipe. The flat sides of this heat pipe allowed the use of high efficiency, gas side, corrugated finning. A modular design of HPHE was also

perfected, each module consisting of a heat pipe sandwiched between
two blocks of corrugated finning. The complete heat exchanger is
constructed by clamping together the required number of modules.
Fig. 28a illustrates the type of structure which they arrived at.
The cross-section of the heat pipe cam be seen and how it is placed
between two corrugated structures. The top end would be in the
cold gas stream and the bottom end in the hot gas stream and heat
would be transferred from the matrix structure to the heat pipe,
through the heat pipe and then out at the other end again through
the matrix structure. These modular units could be combined to
make up a heat exchanger as shown in Fig. 28b.

    Harwell is not of course a manufacturing organisation. It is
a research laboratory, and so it was necessary to find a manufacturer who would take up the heat pipe, if it were successful, and
put it into routine production. Harwell therefore had discussions
with several manufacturers weighing them up according to various
criteria. In the end a licensing arrangement was reached with a
company. Thus, an improved and promising design of heat pipe heat
exchanger was achieved, a licensing arrangement was established
with a manufacturer and four user companies, each representing
significant end markets, had agreed to install prototype systems
for demonstration trials. The appropriate research associations
were also by now familiar with the applications and with the
equipment and so they were in a good position to make a thorough
technical and economic evaluation of the demonstration units once
installed, and to help promote them in their industries.

    One of the demonstration installations is with United Linen
Services Limited, a commercial laundry company. In this laundry
they use a type of drier from which a lot of waste heat is available. Fig. 29a shows one of these driers before the heat pipe had
been installed. Fig. 29b shows the two heat pipe heat exchangers
being fitted at the base of the exhaust ducts. The heat pipes are
vertical and heat is transferred out of the hot air in the exhaust
duct and up into the air intake to the drier. Thus, heat is being
recirculated from the hot exhaust into the fresh air intake. You
can see how compact the whole unit is and the relative ease with
which it has been accommodated in the space available. Production
units will be even more compact because there will be no need for
the bypass tubes shown in Fig. 29b, so that the heat exchangers can
be fitted in without having to deflect the vertical exhaust ducts.

    Well, this system has now been operating successfully for some
time and the Shirley Institute, the RA which is concerned with the
textiles industry, has carried out a detailed assessment of performance. Table 9 shows the results of their energy audit. The first
column gives data on the performance of the drier without the heat
exchanger, in December 1977. The next two columns give the
performance as found on 12 March 1980 and after cleaning on 13

Fig. 28b.   Heat Exchanger Incorpora-
ting Several Modules.

Fig. 28a.   Basic Heat Pipe Module Developed by
Harwell.

Fig. 28.   Source: AERE Harwell.

Fig. 29a.

Fig. 29b.

Fig. 29 a - Laundry Drier at United Linen Services.

b - Fitting the Heat Pipe Demonstration System
to the Drier.   Source: AERE Harwell.

	WITHOUT HEAT EXCHANGER DEC 1977	WITH HEAT-EXCHANGER	
		AS FOUND 12 MARCH	CLEANED 13 MARCH
Av. drying rate, kg/min	0.79	0.51	0.49
Av. air flow rate, kg/min	73.5	33	42.9
Energy consumption and sources MJ per useful MJ			
Consumption:			
water evaporation	1.00	1.00	1.00
hot air (before exchanger)	2.44	1.35	2.46
other losses	0.11	0.20	0.26
total*	3.55	2.55	3.72
Sources:			
steam	3.48	1.86	2.44
electricity	0.07	0.12	0.12
heat recovery	-	0.69	1.10
total*	3.55	2.65	3.66
Rate of heat recovery, kW	-	15	23

* discrepancies between the totals represent measurement errors

Table 9.   Results of Energy Audit by Shirley Institute
on Heat Pipe Demonstration System.

Source: Shirley Institute Report.

March 1980. This table summarises data in the form which I
described earlier when talking about energy audits. At the top we
have the measurement of losses and at the bottom we have the
various energy inputs, steam, electricity and recovered heat. You
can see that the totals above and below don't agree precisely.
This is because of the errors in the measurement which I talked
about before.

You can also see that on 13 March about 30% heat recovery was
being achieved. The difference between the 12 March and 13 March
is partly due to fouling. This is a problem that has cropped up on
all four of the installations. In their efforts to make the heat
exchanger more compact and more efficient Harwell went for a design
involving a close mesh of metal; part of the price that one has to
pay for this is fouling. There are a lot of threads and particles
floating around in a laundry. Some of these get caught in the
filter in front of the heat exchanger, but some of the material
also gets through the filter and fouls up the heat exchanger
surfaces thereby reducing the heat transfer coefficient. In their
report the Shirley Institute recommended:

(i)   the use of an improved filtration material and filter
            housing

(ii)   facilities for easier access to the heat exchangers
            for inspection and cleaning

(iii)   provision of a simple indication of the need for
            inspection, e.g. by fitting manometers upstream of
            the heat exchangers.

Work is now in hand to overcome these problems. It involves
testing different filter materials (e.g. fibreglass and
polyurethane foam) of various porosities in a zig-zag filter frame
which will allow a greater thickness of filter to be used for a
given pressure drop. Tests will also be carried out on a new
filter material which has a graded porosity through its thickness,
i.e. coarser at the front face, becoming progressively finer. We
hope this work will be successful, but it illustrates once again
the need for careful testing of new equipment in situ and in many
cases for adaptation to make it effective and acceptable to the
people who are using it. If the fouling problem cannot be overcome
the system won't perform so well, and its cost effectiveness will
be reduced, so we have to overcome this problem.

I said earlier that the Harwell design also entailed a reduction in size, duct cross-section, etc., at the expense of more
powerful fans than would normally be employed. If we look again at
the Shirley Institute's results in Table 9 we can see that the
electricity consumption by the drier in December 1977 was only 0.07
kW, the power required to drive the existing fan. With the heat
pipe heat exchanger installed there is additional electricity consumption by the larger fans required to drive the air through the
heat exchanger. You can see that the electricity consumption has
now gone up to 0.12 kW, and although this is a significant increase
in electricity use it is still only 10% of the energy which is
being recovered.

Before leaving this project, there is one other point which I
should make. It concerns not so much energy conservation as innovation in general. When coming to a licensing arrangement Harwell
weighed up a great many factors and, of course, the companies with
whom they were having discussions also had to weigh up the pros and
cons from their side. The company which was eventually chosen, was
selected partly because of its technical ability to construct the
enhanced heat transfer surfaces required for the heat pipe heat
exchanger, i.e. its technical ability to make the product. The
company has now decided not to continue with the production of
these systems. It has been taken over by another company which
has concluded that heat pipe heat exchangers don't fit in with its
business. Once again, therefore, it would appear that the

innovators at Harwell, despite a full awareness of the problems of innovation and a very careful attempt to take everything into account, still did not give adequate weight to commercial factors and their plans have been upset by the takeover which, of course, could not be predicted. They now have a promising product, which is proving satisfactory to the demonstration host companies, and which is in demand by other companies, but they don't have a licensee able and willing and committed to producing and selling it routinely in the market place. Hopefully, they will find someone else, but time will have been lost and it will cost more money to get the system into production.

High Temperature Heat Pipes for the Future?

There is one final part to my heat pipe story. I told you first of all about a well conceived demonstration project to promote existing technology, i.e. commercially available heat pipes from the American company Q-Dot. Then I told you about the Harwell project which has led to an improved design of heat pipe, but still basically a conventional heat pipe using conventional fluids. One of the limitations of the heat pipes commonly used for industrial waste heat recovery is their maximum operational temperature. In the BNF project the hot gas coming out of the furnaces is much too hot for the heat pipe and so cold air is bled into the exhaust stream to reduce the gas temperature to one which is suitable for the heat pipe heat exchanger, the whole thing being under the control of a microprocessor. If the heat pipes could accept gas at higher temperature then greater thermodynamic efficiency would be achieved, hence a higher rate of heat recovery and the overall economics would be even more attractive. At present this cannot be done because the organic fluids which go through the evapor-ation/condensation cycle in the heat pipe begin to deteriorate above ~ 300°C (with water as the fluid the internal vapour pressures become dangerously high above 250°C).

We know that there is great potential for heat recovery in the range 200–500°C. A Czechoslovakian paper which was published a few years ago reported that a sulphur/iodine mixture could be used as a working fluid in heat pipes up to 600°C. Harwell has therefore proposed a further research project, to the EEC and the UK Department of Energy, to investigate the feasibility of sulphur iodine and other potential fluids for high temperature heat pipes. Table 10 gives a list of possible fluids and their maximum working temperatures. Harwell propose to investigate the thermal characteristics of the heat pipes and the compatibility of the fluids with various construction materials for the heat pipe itself. If this work goes ahead and the results are promising then a further stage of prototype development would be proposed and at that stage, hopefully, they would be able to attract industrial interest and get a company to invest in the R&D with a view to

MEDIUM	MELTING POINT (°C)	BOILING POINT AT ATMOS. PRESS. (°C)	USEFUL RANGE (°C)
Helium	-272	-269	-271 - -269
Nitrogen	-210	-196	-203 - -180
Ammonia	-78	-33	-60 - 100
Freon 11	-111	24	-40 - 120
Pentane	-130	28	-20 - 120
Freon 113	-35	48	-10 - 100
Acetone	-95	57	0 - 120
Methanol	-98	64	10 - 130
Flutec PP2*	-50	76	10 - 160
Ethanol	-112	78	0 - 130
Heptane	-90	98	0 - 150
Water	0	100	30 - 200
Flutec PP8*	-70	160	0 - 225
Thermex	12	257	150 - 395
Mercury	-39	361	250 - 650
Caesium	29	670	450 - 900
Potassium	62	774	500 - 1000
Sodium	98	892	600 - 1200
Lithium	179	1340	1000 - 1800
Silver	960	2212	1800 - 2300

* Included for cases where electrical insulation is
a requirement

Table 10.   Heat Pipe Working Fluids.

Source: Ref. 25.

marketing a product eventually for operation at temperatures up to
600°C that would make new technology available and expand still
further the scope for cost effective heat recovery in industry.

By these cases I hope I have illustrated how the concepts and
principles which I have discussed earlier in more abstract terms
really do apply to practical situations. In this heat pipe area we
have demonstrations, development and now a further proposed
programme of fairly basic research, all of which may lead to
widespread use of a variety of heat pipes for different
temperatures and applications in a range of industrial sectors.
Perhaps at a future summer school we may learn to what extent these
aspirations have been fulfilled - or not.

REFERENCES

1.   O. Lyle, "The Efficient Use of Steam", HMSO, London (1947)
2.   "Energy Analysis Workshop on Methodology and Conventions",
      International Federation of Institutes for Advanced Study,
      Stockholm (1974).
3.   T. V. Long II, ed., "Energy Analysis Methodology", Technical
      Information Center, US Department of Energy, (1978).

4. "Energy Conservation Research Development and Demonstration", (Energy Paper 32), HMSO, London (1978)

5. "Improved Energy Efficiency," Shell Briefing Service, London, (June 1979).

6. "Exploring Energy Choices", Ford Foundation, Washington (1974).

7. "Technological Innovation in Britain", HMSO, London (1968)

8. "A Framework for Government Research and Development, Cmnd 4814, HMSO, London (1971)

9. "Punch", p.17, Punch Publications Limited, London (2 July 1980)

10. "Management of New Products", Booz Allen & Hamilton, Washington (1968)

11. I. Ansoff, "Corporate Strategy", McGraw Hill, New York (1965)

12. "Perspectives on Experience", Boston Consulting Group Inc., Boston (1968)

13. "Success and Failure in Industrial Innovation", Science Policy Research Unit, University of Sussex, published by the Centre for the Study of Industrial Innovation, London (1972)

14. J. K. Galbraith, "The New Industrial State", Penguin Books, London (1969).

15. A. F. Beijdorff, "Energy Efficiency", Shell International Petroleum Co. Ltd., London (1979)

16. G. N. Hatsopoulos et al, "Capital Investment to Save Energy", Harvard Business Review, p.111, (March/April 1978)

17. W. M. Currie, Energy Conservation R&D - Is it Vital or is it Incidental? in "New Ways to Save Energy", A. S. Strub and H. Ehringer eds., D. Reidel Publishing Co., Dordrect (1980)

18. "Energy Conservation in Industry in IEA Countries", OECD, Paris (1979)

19. H. H. Landsberg et al, "Energy - The Next Twenty Years", Ford Foundation, Cambridge Mass. (1979).

20. D. A. Reay, "Heat Recovery Systems", Spon, London (1979)

21. "Use of Energy in the Iron and Steel Industry", Document IISI/STATS/65, International Iron and Steel Institute, Brussels (1978).

22. D. K. Hale et al, The Industrial Energy Thrift Scheme, in "Energy for Industry", P. W. O'Callaghan ed., Pergamon Press, London (1979)

23. D. A. Reay, "Industrial Energy Conservation", Pergamon Press, London (1977).

24. "Energy Technology for the United Kingdom", (Energy Paper 39), HMSO, London (1979).

25 P. D. Dunn and D. A. Reay, "Heat Pipes", Pergamon Press, London (1976).

APPENDIX

A PROPOSED ENERGY AUDIT PROCEDURE

The suggested procedure is designed to produce all the relevant data necessary for detailed materials and energy balances for the plant under consideration. Measurements are recorded finally in basic S.I. units but it may be advantageous, particularly when dealing with craft industries, to work in British units and convert later. This often makes communication with the client company much more effective.

It will be noted that under ENERGY OUTPUTS, the first requirement for information are not vital to the audit. They are however considered important and deliberately included to show the company being audited that there might well be simple commonsense action that could be taken to conserve energy at very little capital cost to the company.

ENERGY INPUTS

The items of information required vary in number and complexity according to the type of process plant being audited. They can however be simplified to cover the following main sources of energy input to the plant.

(a)   Steam, heat transfer fluid, gas, oil or coal used for direct plant heating.

(b)   Electric power to fans, compressors, pumps etc. and for motivation of plant.

(c)   Heat energy already present in ambient air supplies and the humidity condition of the air.

(d)   Energy present in the product entering the process plant.

(e)   Also required is a record of the Energy Source costs, the product cycle e.g. drying time and the fractional use time of the plant per annum. In the case of intermittent plant usage, it is also necessary to obtain an estimate of the total metal weight of the plant so as to estimate heat required to bring the plant up to operating temperature.

ENERGY OUTPUTS

Items of information required under outputs, like inputs, can vary in complexity but a simplified summary would be as follows:-

(a)   Housekeeping losses – steam leaks, hot air or gas escape from plant.

(b)   Radiation losses from the machine casing.

(c)   Where non continuous operation, transient machine heating requirements.

(d)   Gas to air outflow losses (sensible and latent) from the process to atmosphere and, where direct gas or oil fiting is used, the composition of the waste gases.

(e)   Cooling air losses.

(f)   The condition of the product leaving the plant.

The measurements required under the various INPUT and OUTPUT categories are set out on the specimen Audit Record sheets based on a Loose Stock Wool Fibre Dryer and included in this Appendix as Sheets 1 and 2.

RESULTS

Having outlined the main requirements of the energy audit procedure and assuming that the relevant measurements have been undertaken and all the data recorded on the Audit Record Sheets, two further steps are suggested.

First, the data should be set out in the form of an Audit Summary which comprises seven main sections. These are as follows:

(i)   Details of product throughput

(ii)   Details of heat source, moisture contents of the air in and product in and out. Also required is total moisture removed and total energy input to the plant.

(iii) Details of the exhaust air or gases – velocity, humid volume, equivalent weight of cold dry air at ambient temperature and the total moisture in the exhaust air or gases.

(iv)   A Heat Balance – the total energy output which should, within reasonable limits, agree with the total energy input given above.

(v)   An assessment of possible heat recovery on whatever basis may be chosen for calculating the potential heat recovery in relation to the type of heat exchanger to be used.

(vi)   Also based on cycle time and the fractional use time of the plant, the amount of heat that could be recovered per annum.

(vii)   At the quoted energy source cost, equivalent cost savings per annum.

A typical example of this Audit Summary sheet for the Loose Stock Fibre Dryer is attached as Sheet 3.

The second and final step, having all the relevant data on either the Audit Records or set out on the Audit Summary, is to produce a detailed flowsheet for the materials and energy balance for the plant.

At first sight this might appear to be superfluous but the example I have given is a fairly simple audit. I can assure you that from experience of checking a number of audits carried out by different organisations, when all the information is set out in flowsheet form, on many occasions discrepancies have been found which were serious enough to call for re-examination of certain data recorded on the original audit record sheets. A typical example of such a flowsheet, again for the case of the Loose Stock Fibre Dryer is given in the attached Sheet 4.

ENERGY AUDIT RECORD   -   SHEET 1   -   ENERGY INPUTS

Processing Company ....   Machine Manufacturer ....

Type of Process - Loose Stock Fibre Dryer - 4 Drum.
(All values calculated above Ambient Air Temp of 18.3°C)

ITEM	PARAMETER	VALUE	SOURCE	ENERGY RATE		COST/ANNUM
				Kw	%	
PROCESS STEAM	Flow rate Inlet pressure " temperature " dryness fraction	0.231 Kg/s 6.51 barabs. 162°C 100%	measured measured derived assumed	638	93.1	£12,387
PRODUCT IN.	Flowrate (wool & moisture) Temperature Water	0.21 Kg/s 21°C 0.115 Kg/s	measured measured measured	1.5	0.2	-
INLET AIR	Flowrate (air) Temperature(ambient) Humidity (") Flowrate (Water) Pressure	1.793 Kg/s 18.3°C 75% RH 0.011 Kg/s 1 bar	measured measured measured derived assumed	-	-	-
ELECTRIC POWER Dryer fans	Gross Kw Proportion of energy into air and wool	53.4 Kw 85%	measured estimated	45.4	6.6	£ 3,473
Dive for Drums	Gross Kw Proportion of energy into air and wool	0.75 Kw 60%	measured estimated	0.45	0.1	£ 34
				685.3	100%	£15,894

ENERGY SOURCE COSTS	steam per 1,000 Kg £3.31p Electricity per Kwh 1.7p	
DRYING CYCLE	90 hours per week	
FRACTIONAL USE TIME PER ANNUM	50 weeks	

ENERGY AUDIT RECORD - SHEET 2 - ENERGY OUTPUTS

(All values calculated above Ambient Air Temp. of 18.3°C)

ITEM	PARAMETER	VALUE	SOURCE	ENERGY RATE		COST/ANNUM
				Kw	%	
HOUSEKEEPING LOSSES Steam Leaks Hot Air Escapes Plant Control	Flanges-Flowrate- Flowrate- Temperature Under utilisation, poor temp.control etc.	0.0014 Kg/s 0.205 Kg/s 120%	Estimated Estimates & measurement Control satisfactory	4.0 20.9	0.6 3.0	£ 75 £407
MACHINE CASE LOSSES	Temperature-Top Temperature-sides	44°C 41°C	measured measured	10.7	1.6	£208
TRANSIENT MACHINE CYCLES	Temperature-Cold- " at Working Condition Fractional Down Time	25°C 162°C Once per week	Estimated from Total Weight of Dryer	6.4	0.9	£124
CONDENSATE RETURN	Flowrate Temperature	0.102 Kg/s 115°C	measured estimated	93.5	13.6	£1,816
PRODUCT OUT	Flowrate (Wool & Moisture) Temperature Water	0.114 Kg/s 43°C 0.019 Kg/s	measured measured measured	5.2	0.9	£101
EXHAUST AIR OUTFLOW LOSSES	Flowrate (air) Temperature Relative Humidity Flowrate (water) Pressure	1.793 Kg/s 92°C 0.125 Kg/s 1 bar	measured measured measured derived assumed	538.2	78.5	£10,453
			Total	678.9		
			Unaccounted losses	6.4	0.9	£124
				685.3	100.0	£13,308

<u>ENERGY AUDIT SUMMARY SHEET 3</u>

Work throughput	410 Kg/hr @ 20% regain 342 Kg/hr bone dry weight

Steam pressure and temperature	6.51 bar abs. - 162°C

Initial moisture content of wool in Final " " " " out Weight of water evaporated from wool " " " " " incoming air	121% 20% 345 Kg per hr. 39 Kg per hr. 384 Kg per hr. (0.06 Kg/Kg dry air)

Total heat input from steam consumption		2,296 Mj/hr
Heat input via product		5.4 Mj/hr
Heat input via fan energy		163.4 Mj/hr
Heat input via drum drive energy		1.6 Mj/hr
	Total	2,466.4 Mj/hr

Volume of exhaust gases out (Sp.Vol 1.19 -psychrometric chart)	7,683 M³/hr.

Temperature of exhaust gases out	92°C

Equivalent weight of dry air supplied (ambient temp. 18.3°C)	6,456 Kg/hr

HEAT OUTPUTS FROM PLANT

Total heat content of exhaust gases (≲300 Kj/Kg psychrometric chart)		1,936.8 Mj/hr
Condensate return to boiler		336.5 Mj/hr
Heat losses from dryer casing		38.5 Mj/hr
Heat losses in product out		18.7 Mj/hr
Transient machine heating up and cooling losses		
Down time average once per week - equivalent to loss per hr.		23.2 Mj/h
Housekeeping losses - steam leaks, hot air escapes from machine		89.6 Mj/hr
Unaccounted losses		23.1 Mj/hr
	Total	2,466.4 Mj/hr

Possible heat recovery based on increasing the temperature of the incoming cold air to a heat exchanger up to a temperature equal to 70% of the temperature of the hot exhaust air - 64.4°C	299 Mj/hr

Equivalent amount of heat recovered per annum based on the dryer running for 90 hrs. per week - 50 weeks per annum.	1.3455×10^6 Mj/annum

At the reported steam cost of £3.31p/1,000 Kg - equivalent to a possible saving of:	£1,613 per annum

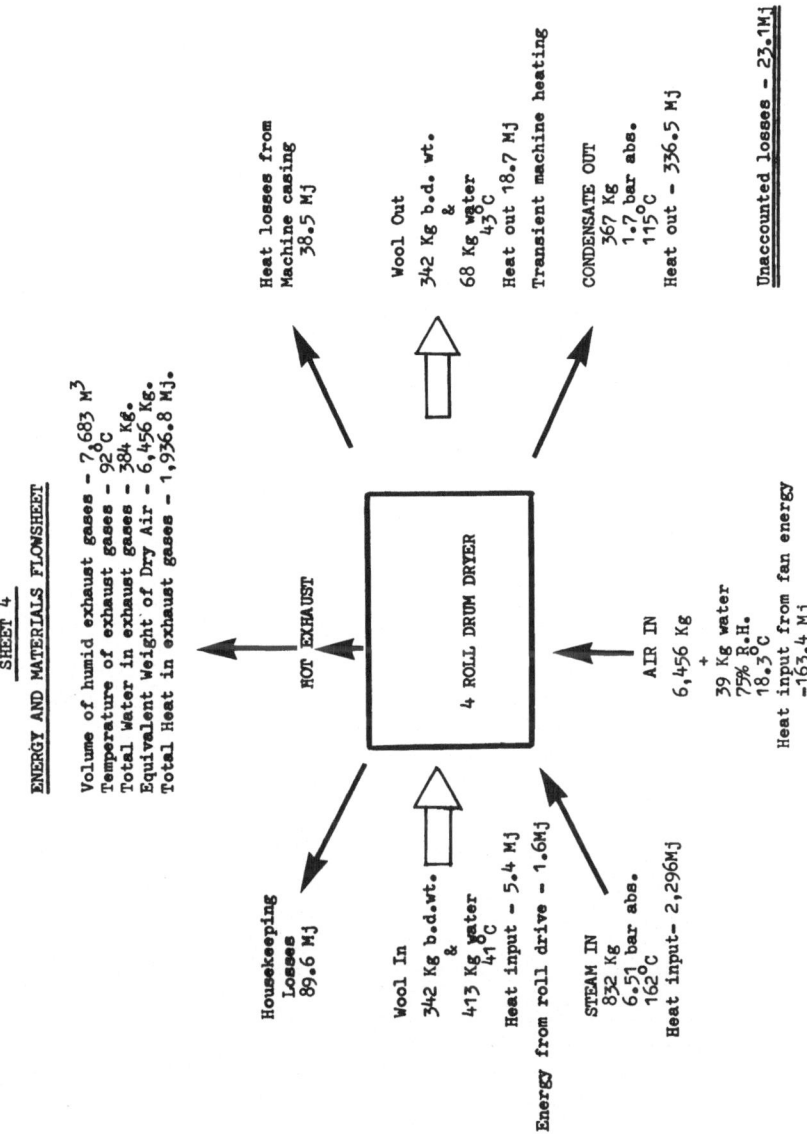

SHEET 4

ENERGY AND MATERIALS FLOWSHEET

Volume of humid exhaust gases - 7,683 M³
Temperature of exhaust gases - 92°C
Total Water in exhaust gases - 384 Kg.
Equivalent Weight of Dry Air - 6,456 Kg.
Total Heat in exhaust gases - 1,936.8 Mj.

4 ROLL DRUM DRYER

HOT EXHAUST

Heat losses from
Machine casing
38.5 Mj

Wool Out
342 Kg b.d. wt.
&
68 Kg water
43°C
Heat out 18.7 Mj

Transient machine heating

CONDENSATE OUT
367 Kg
1.7 bar abs.
115°C
Heat out - 336.5 Mj

Unaccounted losses - 23.1Mj

Housekeeping
Losses
89.6 Mj

Wool In
342 Kg b.d.wt.
&
413 Kg water
41°C
Heat input - 5.4 Mj

Energy from roll drive - 1.6Mj

STEAM IN
832 Kg
6.51 bar abs.
162°C
Heat input- 2,296Mj

AIR IN
6,456 Kg
+
39 Kg water
75% R.H.
18.3°C
Heat input from fan energy
-163.4 Mj

ENERGY NEEDS OF THE LESS DEVELOPED COUNTRIES (LDCs)

Adnan Shihab-Eldin and Sulayman S. Al-Qudsi

The Kuwait Institute for Scientific Research

Post Office Box 24885, Safat, Kuwait

INTRODUCTION

The world has always used large quantities of energy in its production processes and its consumption of goods and services. In the early years, a plentiful supply of wood provided cheap fuel for home and commerce, and coal was sufficiently plentiful to be easily substituted for wood as wood supplies close to points of demand depleted. Kerosene quickly took over from whale oil for lighting, with petroleum becoming increasingly important as the automobile came upon the scene. If anyone has ever doubted that energy is the foundation of modern industrial economies, the "energy crisis" of the mid-1970s dramatically proved the point. The oil price increase demonstrated the vulnerability of the world economies to interruptions in the smooth flow of energy supply. Japan, for example, depends on imports for 99% of its oil needs, while Western Europe imports 96% and the USA about 45% (Todaro, 1977). The oil-importing developing countries, on the other hand, obtain about two-thirds of their energy from oil and, consequently, were hard hit too in the mid-1970s (ICIDI, 1979).

Economic and population growth have required an increasing amount of energy and recently have put tangible pressure on the world supply of its sources. Since 1900, world consumption of energy has increased about twelve-fold. In the same period, the world population has almost tripled, so per capita energy use has increased more than four times. Energy production has been increasing at an average rate of 3.5% per year for the past 50 years, on an absolute basis, or 2% on a per capita basis. Growth in energy production to meet world demand for this time-space has not been uniform. It slackened during both world wars and the depression

of the 1930s, as well as during the events of the 1970s. From
1950 until recently, absolute growth has been averaging 4.8% per
year, but in 1977 it was close to 1% and even negative in 1975
(Kouzminov, 1979).

In the wake of the 1973 "energy crisis" worldwide attention was
given to the future energy needs of mankind and to the world's
future energy supply potential. The cost and availability of
energy has become a worldwide concern. Countries of the world
have become so dependent upon energy (particularly oil) to carry
out their daily socioeconomic functions that any disruption in the
supply of oil is apt to affect the very essence of progress and
continuity in transportation, industrial and agricultural production,
national defense and other vital functions.

In these lectures we will be examining the energy needs of
the LDCs in Africa and Asia. The development of these lectures
will be as follows.

First we discuss the factors that influence energy needs and
their growth. Then we examine the projected needs of LDCs (less
developed countries) in Africa and Asia as predicted by the recent
major international energy studies. The assumptions underlying
these projections will be assessed from the point of view of the
LDCs. In the third part we develop counter-scenarios for the
energy needs of LDCs in Africa and Asia. Projected LDC's future
energy supplies are discussed in part four. And finally, we take
up the projected future energy supplies and contrast them with
the projected energy demands of the major energy scenarios and
of our counter-scenarios. The implications of these balances
for the LDCs and the world energy picture in general are brought
to light.

Factors Affecting Energy Demand

There are three primary factors that determine to a large
extent growth in energy demand. These are (Hafele, 1980): (1)
population growth, (2) economic growth, (3) technological progress.
Associated with these are other factors, including: (4) energy
resources availability and prices, (5) urbanization, (6) climate,
(7) life-style and (8) political will.

Population Growth. A country's demand for energy is typically
a function of (among other variables) its population size; a
growth in the population leads, other things remaining equal, to a
growth in its energy demand.

Fig. 1 shows the historical and projected population trends

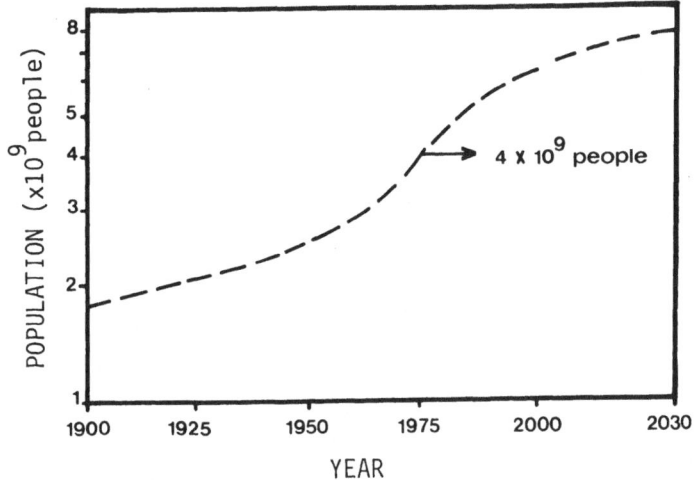

Fig. 1. World Population

Source: W. Hafele, 1980, Energy in a Finite World, A Global Energy
Systems Analysis, IIASA, Laxenburg.

of the world. At present, about 75% of the world's population
live in LDCs. Population growth rates are given Table la.

The present and projected population growth rates are highest
for LDCs in Latin America, Africa and Asia. This is primarily due
to reduced death rates as a result of the application of highly
effective imported modern medical and public health technologies
with no corresponding decline in birth rates. As this table
shows, Africa and Asia made up 38% of the globe's four billion
people in 1975 and it is estimated that nearly half of the world's
estimated population of six billion people will be living in
Africa and Asia by the year 2000 (see Table 1b).

Economic Growth. It has been historically illustrated that
as an economy grows so does its energy demand. For example, a
distinguishing element in the Industrial Revolution was the unprece-
dented increase in output per worker as the substitution of energy-
using steam power for physical labor lessened the labor force constraint. Automobiles, at a later stage, gave the entire population
an enhanced mobility, increased its productivity, made an impact
on its locational residential patterns and increased its consumption
of liquid fuels. Subsequently, derived demands for highways,
bridges, and infrastructure in general were increased. The very
expansion of employment and output that followed increased income

Table 1a. Population Growth Rates, High and Low Scenarios (%/yr)

Region	Average Annual Growth Rate				
	1950-1975	1975-1985	1985-2000	2000-2015	2015-2030
I (NA) + II (SU/EE) + III (WE/JANZ)	1.2	0.8	0.7	0.4	0.3
IV (LA) + V (Af/SEA) + VI (ME/NAf)	2.4	2.8	2.1	1.3	1.0
VII (C/CPA)	1.7	1.9	1.3	1.0	0.7
World	1.9	2.0	1.6	1.1	0.8

Table 1b. Population Projections by Region, High and Low Scenarios (10^6 people)

Region	Population	Projection	
	Base Year 1975	2000	2030
I (NA)	237	284	315
II (SU/EE)	363	436	480
III (WE/JANZ)	560	680	767
IV (LA)	319	575	797
V (Af/SEA)	1,422	2,528	3,550
VI (ME/NAf)	133	247	353
VII (C/CPA)	912	1,330	1,714
World	3,946	6,080	7,976

Source: W. Hafele, Energy in a Finite World, in "A Global Energy Systems Analysis" IIASA, Laxenburg.

I	(NA)	North America
II	(SU/EE)	Soviet Union and Eastern Europe
III	(WE/JANZ)	Western Europe, Japan, Australia, New Zealand, S. Africa, and Israel.
IV	(LA)	Latin America
V	(Af/SEA)	Africa/Southeast Asia (Except Northern Africa and S. Africa)
VI	(ME/NAf)	Middle East and Northern Africa
VII	(C/CPA)	China and Centrally Planned Asian Economies

levels of the people and subsequently increased their demand for
energy-using gadgetry (e.g., household appliances, heating and
cooling systems). The mutual interdependence of energy consumption
on economic growth and vice-versa can indeed be looked upon as a
stylized fact of economic progress (Griffin and Steele, 1980).

Empirically, Fremont Felix, using 1961 data on 153 countries,
showed that per capita national income and per capita energy
consumption are highly correlated. In 1971, Darmstadter, et al.,
investigated the correlation between per capita income growth and
energy consumption and found that as a country's GNP rises over
time so does its energy consumption, in close, even if not propor-
tionate, conformity. Such statistical relationships indicate that
energy does matter, that inadequacy of energy supplies can bring
the development process to a halt, and that assurances of an
adequate supply and mix of energy inputs can be a great stimulant
to development (Tyner, 1978).

Technical Progress. Technological progress can lead to (i)
increased energy availability, and (ii) moderate energy demand
growth. Technical advancements can result in increased interfuel
substitution possibilities in both production and consumption
activities. It can improve modes of transportation in such a
manner that remote energy resources become economically competitive.
In this role, technology can create or expand energy demand (Howe,
1979), yet in another role technology acts to moderate energy
demand growth. This could happen through recovery of waste heat,
recycling and redesign of products in a manner that improves their
energy efficiency (e.g., improving car mileage, insulation of
homes) (Stobaugh and Yergin, 1979).

Urbanization. A key factor affecting energy demand growth,
especially in developing countries, is urbanization. In modern
developed societies the bulk of energy is consumed in the urban
sector. This is also the case in the LDCs. In fact the average
per capita energy consumption in developing countries is not substantially smaller than that in the MDCs (most developed countries).
For example, India's urban population already has an energy consumption of 2.1 kW per capita, while the per capita energy consumption of
its rural and urban populations combined is about 0.33 kW. "As
wide as it might be, the real gap in the development process is
not simply between developed and developing nations. It is
(primarily) between developed nations and the rural areas of
developing nations" (Studsvik, 1979). Similarly, the rural share
of commercial energy consumption in Asia and Latin America is
about 23% and 4% in Africa (Hafele and Sassin, 1977).

Other Factors. Availability of energy at prices that do not
reflect its social costs would tend to encourage inefficient and
wasteful consumption habits, for example, some reduction in energy

consumption growth would result from rising energy prices. Climatic conditions have an impact on energy demand, too. It is generally observed that households' energy consumption in the Western countries peaks during the cold winter season while a peak is reached during the hot summer in many other countries (e.g., Kuwait).

To appreciate the impact of lifestyle upon energy demand, compare the American way of living with that of Sweden. Per capita GNP in both countries is similar yet Sweden consumes less energy per capita than the US. This is generally due to the high locational concentration pattern of the Swedish population, their greater use of multifamily dwellings and central heating plants as well as their relatively short shopping trips. In the US by contrast, households' residential locations are more dispersed and electricity is used for such things as swimming pools (Howe, 1979)

Finally, growth in energy demand can be checked by the political will of a country. Explicit government policies such as taxes and tariffs on energy and standards for the performance of energy-using equipment can curtail some of the growth in energy demand (The National Research Council, 1979).

World Energy Demand Models

In this section a selective and critical review of three world energy models is carried out. These models investigate energy demand and supply and their interaction through the energy market, which is subject to national and international policies and conflicts and to environmental restrictions and legislation.

The literature is extensive and this study does not attempt a complete review of all the world energy models, available techniques and theoretical approaches, but focuses selectively on three major works:
1. The World Energy Conference (WEC)
2. Workshop on Alternative Energy Strategies (WAES)
3. The model of the International Institute for Applied Systems Analysis (IIASA).

WEC. In the WEC model, the time period considered is 1972-2020. The world is broken down into three regions: OECD (Organization for Economic Cooperation and Development), Centrally-Planned Economies and Less Developed Countries.

Total primary demand is projected using a simple constant elasticity demand function linking energy demand to income and energy price. Income elasticities are .95 in the OECD region and 1.1 elsewhere. The overall price elasticity with respect to energy demand is assumed to be −0.3. In accordance with historical

growth performance, annual world economic growth rate is assumed
to be in the neighborhood of 3-4%. Several scenarios are considered.
The first is the Cavendish report in which world demand projections
are made using two alternative growth assumptions: (a) low (3%
per year) and (b) high (4.1% per year). The high scenario assumptions in this report produced an energy demand of 840 EJ (IEJ =
10^{18} J), which is about 2.5 times greater than today's demand.
The second scenario is based on a combination of historical and
extrapolation trends. Depending on the benchmark period, the
projected demand was reported to be too small to meet the needs
of the world's population growth until 2020 or too large to be
reasonably balanced by the anticipated supply level. The third
scenario is based on declining income elasticities for energy and
increasing energy prices, a decline envisioned to occur mainly in
the industrialized nations. By the year 2020, the WEC estimates
that income elasticity would go down to one-half its present level.
Conservation in the industrialized countries occurs as a result
of more efficient production techniques and more restrained
consumption habits due to saturation of demand for certain consumption items, notably passenger cars (World Energy Conference,
1978).

   WAES. As with WEC, WAES is based on alternative scenarios.
The world's future energy needs are estimated from regional
economic growth rates coupled with hypothesized future pricing
patterns of oil and given energy coefficients. National policy
response receives more explicit treatment here than under the
WEC. The 1972-2020 period is broken down into two sub-periods
(until 1985 and from 1985 to 2000). High (H) and low (L) growth
rates are assumed for each sub-period--(1972-85: H = 6%, L =
3.5%) and (1985-2000: H = 5% and L = 3%). Three real pricing
patterns are considered:
   1. Constant (oil price being fixed at its real 1976 level--
       $11.50.)
   2. Increasing by 50% of its 1976 level--$17.25.
   3. Decreasing by 50% of its 1976 level--$7.66.

   For the period 1985-2000, WAES makes two alternative price
assumptions. In the case of high energy prices, the real price
of energy is assumed to increase 2-5% per annum. Yet WAES expects
a gap between demand and potential supply of oil to emerge after
1985. Various ways of closing this gap are discussed. According
to this analysis, the above mentioned price assumption, in conjunc-
tion with relatively slow economic growth (3% per annum), is consistent with a balanced development of supply and demand in the
international oil market.

---

*About 26.6 TWY/yr.

National energy response is assumed to be restrained up until 1985 and vigorous beyond that date. The implication of the assumption is that, in the long run, governments are more willing and/or decided to improve conservation measures in the production and consumption sectors than in the short run (World Energy Conference, 1978).

As in the WEC, growth rates in WAES are assumed to differ among the different regions. LDC rates are slightly higher than those particulated for the industrialized countries (Ulph, 1980). To check the energy supply-demand balance, WAES uses two procedures. In the first, the projected demand is not constrained (the sum of national desires for fuel are compared to the global maximum supply potential). In the second, projected demand is constrained so as to balance its total with the supply available. Preferences are honored to the extent possible, which implies that interfuel substitution possibilities exist.

IIASA. Again the model here is that of scenario-building. Two basic scenarios having upper and lower bounds are considered. The scenarios are defined by two basic variables--population and gross domestic product. Income and price elasticities play as important a role in IIASA's model as they do in the WEC and WAES scenarios. Table 2 shows IIASA's projected regional GDP and population growth rates. The primary energy-GDP elasticities, ϵ_p, are assumed to be different for different regions--reflecting differences in their stages of economic growth (see Table 3). These differences demonstrate that economic growth in the LDCs depends on the building up of their infrastructures, which are typically energy intensive. The final energy-GDP coefficient is assumed to go as low as 0.3 for industrialized countries but to only slightly less than 0.8 for developing countries. As with the above models, IIASA's high and low scenarios are conservative, representing departures from observed trends. In each scenario, world population is assumed to grow to eight billion by 2030. In one scenario, IIASA assumes a relatively low economic growth, fairly large advances in energy end-use technology, and a rather positive attitude towards energy saving. The other scenario assumes a modestly high growth. In both scenarios optimistic assumptions are made about energy supply availability, which include effective and timely decision making and implementation, as well as due regard for the needs of the developing countries (IIASA, 1980).

Assessment of World Energy Models: The LDC's Perspective. In assessing world energy models from the perspective of LDCs, attention should be focused on the crucial fact of present maldistribution of energy among the world's population. World energy use is greatly unbalanced as Fig. 2 illustrates. The consumption of energy per head in industrialized countries

Table 2. 1975 Per Capita GDP, Historical and Projected Real Growth Rates for Two Scenarios to 2030

Region	Historical Growth Rate of Per Capita GDP (%/yr) 1950-1975	GDP Per Capita (dollars) 1975	Projected Growth Rate of Per Capita GDP (%/yr)			
			High Scenario		Low Scenario	
			1975-2000	2000-2030	1975-2000	2000-2030
I (NA)	1.9	7,046	2.9	1.8	1.7	0.7
II (SU/EE)	6.7	2,562	3.6	3.2	3.1	1.9
III (WE/JANZ)	4.0	4,259	3.0	1.8	1.7	0.9
IV (LA)	2.9	1,066	3.0	2.4	1.6	1.9
V (Af/SEA)	2.5	239	2.8	2.4	1.7	1.4
VI (ME/NAf)	5.7	1,429	3.8	2.8	2.4	1.2
VII (C/CPA)	5.1	352	2.8	2.4	1.6	1.4
World[a]	3.1	1,565	2.4	1.9	1.3	0.9

[a]These global per capita growth rates may appear inconsistent with the values for the regions but they are correct. Growth rates of aggregate ratios (e.g., GDP per capita) need not always be within the range of their component parts when both numerator and denominator grow at different rates.

Source: W. Hafele, 1980, Energy in a Finite World, in "Global Energy Systems Analysis", IIASA, Laxenburg.

Table 3. Primary Energy–GDP Elasticities, ε_p, 1950–2030

Region	Historical 1950–1975	High Scenario		Low Scenario	
		1975–2000	2000–2030	1975–2000	2000–2030
I (NA)	1.03	0.42	0.67	0.36	0.89[a]
II (SU/EE)	0.77	0.65	0.67	0.62	0.62
III (WE/JANZ)	0.96	0.70	0.77	0.65	0.73
IV (LA)	1.28	1.04	0.98	1.06	0.97
V (Af/SEA)	1.52	1.15	1.11	1.18	1.19
VI (ME/NAf)	1.20	1.16	0.96	1.23	1.10
VII (C/CPA)	1.57	1.06	1.17	0.98	1.27[a]
World	0.99	0.70	0.90	0.67	0.93

[a]The primary energy–GDP elasticity is unusually high for Regions I and VII in the Low scenario. In the later time period in these regions, demand for liquids must be met from coal liquefaction which has significant conversion losses, thus adding to primary energy use. Since the GDP growth is small in the Low scenario, the elasticity of primary energy use with GDP is increased. If these losses are subtracted from primary energy consumption in 2030, the resulting elasticities are 0.53 and 0.94 for Regions I and VII, respectively. The same effect is present in the High scenario for Regions I, II, III, and VII but is less pronounced in the elasticity because GDP growth is higher.

Source: W. Hafele, 1980, Energy in a Finite World, in "A Global Energy Systems Analysis", IIASA, Laxenburg.

compared to middle-income and low-income countries is in the
proportion of 100:10:1. One American uses as much commercial
energy as two Germans or Australians, three Swiss or Japanese,
six Yugoslavs, nine Mexicans or Cubans, 16 Chinese, 19 Malaysians,
53 Indians or Indonesians, 105 Sri-Lankans, 438 Malians, or 1072
Nepalese. All the fuel used by the Third World for all purposes
is only slightly more than the amount of gasoline the developed
countries burn to move their automobiles.

Fig. 3 shows the distribution of the per capita commercial
energy consumption for the countries of the world in the year
1971. It illustrates the dichotomy between per capita energy
consumption in the LDCs and in developed countries; while it
hovers around 0.2 kW in the former, it is substantially greater
than 2kW in the latter.

World energy models do not pay enough attention to the
existing energy maldistributional pattern, perhaps because they
address the energy question from the perspective of developed
countries. They proceed to project LDC's future energy requirements
as though the prevailing pattern does not call for reversal. The
rest of this section is devoted to a critical analysis of the
assumptions upon which these models base their projections of
future energy requirements of the LDCs.

First, the above surveyed scenarios assume growth rates for
LDCs that, while higher than those assumed for MDCs, fall short
of the aspirations of the LDCs' people. In other words, such
relative rates of growth almost invariably keep the gap between
the rich and poor countries in existence for the projected period.
The above models use growth rates derived (or slightly modified)
from the past growth record. However, while it is true that
during the 1950-75 period per capita income of the developing
countries was growing at a high rate (Table 2), it is equally
true that per capita income in developed countries was growing at
a higher rate. As a result, the gap between the rich and poor
countries, which had been increasing for 100 to 150 years, continued
to widen (Lewis, 1980).

In 1950, the average per capita GNP in the industrial countries
(in 1974 dollars) was $2191 greater than that of the developing
countries. By 1975, this difference had more than doubled, to
$4,838 (Morawetz, 1977). If we were to adopt the growth assumptions
that those models make, it would be tantamount to allowing historical growth rates to continue in the future. However, it is an
easy exercise to show that such an assumption means that the
number of years it would take to close the absolute gap between
the LDCs and MDCs is ridiculously long. For the large majority
of the LDCs the gap would never be closed since their historical
growth rate has been slower than those of industrialized countries.

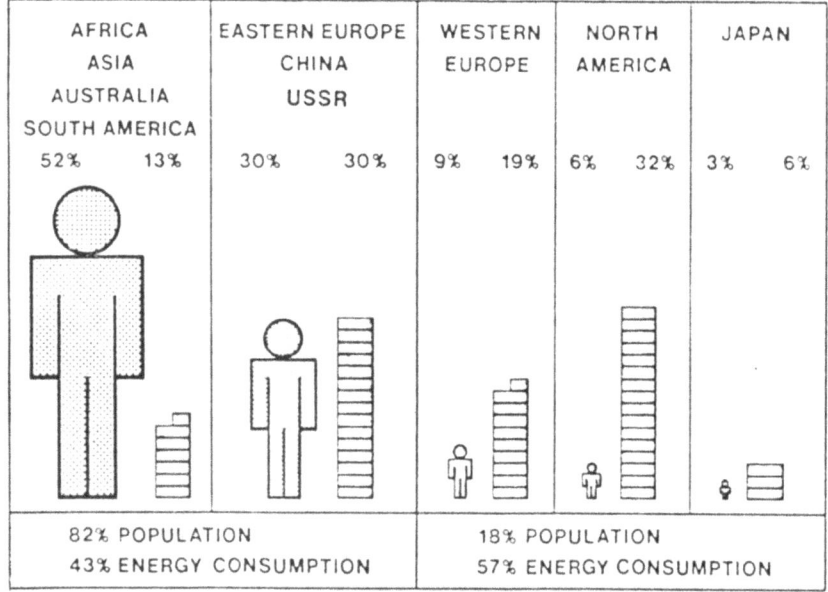

Fig. 2.   Energy Consumption in Comparison to the Population.

Source: Kraftwerk Union A. G., 1980, "Nuclear Energy for
          Developing Countries."

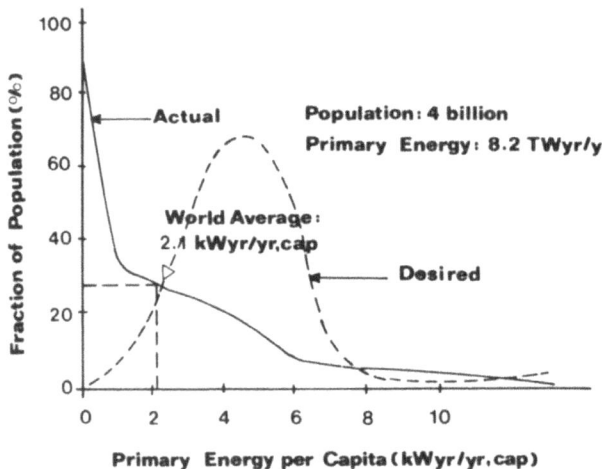

Fig. 3.   Global per capita commercial primary energy consumption,
          1975. Values show fraction of population whose primary
          commercial energy consumption is smaller than the corres-
          ponding value on the horizontal axis.

Even among the fastest-growing LDCs, only eight would close the
gap in 100 years (Morawetz, 1977).

Second, the above models treat economic growth as the main
determinant of energy demand and neglect income distribution
within countries. Such low emphasis on income distributional
patterns is understandable in view of the desire on the part of
model builders to keep their model mathematically manageable.
However, from a practical point of view, the way in which income
is distributed may very well have an important impact on energy
demand. As the consumption pattern of the rich is different from
that of the poor, any redistribution of income would have an
impact on the vector of final demand for goods and services,
including the final demand for energy use. The above scenarios
seem to continue the tradition of economists in the 1950s and
1960s in looking upon growth as the prime objective of LDCs. The
"trickle-down" mechanism was supposed to solve poverty and income
distribution problems if only growth were fast enough. But,
after two decades of concern with the problem of raising per
capita GNP in low income, developing countries, the development
community has shifted its focus to the challenge of increasing
the equity of income distribution. "This shift proved to be
dramatically needed as the empirical studies of the distribution
of the benefits of economic growth showed that the expected
trickle-down was not taking place. More serious, a number of
studies indicated a systematic worsening, both relative and
absolute, in the position of the poorest stratum of income reci-
pients" (Adelman, et al., 1976).

The Santiago Declaration of Third World economists (1973)
summarizes the problem lucidly: "The Third World, with 70 percent
of the world population subsists on only 20 percent of the world
income--and even this meagre income is so maldistributed internally
as to leave the bulk of its population in abject poverty" (Baer,
1974). In response to these unequal distributional patterns came
the call that "distributional objectives should be treated as an
integral part of development" (Chenery, et al., 1974). We
suspect that as the distributional pattern in LDCs improves, per
capita energy consumption will increase. The logic is simple:
energy is an indispensable commodity for all sectors in the
economy. As the masses of the LDCs escape the poverty trap, and
as their incomes increase relative to the income of the rich
class, their demand for energy-using gadgetry would increase at a
much faster rate than the decline (if any) in energy consumption
by the rich. To put this hypothesis to the test, we ran a multiple
regression equation in which we specified per capita energy consumption as a function of per capita income and a measure of income
inequality (the Gini Coefficient). A cross section of data comprising 14 developed and developing countries were employed in the test.
The results obtained were:

$$(E/P) = 3852 - 7667(G) + .8 (GNP/P) \quad . \quad . \quad . \quad . \quad . \quad 2.4.1$$

t-value   (1.9)   (8.1)

where

(E/P) = per capita energy consumption,

G   = Gini Coefficient, and

(CNP/P) = per capita GNP (see Table 4).

The two coefficients are statistically significant at the 10% confidence level. These results indicate that at the mean values of G and (E/P) a one percent increase in inequality is likely to bring about nearly one-half of one percent reduction in per capita energy consumption. By contrast, at the mean values of (E/P) and (GNP/P), a one percent increase in per capita income is likely to lead to about nine-tenths of one percent increase in per capita energy consumption. While this test does not claim to be overwhelmingly conclusive in view of the usual problems that plague cross-sectional data, the statistically negative causal relationship between energy consumption and inequality should not be brushed aside.

The above two critical comments pertain to the broad assumptions of the surveyed energy models. Further comments on the implications of two technical concepts employed in these models are in order: (i) the price elasticity of energy demand, and (ii) the energy-supply constraints in the LDCs.

Price Elasticity of Demand

We have seen that all the world energy models have assumed a fairly price-inelastic world energy demand. Such an assumption is based on an increasing number of empirical works (Manne, et al., 1979). The assumed price elasticity of demand is about -0.3. This implies that any future price increases of energy would be translated into an increase in the total outlay on energy. For the less-developed countries, such an outcome is likely to set in motion two effects:

1. The balance of payments effect, as increased energy prices, would inflate an already high import bill. In 1975 the LDCs as a whole spent nearly $30 billion for energy imports. This amounted to 15% of the total imports of these countries combined and was second only to imports of machinery and transport equipment ($70 billion) and more than those of all food items ($26 billion).

2. The employment effect: the increase of energy prices is likely to affect the level of employment in the LDCs through its impact on: (a) industrial competitiveness due to rising costs of production, (b) curtailment of some developmental projects and the slower economic growth rates in general, and (c) consumers' spending patterns. As the relative price of energy increases, consumers' spending on it would increase because its demand function is thought to be inelastic. This would induce consumers

Table 4. Data Used to Estimate (2.4.1)

Country	Per Capita Energy Consumption	Per Capita GNP	Ginni Coefficient
Brazil	731	1,360	.61
Colombia	685	720	.54
Peru	642	840	.57
Mexico	1,222	1,120	.58
India	218	150	.46
Philippines	329	450	.50
Sudan	143	290	.43
Chile	987	1,160	.49
Ceylon	106	200	.37
Denmark	5,320	8,040	.42
New Zealand	3,617	4,380	.37
United Kingdom	5,268	4,420	.32
U.S.	11,554	8,520	.31
Kuwait	9,198	12,270	.47

Source: Gini Coefficient data from Ahlwalia, M. "Dimensions of
the Problem" in Chenery H. Op. Cit. Data on per capita
energy consumption and GDP from World Bank. World
Development Report, 1979, Oxford University Press,
Washington D.C.

to spend less on other goods and services, thereby reducing
employment in a number of sectors. As most LDCs are projected to
be net energy importers, the additional spending (due to higher
energy prices) is likely to accrue to the world outside the
energy-importing LDCs, thus depriving them of possible positive
income and employment linkages (Dacy, et al., 1978).

Supply Constraints in LDCs

     Planned and assumed growth rates may not materialize if
energy supplies are inelastic and/or not forthcoming in the LDCs.
This may create uncertainty about the future regarding energy
availability. Such uncertainty could be a deterrent to investment
outlets in both public and private sectors. Investors and consumers'
expectations about future energy availability are of critical
importance for the implementation of the designed development
plans when the role of energy input dominates that of other
inputs, as the LDCs experience since the mid-1970s indicates.
The point we want to make is that expectations about future
energy prices and availability are likely to affect growth performance in these countries. Projecting LDC's growth rates in abstrac-
tion from these psychological expectations (as the above surveyed
models do) is likely to increase the variance between expected
and actual future economic and energy demand growth rates.

     Output supply is typically more inelastic the lesser the
substitution possibilities among factor inputs and the lower the
input supply elasticity. Output supply in LDCs is, by and large,
more inelastic than the industrialized countries' output supply,
generally due to relative factor input inelasticity in the former
vis-a-vis the latter countries. As future energy supplies become
scarcer, the supply elasticity would dwindle. Competition among
various demanders would shoot future energy prices up. Given the
reasonable assumption of a low substitution parameter between
energy and other factor inputs, energy's share in the production
costs would increase and output supply would consequently fall.
While it is true that energy is still a relatively minor element
of production costs in the LDCs, it can be expected to increase
as these countries become more industrialized (Makhijani, 1975).

     Some may argue that as present energy forms become more
scarce and their prices more expensive, LDCs could resort to
other energy forms that are more plentiful and less expensive.
But, pragmatically speaking, LDCs are less free to choose between
energy alternatives than industrial countries, and are thus once
more constrained in their choices. Currently they depend on oil
to satisfy the bulk of their needs. Their dependence on oil is
more than the developed countries' dependence since LDCs started
their industrialization and development efforts after the coal-oil

shift had been well established. Not only were oil-dependent
structures built up in these countries through imports of oil-
geared capital goods from developed countries, but also, the oil-
based pattern of life of the latter was taken over to a significant
degree.

Even if all the technological problems associated with supply
constraints from the technological point of view are somehow resolv-
ed, there remain problems of financing future energy needs in the
LDCs. World energy models project a tightened future energy supply
and higher real costs of energy; the scenarios carry undesirable
future effects to the LDCs. With the increase in the energy prices
during the 1970s, power generation and expansion represented, on
the average, about 7-8% of the gross fixed capital formation in
these countries. The shift toward plants with high capital costs
(e.g., nuclear-based plants) coupled with rising prices of imported
equipment is estimated by some to raise the power investment share
of gross fixed capital formation in LDCs to about 12% (Friedman,
1976). This heavy investment has a large foreign exchange component
(about 50%) and as such is likely to lead to reduction of other
import categories--both consumption and investment goods--that are
essential to the smooth operation of the developmental process.

Before proceeding to examine the projected energy needs of
Africa and Asia in the above mentioned world energy demand models
and to assess their impact, we will present some counterfactual
scenarios that better reflect the actual needs and aspirations of
the LDCs.

Counterfactual Analysis

It was observed that present energy models are content to
project a future world energy consumption of about 2-3 kWY on a per
capita basis by the year 2000. The statistical problems that mask
such aggregated averages come to mind. Upon closer examination, it
turns out that the uneven future energy consumption is embodied in
those projections. Per capita commercial energy consumption in the
LDCs, according to the WAES high scenario, is projected to be about
1.02 kWY/yr. IIASA's high scenario produces a per capita commercial
energy consumption of about 1.1 kWY/yr. These fall well below the
present consumption levels prevailing worldwide and, of course, far
lower than those of the industrialized countries. Too, they fall
well below IIASA's worldwide per capita average consumption targets
of 3 and 5 kWY/yr in the low and high scenarios, respectively.

It is believed by many that two-thirds of the globe's population in the LDCs will continue to depend on non-commercial energy
sources to a significant degree. Such an assumption is both unfair
and unrealistic. It is unfair because it implies that a substantial

segment of the population living in LDCs will continue semi-primitive ways of living. Furthermore, it implies that the per capita incomes of these people would still be far behind their "modernized" fellow human beings living in the industrialized world and, to a lesser extent, those in urban regions of the developing nations. In IIASA's high scenario (1975-2000), per capita income is assumed to grow by 2.9% in North America, 3% in Latin America, 2.8% in Asia and 3.8% in the Middle East and North Africa. Such growth rates would mean, in practice, that the absolute income and consumption gap between rich and poor countries will prevail and, probably, widen.

The viability of non-commercial energy sources to meeting future energy requirements in the LDCs is questionable and would be costly to achieve. An increase in per capita income in the LDCs is expected to fuel a process in which commercial energy is substituted for non-commercial energy. Future needs of energy in LDCs result from modernization and development. Modernization implies an introduction of new methods of production, for example, draft animals replaced by diesel pumps, animal transportation by trucks, and cow dung as a fuel by kerosene.

It is costly to achieve because dependence on wood, dung and straw has created the so-called "second energy crisis." Increasing deforestation and the diversion of fuel, wood and charcoal to urban areas are threatening to break down the ecosystem that supports village life in many LDCs. It has been estimated that, by present trends, nearly all humid tropical forests will have been transformed into desert or unproductive wasteland by the year 2000 (Larsson, 1979). Such projections are against the pronounced developmental goals of the UN, namely, that land productivity and food supplied in LDCs should increase to meet the increasing demand for food.

The principle of opportunity cost operates, as expected, if the projected non-commercial energy use in the LDCs materializes. Animal dung is used in these countries as an alternative cooking fuel. But every cowpat that is burned as cooking fuel means that much less nitrogen is available as fertilizer. Three heads of cattle or buffalo produce 22,400 kg of manure a year, yielding 125 kg of nitrogen or enough to fertilize one hectare of corn. Using cattle manure to produce biogas as an alternative domestic fuel in LDCs preserves nitrogen and other mineral nutrients for use in agriculture, if the biogas residue is returned to the land. If it is not returned, agricultural productivity is decreased (if no fertilizer is applied) and import bills are increased (if resort to food and/or fertilizer importation is made).

We believe that improving standard of living in the LDCs would entail more commercial energy consumption on a per capita

basis than is generally assumed by world energy models. The fact
that 70% of the world's 4.4 billion people will double in the
coming three to four decades accentuates our belief that the
projected economic growth and energy demand for LDCs is too
small. Such heavy population concentrations will increase the
demand for food, which might require more cooking fuels, and for
services such as improved water and housing which embody significant
amounts of energy.

Below we discuss three counter energy scenarios and assess
their impact on the future world energy balance. The basic
thesis of our present analysis is that world energy studies
underestimate future energy needs of the LDCs. To be just to the
scholars who carried out these studies, we must acknowledge that
their "projections were not meant to predict most likely (or
desirable) rates (of income and hence energy consumption), but to
test a plausible range and to evaluate their implications" (Hafele,
1980). Given a range of projected world energy supplies, what is
the likely impact on the energy supply-demand balance if the LDCs'
energy demand patterns were more pronounced, for their own benefit,
than those assumed by global energy studies?

The three counter-scenarios presented below are simple-minded
and straightforward. They are not the output of sophisticated
econometric simulation or input/output models. They simply are
based on historical and/or target growth rates. The counter-
scenarios below do not claim to predict the future energy consumption
in the LDCs. Rather, they demonstrate what happens if higher
(than assumed by global energy studies) demands for energy were to
prevail in future LDC markets. Although a crude method is employed
in our scenarios, our confidence in the assumptions that underlie
them is extremely high. We assume that a minimum level of per
capita energy consumption is required in order for these countries
to achieve the stage of developmental take-off. This assumption
stems from our belief that economic development cannot be success
fully launched and sustained unless an adequate and viable flow of
commercial energy input is assured. We distinguish two components
that make up the rate of growth of per capita energy consumption
(Banks, 1976). The first might be called the "energy widening"
component; it involves providing each new addition to the population
with the same amount of energy as each present member. The other
component can be called "energy deepening," and has to do with
increasing per capita energy consumption for new and old members
alike. In the United States from 1960 to 1968, energy consumption
increased at an annual compounded rate of 3.2, compared to a
population growth of about 0.9% per year. Thus, energy was
deepened 2.3% annually. In contrast, energy consumption in many
LDCs has been much smaller on a per capita basis and hence energy
was not deepening at high rates. If the LDCs are to develop at
all, an important variable in the development equation "energy"
must deepen on a per capita basis.

Having said that, we acknowledge that the future energy demand and supply is uncertain and that such uncertainty may prove many or all predictions incorrect. "Those who foretell the future lie, even if they tell the truth" is an Arab proverb that nearly summarizes the caution with which our counter-scenario projections (DuMoulin and Eyre, 1979) are presented. The three projections are based on: (1) the South Korean development experience, (2) 4 kWY/yr per capita consumption target, and the UN study of 1977 (Leontief, 1977).

In economic development literature, South Korea is cited as one of the "successful" developmental stories (Reynolds, 1977) 1977). In the past two decades, the country's real annual GNP was over 9%. To achieve such a respectable growth rate, its energy consumption (1960-1973) grew by an average annual rate of 12.5%. On a per capita basis, energy consumption increased from .25 kWY/yr in 1960 to about 1 kWY/yr. While 1 kWY/yr is still below the world's per capita average energy consumption, what is remarkable is its annual growth rate. Over the period 1960-1975, the annual growth rate of per capita energy consumption was 9.5%. Such a growth rate is well above annual rates for other non-OPEC LDCs. While the growth rate of per capita energy consumption in South Korea has been remarkable, the actual per capita energy consumption has been higher than its counterpart in many LDCs as well. This is shown in Table 5.

Table 5

Per Capita Energy Consumption in South Korea as a Ratio of Per Capita Energy Consumption in Selected LDCs

Year	1960	1970	1973	1975
South Korea/Egypt	.88	2.47	2.64	2.65
South Korea/India	1.79	4.16	4.29	4.65
South Korea/Nigeria	10	12	10	9
South Korea/Brazil	.69	1.3	1.15	1.15

Source: Dervied from Predicasts, op. cit., various tables.

Table 6. Per Capita Energy Consumption in Selected Countries

Country		Year			
		1960	1970	1973	1975
South Korea	E*	.19	.73	.88	1
	P	26.8	32.2	34	36
	E/P	.25	.79	.90	.98
India	E	1.79	3.0	3.4	3.73
	P	437	540	575	610
	E/P	.14	.19	.21	.21
Japan	E	3.76	11.01	13.79	12.87
	P	91	104	108	113
	E/P	1.43	3.7	4.42	3.97
China	E	12.19	11.85		
	P	551	771		
	E/P	.77	.53	.60	.68
USA	E	44.57	67.14	74.56	70.56
	P	190.8	207	260.4	213
	E/P	8.1	11.2	12.3	11.5
Sweden	E	1.03	2.02	2.17	2.13
	P	7.7	8.04	8.14	8.20
	E/P	4.62	8.69	9.23	8.99
UK	E	7.55	9.48	10.02	9.78
	P	54.5	55.4	55.93	55.96
	E/P	4.8	5.92	6.2	6.04
USSR	E	17.26	30.72	35.99	39.8
	P	214	243	250	256
	E/P	2.79	4.37	4.98	5.38
Egypt	E	.22	.31	.35	.40
	P	26.8	33.3	35.6	37.2
	E/P	.28	.32	.34	.37

Source: Computed from Predicasts, Inc., World Energy Supply and
         Demand 140, March 1977, various tables.
E* = Total Energy in Quads (=10^{15}BTU)
P = Population in Millions
E/P= Per Capita kWY/yr Energy Consumption.

By contrast, the compounded annual growth rate of the per
capita Indian energy consumption for the same period was about
1.7%. The corresponding figure for Egypt was 1.5%. Table 6
presents per capita energy-consumption in selected countries.

If per capita energy consumption in the LDCs were to match
the South Korean growth rate, then by the year 2000 per capita
energy consumption of 70% of the world population would reach
about 5 kWY/yr. This means that energy needs in the LDCs are
going to be on the order of 15.1 TWY/yr around the year 2000.

Our second counter factual analysis postulates a specific
per capita energy consumption target in the LDCs over the next
two decades or so. We propose, as Hafele did in earlier works,
that a 5 kWY/yr of energy consumption is the minimum amount
needed for the average person to energize a productive and "humane"
standard or living (Hafele and Sassin, 1977). Currently, several
countries of the western world enjoy such a per capita consumption
level. Japan had a per capita energy consumption of about 4
kWY/yr (1975). Students of economic development have been told
over and over again by western scholars that Japan provides an
exciting and illuminating developmental lesson to follow (Kelley
and Williamson, 1974). Nearly 100 years ago the Japanese economic
structure was similar to the economic structure of many countries
in the Third World. Through a set of fiscal, monetary and foreign
policies, together with the sociological characteristics of the
population and the national will, the country was able to launch
a successful development strategy. By the middle of this century,
the country had escaped the poverty trap and was well established
as an industrialized economy. From then until the 1970s, its
real GNP's annual growth rate was well over 10%. Per capita
energy consumption grew at a compounded annual rate of about 7%.
If there are indeed lessons to be derived from the Japanese
economic development, its energy deepening historical pattern is
one of these important lessons. What would happen then if the
LDCs aimed at achieving a per capita energy consumption level
similar to that prevailing in today's Japan by the year 2000?

Given the projected population estimates of the LDCs (Regions
IV + V + VI only) and given that their per capita energy consumption
approaches 4 kWY/yr, then their total energy requirements would
be in the order of 13.4 TWY/yr by the turn of this century. The
magnitude of this figure should become clear when we recall that
it represents about 80% of IIASA's projected demand for the whole
world (not merely the LDCs). On the other hand, if we were to
aim at 5 kWY/yr as our target for the year 2000, then total
primary energy needs of the LDCs would be about 16.75 TWY/yr,
which is just about the projected total world energy demand by
that year.

Even if we were to lower the future per capita energy consumption target from 4-5 kWY/yr to about 3.2 kWY/yr (which is IIASA's target for the LDCs in the year 2030), then total energy demand in these countries would be about 10.7 TWY/yr--nearly two-thirds of the world demand in the year 2000. This would be a sizeable jump from the current LDC's consumption level of about .84 TWY/yr.

In our third counter-scenario, we postulate that the rate of growth of energy consumption in the LDCs is given by the following simple equation:

$$\dot{D}/D = \alpha \dot{Y}/Y + \beta \dot{P}/P, \qquad \qquad \cdots \cdots \; 3.4.1$$

where,

$\dot{D}/D$ = Percentage rate of growth of energy demand
$\dot{Y}/Y$ = Percentage rate of growth of Gross Domestic Product (GDP)
$\dot{P}/P$ = Percentage rate of growth of energy prices
α = Income elasticity and
β = Price elasticity.

Below, we assume that α takes a value of 1.3 and β of -0.3. These values are in line with those proposed by world energy models. Prices in 2000 are assumed to grow by 50% above their 1975 level, which is again similar to projected energy pricing in these models. The point of departure between our scenario here and the surveyed models is that we propose a fairly high income growth rate for the LDCs via à vis these models. Following W. Leontief, we suggest that the annual growth rate in these countries may not go below 6.9%. Such a growth rate would cut the income gap between industrialized countries and LDCs in half. Using such a scenario would produce a compounded annual energy consumption growth rate of about 7.5%. By the year 2000, the average per capita energy consumption would be about 2 kWY/yr and the total primary energy consumption in the LDCs (China excluded) would reach about 6.7 TWY/yr.

One may wonder about the realism of such counter-scenarios. These are targets that may or may not materialize due to a host of domestic and international socioeconomic factors. But the point that we want to make is that they are not out of line with the recommendations of some scholars and international bodies. For example, in a declaration issued in 1975, the LDCs set a goal for themselves of providing at least 25% of the world's industrial production by the year 2000. Subsequently, in May 1975, the UN Conference on Trade and Development (UNCTAD) declared that this goal would mean an 11% annual growth rate in manufacturing output for each of the intervening years (Schurr, 1978). Now the industrial sector of the LDCs accounts for about 40-50% of the total sectoral energy consumption. It can then be readily agreed

that meeting the UNCTAD target would require an annual energy
growth rate in these countries of at least 6.5% and as high as
8%. More specifically:

$$\dot{D}_E = w_1 \dot{O}_E + w_2 \dot{I} \qquad \qquad \ldots \ldots 3.5.1$$

where,

$\dot{D}_E$ = total growth rate of energy demand

$\dot{O}_E$ = other sectors' growth rate

$\dot{I}$ = industrial sector's growth rate

     w_1 and w_2 are the weights (relative energy use) of the other
sectors and the industrial sector, respectively. If the industrial
sector is to grow by an annual average rate of 11%, then we
propose upper and lower limits on the annual growth rates of
other sectors as follows: (i) 5% lower limit, and (ii) 6.9%
upper limit.

     The lower limit assumes a minimum amount of growth in the
other sectors if the industrialization process is to be successfully launched. In other words, it guards against possible
supply bottlenecks in the other sectors and provides growth in
employment and incomes in these sectors in a manner that would
strengthen the intersectoral linkages. The upper limit was
chosen on the assumption that other sectors' growth rates would
proceed along the lines of aggregate growth rate recommended by
W. Leontief et al. (1977).

     Total sectoral growth rate of energy demand would be about
8% under the lower limit and 9% under the upper limit. A lower
limit of 8% would imply a per capita energy consumption in the
LDCs of 3.2 kWY/yr and a total primary energy demand of 10.8
TWY/yr. In the upper limit, the figures are nearly 4 kWY/yr
and 13.4 kWY/yr, respectively.

     Such counter-scenarios produce potential energy needs in the
LDCs that are alarmingly high. Yet they are no more alarming
than estimates made by others. Robert Heilbroner, for example,
estimates that raising per capita energy consumption in the LDCs
to OECD level would mean a staggering increase (at least 20 to 30
times) in these countries (Heilbroner, 1974). It is our belief
that international energy models have been satisfied to imply
that the gap between LDCs and industrialized countries would
persist and perhaps widen in absolute terms by the years 2000 and
2030. Our counter-scenarios, on the other hand, propose a different
thesis: Strive for a future world in which the current extreme
income (and hence energy consumption) inequalities are significantly
reduced, not by a worldwide return to primitive living, but by

increasing the living standards (and energy consumption) in LDCs
much more rapidly than in the industrialized countries. This
prompted us to use the South Korean income and energy growth
patterns as a possible future guideline for other LDCs to follow.
While such a target may appear, on historical grounds, too ambitious
for all other LDCs to follow over the next two decades or so, we
think that the future is not irrevocably determined by historical
and current trends, but that choices exist that, when implemented
in policy and action, can modify trends, sometimes slightly and
other times profoundly (Chesshire and Paritt, 1978).

    In Table 7 we contrast energy demand projections for the LDCs
and the world using the targets of 2 and 5 kW per capita with
IIASA's high scenario. For regions V and VI, energy demand projections call for 5.6 and 13.8 TWY/yr when 2 and 5 kW per capita
targets are employed (year 2000). These are substantially higher
than IIASA's high scenario projections of 2.2 and 7 TWY/yr for the
years 2000 and 2030, respectively. It is interesting to note that
the ratio of per capita energy consumption in region V to that in
region III is not projected to increase appreciably in IIASA's
high scenario even for the year 2030. In fact, in 2030, the per
capita energy consumption in Africa and Asia is projected by IIASA
to remain below the 1975 world average. Basing future energy
needs in LDCs on specific per capita energy consumption targets
produces global energy requirements that surpass those projected
by IIASA. For the 2 kW per capita, the total world energy demand
amounts to 21.3 TWY/yr or 4.4 TWY/yr higher than IIASA's high
scenario for the year 2030. A target of 5 kW per capita yields a
world total of 35.3 TWY/yr, more than double IIASA's high scenario
projections for the year 2000 and equal to their projections for the
year 2030.

    It can also be observed from Table 8 that North America's per
capita energy consumption in IIASA's high scenario continues to
grow to some staggering figures: 14 and 19 kW in the years 2000
and 2030, respectively. Even in the IIASA low scenario, these
figures are 12 and 14 kW. The absolute gap between North America
and Africa and Asia remains very high (13-17.5 kW per capita for
the high scenario and 11-13 kW per capita for the low).

    It is striking that the region of the Middle East and North
Africa, endowed with huge energy resources supporting a substantial
fraction of the energy needs of the Western developed regions and
having sufficient financial resources to support a high economic
growth, is projected to have per capita energy consumption that
would lag far behind North America and other developed regions,
even in the year 2030.

Table 7. Per Capita Energy Consumption Projections

Year / Region	Ratio Region/Region III IIASA's High Scenario			Targets	
				-A- (kW/yr)	-B- (kW/yr)
	1975	2000	2030	2000	2030
I	11.2	13.7	19		
II .	5.1	8.5	15.2		
III	4.0	6.3	9.3		
IV	1.1 (0.275)	2.3 (0.36)	4.6 (0.49)	2	5
V	0.23 (0.058)	0.56 (0.089)	1.3 (0.14)	2	5
VI	0.98 (0.245)	3.11 (0.5)	6.7 (0.72)	2	5
VII	0.5 (0.125)	1.1 (0.17)	2.6 (0.28)	2	5
World	2.1	2.8	4.5		

NOTE: (A) = Per Capita Target of 2 kW/yr.
         (B) = Per Capita Target of 5 kW/yr.
         Figures between parantheses indicate the projected per capita
         energy consumption in the ith Region as a fraction of per
         capita energy consumption in Region III.

Table 8. Energy Requirements: IIASA's Projections vs. Our Targets*
(TWY/Yr)

Year / Region	1975	Target (kWY/Yr)		IIASA's High	
		2	5		
		2000		2000	2030
I	2.65	3.89	3.89	3.89	6.02
II	1.84	3.69	3.69	3.69	7.33
III	2.26	4.29	4.29	4.29	7.14
IV	0.34	1.15	2.9	1.34	3.68
V	0.33	5.06	12.6	1.43	4.65
VI	0.13	0.5	1.2	0.77	2.38
VII	0.46	2.7	6.7	1.44	4.46
Total	8.01	21.3	35.3	16.9	35.7

*Targets pertain to regions IV, V and VI.
**Figures for total regional energy requirements were rounded up.

Energy Supply

Some futurists are optimistic about future energy supplies in our globe. Their source of optimism is based on the hypothesis that "a sufficient research and development effort will make available before the year 2000 several new technologies that can provide the world with nearly unlimited and economical quantities of clean energy from renewable inexhaustible resources" (Kahn, 1979).

Since 1900, world production of energy has increased more than tenfold. Since 1925, world population has almost doubled while energy production increased fivefold, so per capita energy consumption has almost tripled. World energy production has been increasing at an average rate of 3.5% per year or 2% per year on a per capita basis (WEC, 1974). In the third quarter of this century (1950-1975) world energy supply increased at a compounded annual rate of 4.8%. However, geographical distribution of the major supplying sources shifted significantly as demand in the industrialized countries outstripped indigenous energy supplies. This was accompanied by a shift away from coal to crude petroleum as the principal form of primary energy. While consuming about 10% of the world's total energy the LDC's share of world primary energy production reached 30.3% in 1975. In 1950, they produced 13% and consumed 5.6% of the world's total energy. Thus, by 1975, their share of the world's energy production had increased about 228% while their share of energy consumption increased by only 73%. Table 9 presents these trends over the period 1950-1975.

According to the studies reviewed above, energy production in the LDCs is expected to increase significantly by the year 2000. Such an expectation is generally based on the hypothesis that "higher energy prices will encourage the exploration for and production of fuel in many developing countries where such activities were seen as uneconomic before. This is particularly true for oil, natural gas and coal. Developing countries with adequate reserves of coal and natural gas may choose to develop these rather than risk further deterioration in their balance of payments because of higher oil import bills" (WAES, 1977). However, the WEC projects that the LDCs as a group would go into energy deficits by the year 2020. But these projected deficits conceal the substantial export potential of OPEC, which remains until 2020, while the non-OPEC developing countries continue to experience an increase in their energy import bills (WEC, 1978). In IIASA's high scenarios the LDCs (China and Centrally-Planned Asian economies excluded) would produce about 21% of the world's total energy supply in the year 2000 and about 30% in the year 2030.

Latin America is projected to supply 7.9% of the world total by the end of this century. Africa and Southeast Asia would

provide around 8.5%, while the Middle East and North African
region is projected to provide about 5% of the world's total
supply in the year 2000. Latin America's share is projected to
increase to 10.3%, Af/SEA would produce 13% while the Middle East
and North Africa region would provide a 6.6% of the world's total
supply in 2030. These projections and estimates of ultimately
recoverable resources are given in Tables 10 and 11.

Table 10. Primary Energy by Region
1975-2030 (TWyr/yr)
Total Supply: Two Scenarios

Region		High Scenario			Low Scenario	
	1975	2000	2030	2000	2030
I. NA	2.65	3.89	6.02	3.31	4.37
II. SU/EE	1.84	3.69	7.33	3.31	5.0
III. WE/JANZ	2.26	4.29	7.14	3.39	4.54
IV. LA	0.34	1.34	3.68	0.97	2.31
V. AF/SEA	0.33	1.43	4.65	1.07	2.66
VI. ME/NAF	0.13	0.77	2.38	0.56	1.23
VII. C/CPA	0.46	1.44	4.45	0.98	2.29
TOTAL	8.01	16.85	35.65	13.59	22.40

Source: W. Hafele, 1980, Energy in a Finite World : A Global Energy
Systems Analysis, IIASA, Laxenburg, Austria, Table 17.3.

In the mid-1970s the non-OPEC developing countries produced
primary energy equivalent to 6.1 bld of oil, which supplied a high
proportion (72%) of their total energy requirements (Lambertini,
1976). According to some estimates, Latin American countries
depend on oil to satisfy 70.4% of their total commercial energy
needs. The corresponding figures for Asia and Africa are 56.3 and
74.5, respectively. Taken as a group, the LDC's dependence on oil
in 1978 was as high as 65.3% (Oil and Energy Trends, 1980).
Currently, OPEC countries account for about 80% of oil reserves in
the world outside of Communist countries.

Table 11. Summary of Estimates of Ultimately Recoverable Resources, by Category

RESOURCE	Coal (TWyr)		Oil (TWyr)			Natural Gas (TWyr)			Uranium (10³ tU)	
Cost Category*	1	2	1	2	3	1	2	3	1	2
REGION										
I. NA	174	232	23	26	125	34	40	29	1,920	1,500
II. SU/EE	136	448	37	45	69	66	51	31		4,140
III. WE/JANZ	93	151	17	3	21	19	5	14	793	2,087
IV. LA	10	11	19	81	110	17	12	14	56	3,544
V. Af/SEA	56	52	25	5	33	16	10	14	311	5,269
VI. ME/NAf	<1	<1	132	27	15	108	10	14	78	1,524
VII. C/CPA	92	124	11	13		7	13	14		1,980
World	560	1,019	264	200	373	267	141	130	3,158	20,044

*Cost Categories represent estimates of costs either at or below the stated volume of recoverable resources (in constant 1975 $).

For oil and natural gas:	Cat. 1: 12$/boe	For coal:	Cat. 1: 25$/tce
	Cat. 2: 12-20$/boe		Cat. 2: 25-50$/tce
	Cat. 3: 20-25$/boe	For Uranium:	Cat. 1: 80$/kgU
			Cat. 2: 80-130$/kgU

Source: W. Hafele, 1980, Energy in a Finite World: In a Global Energy Systems Analysis, IIASA, Laxenburgh, Austria, Table 17.6.9.

WAES made three assumptions about future oil production in OPEC countries--33, 40 and 45 mbd by the year 2000. These three assumptions correspond to three governmental policies with regard to production programming, particularly in the oil-producing countries of the Middle East. The higher assumptions correspond to higher production ceilings in these countries, high additions to reserves (10 billion barrels) and rising oil prices (WAES, 1977).

The picture is different in non-OPEC developing countries. Much of the future production will probably be limited to a few countries like Egypt, Mexico and Brazil. But the countries will continue to be oil-importers. Table 12 summarizes WAES's projection.

It is assumed that total oil production in the Middle East and North Africa will rise to meet demand until a level that is no more than 50% higher than the region's 1973 production is reached (i.e., 33.6 mb/d). Ultimately, however, this oil rich region would have to decelerate its production rates as the limit of its recoverable oil resources is reached (Hafele, 1980).

Table 12.   OPEC and NON-OPEC Production Rates
2000-2025
(in mbd)

YEAR	OPEC		NON-OPEC	
	Scenario C-1	Scenario D-8	Scenario C-1	Scenario D-8
2000	33	33	25	19
2005	33	33	23	17
2010	33	33	21	15
2015	33	32	19	13
2020	33	25	16	11
2025	33	20	13	9

Source: WAES, 1977, Energy, in Global Prospects, McGraw Hill, New York, pp.134-139.

NOTE: C-1 = High Additions to Reserves

D-8 = Low Additions to Reserves

Natural gas has become important as an energy source relatively recently. Due to the difficulty of transporting it in gaseous form, it was not utilized but was flared off at the wellhead. Estimates of proved reserves indicate that the largest concentrations are found in the Middle East, the USSR and the United States. About 60% of the world's (outside Communist countries) proven reserves of natural gas are in the OPEC region.

Non-oil LDCs depend only marginally on natural gas to satisfy their energy needs. However, its share is increasing: The World Bank estimates that the share of natural gas in total energy consumption has increased from 4% in the mid-1960s to 8% in the mid-1970s (World Bank Staff, 1976) WAES projects that about 17% of the LDC's energy production in the year 2000 will take the form of natural gas.

As LDCs started their development process after the coal-to-oil transition took place, they have in general not relied on coal as extensively as they did on oil and, consequently, have not explored nor invested substantially in coal exploration activities. Rising oil prices may encourage non-OPEC developing countries to escalate their exploratory activities in the field of solid fuels to help meet domestic energy needs. However, known world coal reserves are distributed very unevenly, with the US, USSR and China accounting for 75% of the total, and Europe (West and East) an additional 17% (Predicasts, 1977). According to some estimates, LDCs coal reserves are about 8-9% of the world total, Asia's potential is about 5%, Africa (including the Middle East) has about 3%, while Latin America has less than 1%. In the WAES high scenarios, about 6% of the total energy production in the LDCs in the year 2000 will come from coal. In 1975, coal production in the LDCs was about 21% of the total commercial energy produced in the region. Coal's share of total production was 62% in Asia, 4.2% in Africa and about 2.7% in Latin America (Predicasts, 1977).

Electrical power generation is the largest energy-related industry in non-OPEC developing countries. Over a 15-year period (1960-1974), installed power capacity in these countries tripled (World Bank, 1976). Electrical power generation derived from waterfalls is becoming increasingly attractive in the LDCs as the cost of fuel for heat-power generation increases, and the limitations of supply become more starkly apparent. World Bank estimates indicate that reserves of hydroelectric power are relatively abundant in the LDCs and that a marginal utilization rate of 4% of the potential is currently being achieved in these countries. Besides hydropower, other sources of primary electricity are currently being developed in the non-OPEC developing countries. It is expected that between 10 and 14% of the primary energy generated in these countries in the early 1980s will come from nuclear plants (World Bank, 1976). However, due to the staggering

investment costs of nuclear plants and technological, environmental
and technology-transfer problems, actual figures of nuclear-
generated electricity in these countries may lag behind these
estimates. Along with the World Bank's projection are those made
by WAES, which estimates that nuclear electricity will supply
about 10% of the total energy production in the LDCs (9% in the
non-OPEC and 1% in the OPEC) (WAES, 1977). Table 13 gives a
breakdown of energy supply in Asia, Africa, the Middle East and
Latin America by energy source for the period 1960-1990.

Supply-Demand Balance

All surveyed energy models assume that future energy supply
would more or less balance demand in the LDCs. Surplus is projected
to occur in the OPEC producing countries, particularly in the form
of oil and natural gas. NON-OPEC LDCs are projected to run an
energy deficit by the year 2000. However, through energy trade
such a potential deficit is assumed to be covered. WAES estimates
that total OPEC oil exports would range between 35 and 38 mbd in
2000 (WAES, 1977). It is estimated that if all import demands of
the major consuming countries for natural gas are to be met in the
year 2000, OPEC countries would need to produce about 10 MBDOE (=
.70 TW) for export in addition to their projected domestic demand
of 5 MBDOE. On the other hand, it is projected that between 15
and 25% of non-OPEC energy needs have to be imported in the year
2000. However, substantial coal and gas production are expected
to be forthcoming in these countries once thorough geological and
exploratory activities take place. Therefore, the non-OPEC LDCs
as a group will require no imports of coal or gas, and, in fact,
may be modest exporters (WAES, 1977). These energy supply-demand
projections are portrayed in Table 14.

According to the World Energy Conference, the LDCs would
remain energy exporters beyond the year 2000, but by 2020 the
potential for exports is likely to be reduced. If development
proceeds at WEC's high scenario, the LDCs group as a whole would go
into energy deficit by 2020. However, these overall figures mask
the substantial export potential of OPEC, which remains until
2020, while the non-OPEC regions continue to increase their
import requirements (WEC, 1980).

But, as expected, energy supply and demand do not balance in
each of the three continents separately. According to Predicasts,
supply in the Middle East and Africa would outstrip domestic
demand due to the geographical location of huge oil reserves in
the Middle East. Asis has been suffering from an energy deficit
for the past two decades and the pattern is projected to continue.
And, while Latin America has had energy output sufficient to meet
domestic demand and export the residual, such a pattern is expected

Table 13. Energy Supply
(quad BTUs)

ITEM	1960	1970	1973	1975	1980	1985	1990
			ASIA				
PRODUCTION							
Coal	15.17	15.08	16.21	17.81	22.01	27.32	32.60
Crude Petroleum	1.33	3.26	5.81	6.73	13.75	20.50	26.90
Natural Gas	.17	.55	.87	1.07	1.43	2.11	3.36
Nonhydrocarbons	1.09	2.01	2.23	2.66	4.17	7.77	12.45
Total Production	17.76	20.90	25.12	28.27	41.36	57.70	75.31
% of Domestic Demand	91.0	68.8	68.1	72.4	74.4	77.6	76.8
			AFRICA-MIDEAST				
PRODUCTION							
Coal	1.33	1.83	2.09	2.35	2.93	3.63	4.50
Crude Petroleum	11.55	42.00	56.40	51.11	67.57	80.33	95.47
Natural Gas	.08	.87	1.49	1.72	4.47	7.80	10.67
Nonhydrocarbons	.11	.43	.55	.64	1.04	1.61	2.34
Total Production	13.07	45.13	60.53	55.82	76.01	93.37	112.98
% of Domestic Demand	475	830	853	675	609	511	450
			LATIN AMERICA				
PRODUCTION							
Coal	.21	.29	.33	.36	.55	.74	.91
Crude Petroleum	7.68	10.72	10.64	9.13	11.32	12.97	15.47
Natural Gas	.50	1.23	1.54	1.63	2.31	3.05	3.89
Nonhydrocarbons	.48	1.13	1.52	1.80	2.54	3.62	5.14
Total Production	8.87	13.37	14.03	12.92	16.72	20.38	25.41
% of Domestic Demand	225	175	147	121	107	92	83.9

Source: Predicasts, Inc., 1977, "World Energy Supply and Demand,
Predicasts, Cleveland.

Table 14.   WEC Supply-Demand Projections (TWY/Y)
(2000)

	L	H	L	H
NON-OPEC				
Demand	2.47	3.3	4.8	8.43
Supply	1.90	3.17	2.85	5.71
OPEC				
Demand	.60	1.5	1.1	2.51
Supply	3.2	5.10	2.85	5.71
TOTAL				
Demand	3.7	4.35	5.81	10.94
Supply	5.10	8.27	5.70	11.42

Source: WEC, 1980, World Energy Resources 1985-2020, IPC Press,
Guildford, pp.245-247.

to be reversed as early as 1985 (see Tables 15, 16 and 17).

Africa and Southeast Asia would stay as net energy exporters
until 1990 but then become rather substantial oil importers. The
Middle East and North African region is expected to continue its
role as the main oil supplier. In 2030, the region's output of
conventional oil is projected to be around 80% of the world's
total. Latin America is projected to meet its domestic needs with
perhaps a marginal surplus for exports.

In Africa and Southeast Asia, the relatively small gas supplies
are thought to be adequate to meet the projected demand for gas.
In the Middle East and North Africa gas supply is projected to be
sufficient to meet domestic demand (Hefele, 1980).

Nuclear power is expected to provide a noticeable but modest
share of electricity needs in the LDCs (about 7%) by the year 2030
(WAES, 1977). The International Atomic Agency estimates that 11%
of total generating capacity in developing countries would be in
OPEC countries and 89% in non-OPEC countries. IN WAES's high

scenario, about 56% of the total demand for electricity, by the year 2000, would come from nuclear plants.

Before we proceed to discussing energy supply-demand balance under our counter-scenarios, we would like to highlight some of the problems that we perceive the energy models have with respect to their assumed future energy supply rates in the LDCs. First, it must never be forgotten that the industry and infrastructures in the LDCs are geared towards consumption of liquid fuels, notice-ably oil. The UNCTAD report makes it abundantly clear that the LDCs dependence on oil is greater than in industrialized countries (UNCTAD, 1978).

Table 15.   Latin America Energy Supply and Demand
(quad BTUs)

Item	1975	1980	1985	1990
Demand				
Motor Vehicle	2.28	3.21	4.41	5.93
Household	.54	.67	.80	.98
Electric Power	3.32	4.90	6.77	9.19
Industrial	4.56	6.79	9.88	13.51
Total Demand	10.70	15.57	21.86	29.61
Production				
Coal	.36	.55	.74	.91
Crude Petroleum	9.13	11.32	12.97	15.47
Natural Gas	1.63	2.31	3.05	3.89
Nonhydrocarbons	1.80	2.54	3.62	5.14
Total Production	12.92	16.72	20.38	25.41
% of Domestic Demand	121	107	92.9	83.9
Surplus	2.22	1.15	-1.56	-4.76
TOTAL SUPPLY	10.70	15.57	21.86	29.61

One TWY/Yr = 29.89 Quads.

Source: Predicasts, Inc., 1977, World Energy Supply and Demand, Predicasts, Inc., Cleveland, Ohio.

Table 16.   Africa-Mideast Energy Supply and Demand
(quad BTUs)

Item	1975	1980	1985	1990
Demand				
Motor Vehicle	.97	1.47	2.15	2.97
Household	.54	.71	.94	1.15
Electric Power	2.55	3.85	5.65	7.69
Industrial	4.21	6.46	9.53	13.31
Total Demand	8.27	14.49	18.27	25.12
Production				
Coal	2.35	2.93	3.63	4.50
Crude Petroleum	51.11	67.57	80.33	95.47
Natural Gas	1.72	4.47	7.80	10.67
Nonhydrocarbons	.64	1.04	1.61	2.34
Total Production	55.82	76.01	93.37	112.98
% of Domestic Demand	675	609	511	450
Surplus	47.55	63.52	75.10	87.86
TOTAL SUPPLY	8.27	12.49	18.27	25.12

One TWY/Yr = 29.89 Quads.

Source: Predicasts, Inc., 1977, World Energy Supply and Demand,
Predicasts, Inc., Clevalend, Ohio.

Table 17.   ASIA Energy Supply and Demand
(quad BTUs)

Item	1975	1980	1985	1990
Demand				
Motor Vehicle	1.90	2.78	3.66	4.67
Household	2.47	2.94	3.60	4.35
Electric Power	9.68	14.36	19.95	27.08
Industrial	24.98	35.48	47.16	61.95
Total Demand	39.03	55.56	74.37	98.05
Production				
Coal	17.81	22.01	27.32	32.60
Crude Petroleum	6.73	13.75	20.50	26.90
Natural Gas	1.07	1.43	2.11	3.36
Nonhydrocarbons	2.66	4.17	7.77	12.45
Total Production	28.27	41.36	57.70	75.31
% of Domestic Demand	72.4	74.4	77.6	76.8
Surplus	-10.75	-14.20	-16.67	-22.74
TOTAL SUPPLY	39.03	55.56	74.37	98.05

One TWY/Yr = 29.89 Quads.

Source: Predicasts, Inc., 1977, World Energy Supply and Demand,
Predicasts, Inc., Cleveland, Ohio.

World energy models do acknowledge that both oil price increase and oil shortage are likely outcomes in the next decade or so. Given the relatively rigid economic structure in the LDCs, any future disruption in the supply of oil would tend to have more pronounced and detrimental impact upon their economies than upon those of the industrialized nations. Thus, while other energy sources may be potentially available and forthcoming, their oil-specific needs require supplies that may not flow at the needed rates.

Second, within the LDCs, OPEC countries are assumed to increase their current production rates of oil by about 25 to 100% by the turn of the century. However, such assumptions may prove to be incorrect as current political and economic thinking in the OPEC countries leans towards conserving energy by production-programming (Fesharaki, 1980). Currency devaluation coupled with escalating inflation rates have been cited as factors calling for lower production levels. The oil-importing countries have been, allegedly, storing an increasing volume of OPEC oil. Expectations of higher future oil prices deem such a policy rational as these countries would buy oil at the cheap price and consume it when the price escalates.

The producing countries, on the other hand, have been trying to program their production levels, since the dollar devaluation and world inflation are eating up rather swiftly the value of their non-renewable assets (i.e., the form of Petro-dollars). Domestically, these countries may cut down their growth rates to avoid abrupt changes in the social infrastructure, which may give rise to sociopolitical elements that are hard to control. If growth rates continue their past records, then energy consumption in these countries would grow too, with the gap between production and consumption (= exports) being quickly bridged (OAPEC, 1977). In Kuwait, for example, electrical energy consumption has been growing in the last few years at the phenominal rate of about about 20%. Primary energy consumption growth rate has also been high. If historical trends continue, by the year 2000 domestic demand for energy would siphon off over 50% (probably 75%) of the projected oil production level (i.e., 1.5 MBO/D) as Fig. 4 illustrates. Other oil producing countries are also witnessing similar energy consumption patterns (Fesharaki, 1980).

Third, there is the argument that the LDCs are almost totally dependent on the external supply of technology for managing their energy sector. For a developing country with a meagre infrastructure for R&D, choices of technology have generally been constrained by what was available in the industrialized West (Pachauri, In Press). Such dependence means, in turn that in order to switch from one energy source to another a relatively long time elapses (i.e., shipping of and training on the new technology.)

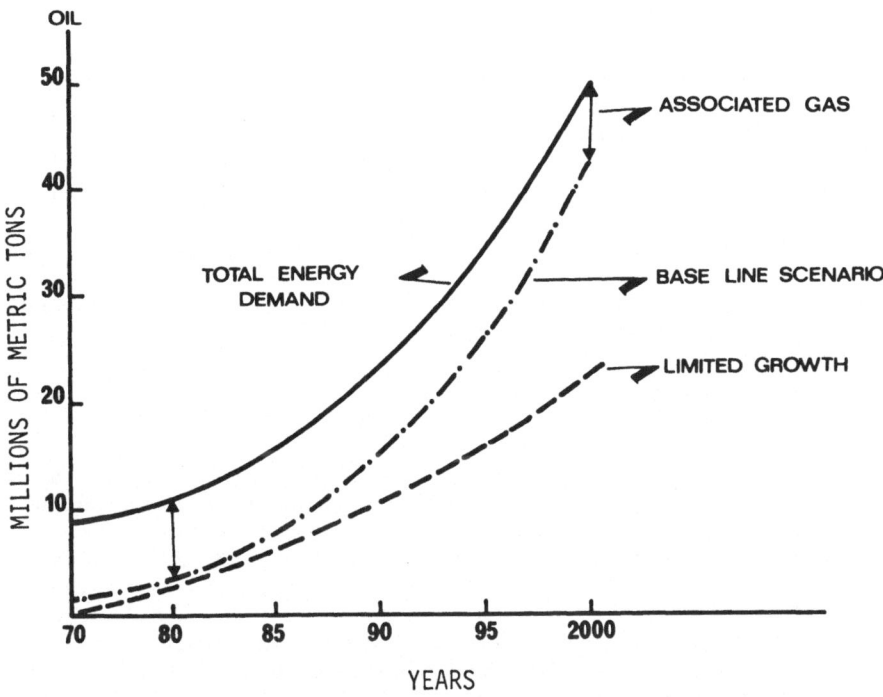

Fig. 4. Kuwait's Local Oil Demand Projections

Finally, the shift toward plants with high capital costs such
as those based on nuclear energy, implies that foreign exchange
bills would be formidable for many LDCs. We saw earlier that
power investment is estimated to absorb about 10-12% of gross
capital formations in the LDCs (Freidman, 1976). Alan Strout has
shown that most developing countries did not develop significant
energy-intensive industries when oil prices were low and the
future higher energy prices assumed by world energy models would,
if anything, discourage them from major investment in energy-
intensive industry (Strout, 1979). The impact of this would again
be smaller future energy production in the LDCs than is generally
projected by the surveyed world energy models.

In what follows, we discuss the supply-demand balance under
our counter-scenarios. As was discussed earlier, the world energy
demand models assumed that, given moderate demand growth rates,
energy supply in the LDCs would be sufficient to meet their
future needs. In the year 2000, the World Energy Conference projects total primary energy demands in the LDCs would be sufficient
to meet their future needs. In the year 2000, the World Energy
Conference projects total primary energy demands in the LDCs to be
within a 4.5-5.5 TWY/yr interval. The lower and upper estimates
for the year 2020 are 12.7 and 17.5 TWY/yr, respectively. Beyond
the year 2000, demand is projected to outstrip energy supply in
the LDCs. On a per capita basis, energy consumption in LDCs is
projected by the Conference to be in the order of 1.7 kWY/yr in
2000 and about 2.78 kWY/yr in the year 2020.

WAES, on the other hand, projects, in its high scenario, a
total energy demand of about 3.06 TWY/yr in the year 2000, which
is equivalent to about .91 kWY/yr on a per capita basis. The
corresponding projections under the low scenario are 2.91 TWY/yr
and .65 kWY/yr, respectively. The projected total energy demand
in the LDCs according to IIASA's high scenario is 3.54 TWY/yr by
the year 2000 and 10.71 TWY/yr in 2030. On a per capita basis,
each person living in the LDCs would consume about 1.05 kWY/yr by
the end of this century and about 2.14 kWY/yr in 2030. Table 18
presents projected total energy needs of LDCs as obtained in
world energy models.

The table conveys an interesting message: To the extent that
future energy needs in the LDCs follow our projections more clearly
than those of world energy models, then the difference in the
projected future needs may lie between 3.2 and 11.5 kWY/yr. These
differential figures represent between 20 and 70% of the projected
future world's (and not the LDCs only) energy demand.

In Table 19 we summarize the regional energy supply-demand
balance. Regions are ranked 1 to 7 according to the size of the
potential excess energy demand, given specific volume of recoverable

Table 18. LDC's Projected Energy Needs, World
Energy Models
(2000)

Total Energy Demand According to World Scenarios	High Scenarios (TWY/yr)	Low Scenarios (TWY/yr)
1A. WEC	4.4	3.1
2A. WAEC	3.05	2.2
3A. IIASA	3.54	2.6

energy at specific prices. Region V (Africa/South East Asia) has potentially recoverable commercial energy sources that would sustain a per capita energy consumption target of 2 kW for about 19 years. Beyond this period the region has no indigenous energy sources that would be economically recoverable at the assumed cost structure. If this region were to depend exclusively upon its recoverable resources, it would run out of this energy form in five years. Its gas reserves would carry it through a period of three years. Its coal potential would tie it over for about a decade, but it has to resolve the technological, environmental and legislative problems associated with coal.

Region III (WE/JANZ) would run out of its indigenous resources within a thirty-year period. However, its relatively abundant source (coal) can be technologically adapted and utilized more readily than coal in Region V due to the lack of technological knowledge available in the latter region. Oil and gas are the most plentiful sources in Region VI (ME/NAF). The region's recoverable coal energy would not sustain it, at a rate of 5 kW per capita energy consumption, more than a few months.

The world at large has enough energy to last for the next 50 years, with coal contributing about 50% of all future energy requirements. If a more optimistic recovery cost structure is assumed, the worldwide energy supply would be sufficient to meet demand for the next 134 years. Notice, however, that the rank ordering of regions change when this more optimistic cost structure of recoverable energy is adopted. North America, Western Europe and the Soviet Block would have much more solid energy form (coal) than the rest of the world. The Middle East and North Africa region, although still ranking first among all regions,

Table 19a.   Total Ultimately Recoverable Resources
at Low Cost Categories

Region	Energy Form				Rank
	Coal	Oil	Gas		
I	174	23	34	231	
(3.89 TWY/yr)	45*	6	8.7	59	[5]
II	136	37	66	239	
(3.69 TWY/yr)	37*	10	18	65	[6]
III	93	17	19	129	
(4.29 TWY/yr)	22*	4	4.5	30	[2]
IV	10	19	17	46	
(1.34 TWY/yr-IIASA's					[3]
2 kW)	7.5*	14	13	34	
V	55	25	16	96	
(5.06 TWY/yr at Target					[1]
of 2 kW)	11*	5	3.2	19.2	
VI	<1	132	108	240	
(1.2 TWY/yr at Target					[7]
of 5 kW	-*	110	90	200	
VII	92	11	7	100	
(2.6 TWY/yr at Target					[4]
of 2 kW	34*	4	2.6	40.6	
WORLD	560	264	267	1091	
22.07	25*	12	12	49	

NOTE: Figures in parentheses in Column 1 indicate projected demand.

*Figures in the second row of each region refer to the projected
number of years that a region's indigenous resource could sustain
it.

Table 19b. Total Ultimately Recoverable Resources
at High Cost Categories

	Energy Form				
Region	Coal	Oil	Gas	Total	Rank
I	406	174	103	683	[4]
(3.89 TWY/yr)	104*	48	26.5	178.5	
II	584	151	148	884	[6]
(3.69 TWY/yr)	158*	41	40	239	
III	244	41	38	323	[2]
(4.29 TWY/yr)	57*	9.6	8.8	75.4	
IV	21	210	43	274	[5]
(1.34 TWY/yr-IIASA's					
2kW)	16*	157	32	204	
V	107	63	40	210	[1]
(5.06 TWY/yr at Target					
of 2 kW)	21*	12.6	8	41.6	
VI	<1	159	132	291	[7]
(1.2 TWY3Yr at Target					
of 5 kW)	-*	132.5	110	242	
VII	216	35	34	285	[3]
(2.6 TWY/yr at Target					
of 2 kW)	83*	13.5	13	110	
WORLD	1579	837	538	2954	
22.07	72*	38	24	134	

NOTE: Figures in parentheses in Column 1 indicate projected demand.

*Figures in the second row of each region refer to the projected number of years that a region's indigenous resource could sustain it.

would not have expanded the contribution of its coal share in the total.

The conclusion that we may derive here is that to the extent that our counterscenario assumption may be borne in the LDC's future energy consumption then a gap of between 3.2 and 11.5 TWY/yr would exist between domestic demand and indigenous energy supply sources in these countries. While these figures are crude approximations to an uncertain energy future, their qualitative implication should be clear: Unless we provide the 70% of the world population with adequate supply of this "fuel of growth" all our egalitarian, humanitarian and peaceful goals for our planet Earth may remain largely unfulfilled. We simply cannot content ourselves with the gross energy consumption inequalities- nicely dressed up by beautiful averages-that have plagued our history.

REFERENCES

Adelman, I. C. Morris, and Robinson, S., 1976, Policies for equitable growth, World Development, 4(3):561.

Baer, W., 1974, The World Bank Group and the process of socio-economic development in the Third World, World Development, 2(6):1.

Chernery, H., et al., 1974, "Redistribution with Growth", Oxford University Press, London.

Chessire J. and Paritt, K., 1978, The Great Debate, "Some Energy Futures" in World Futures., C. Freeman and M. Jahoda (eds.) Universe Books, New York.

Dacy, D. C., et al., 1978, Employment effects of energy conservation in the USA, Energy Economics 1(4):194-202.

Darmstadter, et al., "Energy in the World Economy: A Statistical Review of Trends in Output, Trade and Consumption Since 1925", John Hopkins Press, Baltimore.

DuMoulin, H. and Eyre, J., 1979, Energy scenarios. A learning process, Energy Economics, 1(2):76-86.

Fesharaki, F., 1980, Global petroleum supplies in the 1980s: prospects and problems, OPEC Review, 4(2):27-49.

Friedman, E., 1976, Financing Energy in Developing Countries, Energy Policy, 37-49.

Griffin, J. M., and Steele, H. B., 1980, "Energy Economics and Policy", Academic Press, New York.

Hafele, W., 1980, "Energy in a Finite World: A Global Energy Systems Analysis", IIASA, Laxenburg.

Hafele, W. and Sassin, W., 1977, The Global Energy Systems, in "Annual Review of Energy".

Heilbroner, R., 1974, "An Inquiry into the Human Prospect", Norton, New York.

Howe, C. W. , 1979. "Natural Resources Economics", John Wiley, New York.

Independent Commission on International Development Issues (ICIDI), 1979, "North-South: A Program for Survival", ICIDI,

Kahn, H., 1979, "World Economic Development 1979 and Beyond",
    Marrow Quill, New York.
Kelley, A. and Williamson, G., 1974, "Lessons From Japanese Develop-
    ment", University of Chicago Press, Chicago.
Kouzminov, V. A., 1979, Comment on the world's energy situation,
    Impact of Science on Society, 29(4):289-302.
Lambertini, A., 1976, "Energy and Petroleum in Non-OPEC Developing
    Countries", World Bank Staff Working Paper.
Leontief, W., et al., 1977, "The Future of the World Economy", A
    United Nations Study, Oxford University Press, New York.
Lewis, W. A., 1980, "The Slowing Down of the Engine of Growth,
    American Economic Review, 70(4):555-564.
Makhijani, A., 1975, "Energy and Agriculture in the Third World",
    Ballinger Publishing Co., Cambridge.
Manne, A., et al., 1979, Energy policy modeling: a survey,
    Operations Research, 27(1):Jan/Feb.
Morawetz, D., 1977, "Twenty-five years of Economic Development,
    1950-1975", John Hopkins Press, Baltimore.
The National Research Council, 1979, "Energy in Transition 1985-2010",
    W. H. Freeman & Co., San Francisco.
Pachauri, R. K., (In press), Energy prospects and problems in oil
    importing LDCs, OAPEC Journal.
Reynolds, L., 1977, "Image and Reality in Economic Development",
    Yale University Press, New Haven.
Schurr, S. H., 1978, Energy, economic growth and human welfare,
    Energy Use Management, Pergamon Press, New York, pp.87-92.
Stobaugh, R. and Yergin, D., 1979, "Energy Future: Report on Energy
    Project at the Harvard Business School", Randam House, New
    York.
Strout, A. M., Prospects for Nuclear Power in the Developing Countries,
    in "Advances in the Economics of Energy and Resources", Robert
    Pindyck (ed.), JAI Press, Greenwich.
Studsvik Report, Olof Murelius, Energy in Developing Countries
    EP-79/118, Nykoeping, Sweden.
Tyner, E., 1978, "Energy Resources and Economic Development in India",
    Martinus-Nijhoff Social Sciences Division, Leiden.
Ulph, A. M., 1980, World energy models - a survey and critique,
    Energy Economics, 2(1):46-59.
UNCTAD, 1978, "Energy Supplies for Developing Countries: Issues in
    Transfer and Development of Technology".
WAES, "Energy Supply-Demand Integration for the Year 2000",
    MIT Press, Cambridge, Massachusetts.
WAES, 1977, "Energy: Global Prospects, 1985-2000", MCGraw Hill,
    New York.
World Energy Conference (WEC), 1978, "World Energy Resources", IPC
    Science and Technology Press for the WEC, Guildford.
World Energy Conference (WEC)m 1978, "Energy Resources: Avilability
    and Rational Use," IPC Science & Technology Press, Istanbul.

ENERGY PROBLEMS IN THE THIRD WORLD

José Goldemberg

Institute of Physics
University of São Paulo
São Paulo, Brazil

I. THE NATURE OF THE ENERGY CRISIS IN LDC's

The energy problems of the last decades of the 20th century will probably pass into history as the transitory problems of societies which coupled their growth and development to the consumption of irreplaceable fossil fuels.

Table I shows the current levels of energy consumption for some selected countries and world regions.

As can be seen in this Table yearly per capita consumption ranges from 0.12 TCE in the lower-income countries, to 0.6 in India and to 12.8 in the U.S., with a world average of 2.23 TCE. Of the total world energy budget (8.9×10^9 TCE) 69% is consumed in the developed countries, which account for only 25% of the total population. The U.S., with 5.3% of the world's population, consumes over 30% of the world's energy.

Table 1.   Energy Consumption of World Regions and Some Selected Countries

	Population (billions)	Billions of TCE*/year			Energy/cap.	Energy/capita	Source
		Comm.	Non-Comm.**	Total	TCE/cap.	kcal/day	
World	4.0	7.4	1.5	8.9	2.23	42,500	Ref. 1
Developed Countries	1.05	6.1	--	6.1	5.8	110,000	Ref. 1
Developing Countries	2.95	1.3	1.5	2.8	0.95	18,000	Ref. 1
Middle-Income Countries ***	0.55	0.37	?	0.37	0.67	12,700	Ref. 1
Lower-Income Countries ***	0.89	0.16	?	0.16	0.19	3,400	Ref. 1
Brazil	0.11	0.10	0.03	0.13	1.2	22,800	Ref. 2
India	0.6	0.16	0.2	0.36	0.6	11,000	Ref. 3
China	0.878	0.377	?	0.377	0.384	7,100	Ref. 4
Bangladesh	0.08	0.002	0.007	0.009	0.12	2,300	Ref. 5
U S	0.214	2.7	--	2.7	12.8	243,000	Ref. 1

* 1 kg of coal = 8.6 kwh = 7.4x10^6 cal = 3x10^4 BTU

** Non-commercial sources are mainly fuel, wood, crop wastes and dung.

*** We used a classification of less developed countries as lower-income LDC's (annual per capita income under US $200) and middle income LDC's (income over US $200 and below US $1,000).   See Appendix I.

It is very doubtful that these variations will persist for many decades due to social and political changes around the world. As LDC's develop, their share of the world energy budget tends to grow, which increases the competition for the fossil fuel resources that are not altogether very large. This tendency for an "equalization" of the levels of energy consumption is very strong and unless managed in a satisfactory way will certainly generate conflicts in the rush to gain access to and/or control over fossil fuels.

In addition to that at the current growth rate of 2.1%[1] per year the world's population will reach 10 billion by year 2010; if at that time the "per capita" energy consumption were 12.8 TCE (tons of coal equivalent), the current level of the average U.S. citizen, the total energy consumption per year would be approximately 128×10^9 TCE (3500 Quads).

At this rate of consumption presently known deposits of coal, oil and natural gas would not last more than 60 years, with growing shortages and consequently rising prices occurring much before that. The world's energy problems resides therefore in the fact that the bulk of the energy being consumed comes from fossil fuels which are available in rather limited supplies (Table 2).

In order to understand the differences (and similarities) between developed and LDC's it is useful to review some of the data available for demand of energy in selected countries.

Table 2. World Non-Renewable Energy Resources

	Potential Resource (Billions TCE)
Coal	7,000
Oil	400
Natural gas	400

Source: Ref. 1

Data for developed countries Canada, Denmark, Finland, France, Germany, Italy, Japan, Netherlands, Norway, Sweden, UK, US are available from a number of studies and we used here as a general reference the information given in the WAES study[6]. In this study the economy is divided in a set of separate sectors (transportation, industrial, residential, etc.) and the energy inputs classified as coal, petroleum, etc. It is important to stress that only commercial sources of energy are used. Figure 1 shows the demand of energy for the developed countries and a few additional LDC's: India,[3] Brazil,[2] Bangladesh[5] and China.[4]

What is striking in this figure is that demand profiles do not differ much for all countries considered.

The situation <u>does</u> change however when one takes into account non-commercial sources of energy in the few cases where they are known (India,[3] Brazil,[2] China,[4] Bangladesh[5]), or have been estimated (East Africa[7] and Central American countries[8]) (Table 3).

Table 3.   Commercial and Non Commercial Energy Consumption in LDC's

	Comm.	Non Comm.	Total	Source
India	48%	52%	100%	Ref. 3
Brazil	70%	30%	100%	Ref. 2
China	70%	30%	100%	Ref. 4
East Africa	10%	90%	100%	Ref. 7
Bangladesh	26%	74%	100%	Ref. 5
Costa Rica	69%	31%	100%	Ref. 8
El Salvador	54%	46%	100%	Ref. 8
Guatemala	52%	48%	100%	Ref. 8
Honduras	52%	48%	100%	Ref. 8
Nicaragua	66%	36%	100%	Ref. 8
Panama	81%	19%	100%	Ref. 8

ENERGY DEMAND

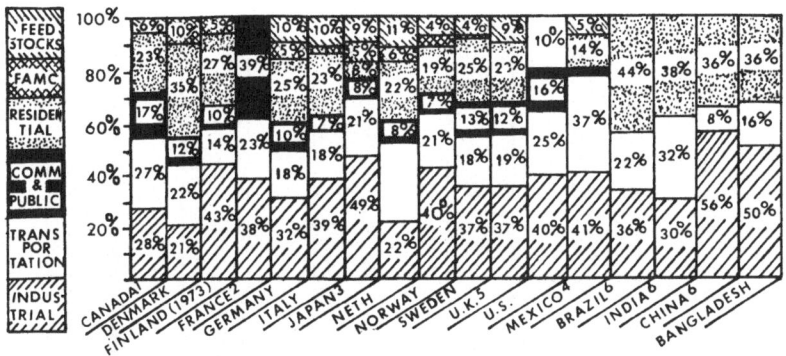

Fig. 1 - Energy demand for developed and LDC's -
                  commercial sources

1-Fishing, agriculture, mining and construction are
  aggregated into other sectors.
2-Only three sectors are shown: domestic (including
  residential and commercial), transportation and industry.
3-Transportation includes bunkers.
4-FAMC is 1% - FAMC: fishing, agriculture, mining and
  construction.
5-Commercial and public also includes fishing and
  agriculture.
6-Commercial public, residential, fishing, agriculture
  lumped together as residential.

        Figure 2 shows the energy demand by different sectors
of the economy including commercial and non-commercial
sources. We have introduced here a model developed
country obtained by averaging the demand patterns of the
developed countries given in Figure 1 (for which non-commercial sources are negligible) and lumped together the
residential, commercial, public, fishing and agriculture
sectors. Although a weighted average (using as weights
the total energy consumed by each country) might be better
this would make the US role too dominant. A simple average
takes more into account diversities of geography and life-
styles within the developed countries.

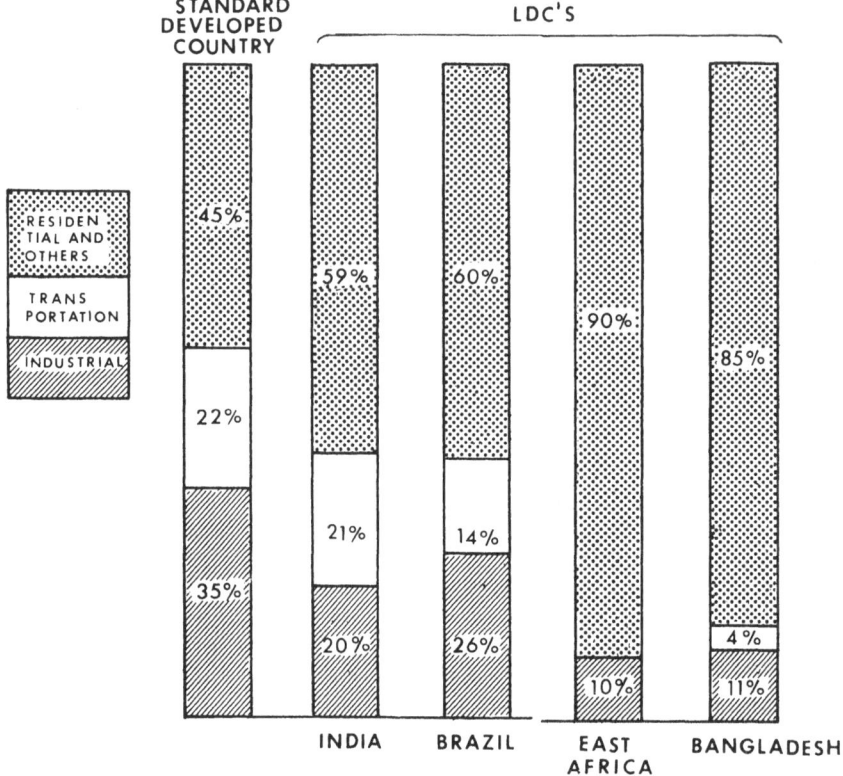

Fig. 2 - Energy demand for developed and LDC's; all
          sources (commercial and non-commercial)

        The major difference between the demand profiles of
the developed countries and the LDC's is the much greater
importance of the domestic sector in the LDC's, where it
accounts for at least 60% of total energy use (90% for
East Africa and 85% for Bangladesh), compared to
approximately 40% in the developed countries. This reflects the demand profile one intuitively expects for
LDC's but which is not apparent in Figure 1, which shows
only commercial sources. Even Bangladesh, which is a
fairly undeveloped country (with an income per capita of
US $110) presents in Figure 1 a consumption profile
similar to that of industrialized countries.

        The reason for the similarity of the profiles based
on commercial energy is the following: most of middle-
income (and some of the lower income) LDC's have a social

structure that is dual in character: 80-90% of the
population live in backward agricultural areas (or in
shacks in the urban areas) and do not really participate
in the economic life of the country;   and 10-20% is quite
affluent, living in big cities with cosmopolitan life-
styles, and accounts for most commercial energy consumption in the country.

The urban fraction of the population (and its
leadership) determine the development policies which
consist in general of pushing the leading industrial
sectors and waiting for the results to "trickle down" to
the people outside of the rapidly expanding economy.

This development model which is widespread in Latin
America and Southeast Asia is often described as the
"Belgium inside India model" for obvious reasons.   There
are effectively two countries in one to deal with in most
cases and average energy consumption and GNP/per capita
have to be analyzed with great caution.

The rural/urban population mix has been changing
rapidly in LDC's.   In general rural life and social
organization in the fields is such that the peasant, own-
ing no land, cannot expect, even working very hard, to
improve his living conditions.   Consequently many migrate
to the large cities whenever possible living in shacks
which might appear unbearable to the well established
urban dwellers but nevertheless constitutes a progress of
sorts for the migrants from rural areas;   they can get in
cities a few things such as medical aid, school for the
children and some amenities such as lighting and TV and
radio entertainment they can't have in the fields.

On a worldwide basis the problem of urbanization
can be seen clearly in Figure 3;   the rural population
which was 80% in 1900 has decreased to 65% in 1975 and
will probably go down to 45% by the year 2000.[9]

In the developed countries less than 35% of the
population lives in rural areas (down from 70% in 1900).
The decrease of rural population has been rapid and
accelerating for these countries.

In the LDC's approximately 90% of the population
was rural in 1900 and this number has decreased slowly
to 75% in 1975.

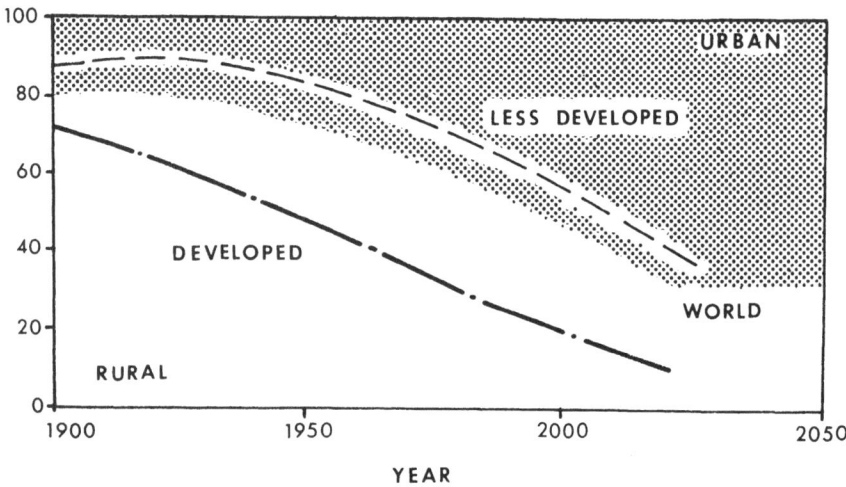

Fig. 3 - Evolution of the world population in rural
and urban areas

As pointed out above the main difference between
developed and LDC's countries lies in the fact that
LDC's depend on non-commercial sources of energy to
account for at least 30% (and generally much more than
that) of their needs.

Table 4 shows how energy use is distributed by
energy consuming sector and by source in the U.S. This
distribution is typical of developed countries.

At the other extreme we show in Table 5 the input-
output energy matrix for a typical Indian village[10]
(population ≤ 500).

Data on villages in Bangladesh[11] and in villages
in China, Tanzania, Northern Nigeria, Northern Mexico
and Bolivia[12] confirm the picture suggested by these
data. These patterns are probably characteristic of a
total population of over 2 billion people in the LDC's
(Table 6).

Table 4. Energy Input-output Matrix for the U.S. (1972)

Energy Source	Energy Consuming Activity (kcal/capita/day)*					
	Agriculture Mining and Others	Commercial	Residential	Transport	Industrial	Total
Oil	5,500	3,300	13,000	59,000	17,000	97,000
Coal	-----	700	-----	------	14,000	15,000
Natural Gas	1,000	6,000	20,000	------	26,000	53,000
Electricity	500	5,000	6,000	------	7,000	17,000
Total	7,000	15,000	39,000	59,000	64,000	184,000

*20,000 kcal/capita/day corresponds to an available power of 1 kw.

Source: Reference 6. Not included in this table is the amount of energy lost in the production of electricity from coal, oil and synthetic gas which is approximately 60,000 kcal/capita/day (25% of the total energy consumption).

Table 5.   Energy Input-output Matrix for Typical Indian Village

Energy Source	Energy Consuming Activity (kcal/capita/day)					
	Agriculture	Domestic Activities	Lighting	Transport	Manufacturing	Total
Human Labor	370	250	------	60	10	690
Animal Work	840	---	------	160	--	1,000
Non-commercial energy (wood, dung, crop residues)	---	4,200	------	---	460	4,660
Oil	270	---	260	---	---	530
Coal	---	100	---	---	---	100
Electricity	90	---	40	---	---	130
Total	1,570	4,500	300	220	470	7,110

Source: Ref. 10

Table 6. Estimated per capita use of energy in rural areas of seven developing countries

	India[*]	China, Hunan[**]	Tanzania[**]	Northern Nigeria[**] (in 10^3 kcal/capita/day)	Northern Mexico[**]	Bolivia[**]	Bangladesh[***]
Human Labor	.67	.64	.64	.61	.75	.71	.67
Animal Work	1.00	.92	---	.13	1.30	1.83	1.00
Fuel Wood	2.86	13.69	15.07	10.27	9.70	22.83	.93
Crop Residues	1.16						1.65
Dung	.67						.57
Total Non-commercial	6.36	15.25	15.71	11.01	11.75	25.37	4.82
Coal, Oil, Gas and Electricity	.53	2.05	---	.02	19.81	---	.27
Chemical Fertilizers	.22	.34	---	.05	5.33	---	.10
Total Commercial	.75	2.39	---	.07	25.14	---	.37
Total All Sources	7.11	17.64	15.71	11.08	36.89	25.37	5.19

[*] Ref. 10
[**] Ref. 12
[***] Ref. 11

What is outstanding in Tables 4 and 5 is not just
the fact that an average U.S. citizen consumes 25 times
as much energy as does a peasant in India but also the
difference in the spending patterns. A full 1/3 of the
energy in the U.S. is spent in transportation and another
1/3 in industrial activities, items that are negligible
in a village. In contrast, agriculture and domestic
activities account for 85% of the energy spent in the
village, items that account for only 25% of the energy
used in the U.S. Cooking by itself represents 61% of
the total in the villages while in the U.S. this item
represents less than 1.5%.

Between these two extremes one has "islands of
prosperity" represented by 10 to 20% of the urban affluent part of the population of almost all countries
outside the developed industrial countries.

One has therefore quite different problems in different parts of the world and strategies to face them are
bound to have many differences.

Since non-commercial sources are renewable and commercial ones, in general, are not, one immediately as-
sociates the first ones with sparsely populated rural
areas (and decentralized uses) and the latter one with
heavily populated urban areas and centralized solutions.

This association is not entirely justified as we
will see later.

II.   SMALL SCALE DISPERSED SOLUTIONS

Big cities have large densities of people, which
consume large amounts of energy "per capita", depend
very heavily on the use of fossil fuels and produce
large amounts of waste and pollution; most of the present
technologies in use in cities are "very hard" (including
the treatment of refuse and sanitation) in the sense of
requiring large and bulky systems for the production and
use of energy; on the other hand in rural areas, in the
developing countries, the energy consumption "per capita"
is much smaller.

As can be seen in Table 7, energy consumption "per
capita" in cities does not change much all over the world.
The energy consumption in rural areas in India and Brazil
however is more than 10 times smaller on a "per capita"
basis than in most cities.

APPENDIX I

Classification of Less Developed Countries

Lower-Income Countries (annual per capita income under
$200) 1972 dollars.

South Asia	Lower-Income Sub-Sahara Africa	
Afghanistan	Burundi	Niger
Bangladesh	Central African Republic	Rwanda
Burma	Chad	Sierra Leone
India	Cahomey	Somalia
Nepal	Ethiopia	Sudan
Pakistan	Guinea	Tanzania
Sri Lanka	Kenya	Togo
	Madagascar	Uganda
	Malawi	Upper Volta
	Mali	Zaire

Middle-Income Countries (annual per capita income over
$200 and under $1000) 1972 dollars

East Asia	Middle-Income Sub-Sahara Africa and West Asia	Caribbean, Central and South America
Fiji	Angola	Argentina
Hong Kong	Bahrein	Barbados
Korea(South)	Cameroon	Bolivia
Malaysia	Congo P.R.	Brazil
Papua New Guinea	Cyprus	Chile
Phillippines	Egypt	Colombia
Singapore	Ghana	Costa Rica
Taiwan	Israel	Dominican Republic
Thailand	Ivory Coast	El Salvador
	Jordon	Guatemala
	Lebanon	Guyana
	Liberia	Haiti
	Mauritania	Honduras
	Morocco	Jamaica
	Mozambique	Mexico
	Oman	Nicaragua
	Rhodesia	Panama
	Senegal	Paraguay
	Syria	Peru
	Tunisia	Trinidad and Tobago
	Turkey	Uruguay
	Yemen AR, DM	
	Zambia	

Source: World Bank

In addition to that the energy consumption density (watt/m^2) in urban areas is also pretty much the same in most countries; in rural areas this density is more than 100 times smaller than in cities.

Table 7.   Urban and Rural Densities

Country or City	Pop. Density (people/km^2)		Energy Consumption (kcal/cap./ day)		Energy consumption density watt/m^2	
	Urban	Rural	Urban	Rural	Urban	Rural
INDIA	6,000	135	41,600	7,200	12	0.04
NEW YORK	560	–	238,000	–	6.4	–
LONDON	1,100	–	108,000	–	5.7	–
TOKYO	980	–	81,000	–	3.8	–
S.PAULO (BRAZIL)	1,260	13	50,000	9,600	3.2	0.006

1 kw of installed power corresponds to a consumption of 20,800 kcal/day.

The association of "renewable" "small scale" and "decentralized" is a fairly natural one because direct solar energy has such low density ($\sim$ 100 W /m^2); biomass, hydropower and wind have even lower densities (Table 8).

The idea is therefore to use solar energy as it comes in small dispersed quantities that do not require much processing ("soft"technologies). The attraction of a long lost bucolic rural life exerts of course a strong attraction in the minds of many proponents of this "soft path".

Table 8.   Energy Densities

Source	watts/m^2
Wind (North Seacost)	~ 4.5
Fuelwood plantation	~ 1
OTEC (tropical oceans)	~ 0,8
Wind (continents)	~ 0,6
Fuelwood (natural forests)	~ 0,2
Biogas	~ 0,18
Hydropower	~ 0,02

We will discuss and evaluate here a number of the technologies used to supply energy from <u>decentralized</u> sources:

1. Biogas production
2. Minihydroelectric stations
3. Direct solar collectors
4. Photovoltaics
5. Wind

1.   Biogas Production

The production of gas by anaerobic conversion of biomass is one of the most promising methods for the solution of the energy problems of villages of the undeveloped world.

The process is quite simple, in principle[13,14,15]: animal dung, pieces of vegetation (crop stalks, straw, grass clipplings and leaves), garbage and waste water are sealed up in insulated containers (digesters) and left to decompose. Digestible organic materials (liquids, proteins and most starches) are broken down by acid-producing bacteria and the resulting volatile acids are in turn converted by anaerobic methanogenic bacteria into a gas that is typically composed of 55 to 70% methane (CH_4), 30 to 45% of carbon dioxide (CO_2) and a trace of hydrogen sulfide and nitrogen. Besides the versatile low-pressure, medium-caloric gas (between 5,300 and 6,300 kcal per cubic meter) the process yields an organic

fertilizer of outstanding quality and improves sanitation
conditions in rural areas. (Figure 4)

   With the help of minor modifications the biogas can
be used to power internal combustion engines and to
substitute for diesel oil in small electricity generators
for lighting and irrigation.

   The burning of biogas for cooking is clearly
advantageous when compared to the burning of animal
manure. Typically the efficiency of biogas digesters is
60%, which means that 1 kg of dry manure produces 400
liters of gas with an energy content of 2,200 kcal; if
the cooking efficiency of this gas is 50%, 1,100 kcal
will be delivered to the cooking pan.

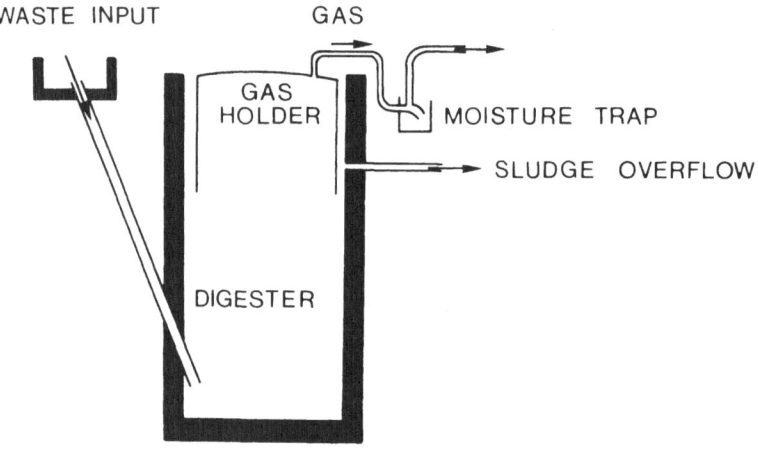

Fig. 4 - Biogas plant schematics

   If 1 kg of dry manure having an energy content of
4,000 kcal/kg (which is probably an overestimate) is burned directly for cooking the amount of heat delivered to
the cooking pan will be 400 kcal in a cooking stove that
is 10% efficient. Biogas is therefore 2.5 times more efficient than manure for cooking purposes.

The method was introduced in China some years ago but picked up momentum only in the last 5 years with 410,000 digesters in use in Szechwan province alone and another 80,000 in Mien-yang country in 1975. Hundreds of thousands of households benefitted from them.[16] It is reported that in the first 6 months of 1976 another 1.3 million digesters were built in China.[17]

The main problem of biogas conversion is that it does not work in cold regions because of the thermal requirements of the fermentation process. In addition to that the pH of the mixture has to be watched and a minimum of maintenance given to the pits.

An estimate made of the potential for biogas generation in China[16] (based on the residues of cattle, horses, pigs, chickens and man) is the energy equivalent of 48 million tons of coal per year (more than 15% of the total consumption of energy in that country). A similar estimate for India[18] gave 60 million tons of coal - enough to satisfy all the rural domestic requirements of energy for the country.

The widespread introduction of biogas generation seems therefore to be feasible in many undeveloped areas of the world.*

2. Minihydroelectric Stations

The technology for generation of hydroelectric power from small stations has been available for many years but its use has always been dwarfed by the construction of gigantic dams and huge hydroelectric power projects.

The definition of minihydroelectric plants is not very clear: in this category one includes, in general, dams less than 100 feet in height, less than 10,000 acre-foot of reservoir storage capacity and with a potential

---

*It is intriguing to observe that biogasefiers have become quite popular in China and are facing many institutional difficulties in India. This is not due to political inducement in China, as one might think. According to Vaclav Smil (private communication) the Chinese farmers use the dung of their domestic animals (pigs and chickens) in their private lots to run the digesters. Community plants on the commune level are rare in China. In India the cultural habits are such that cattle roam around the country making it difficult to collect their dung.

capacity less than 5,000 kilowatts[19]. A dam with 10,000
acre-foot in storage capacity and a height of 100 feet
corresponds to a potential of 5,000 kW.

There are almost 50,000 of these dams in the United
States with a total capacity of 27,000 Megawatts, i.e.,
an average of 500 kW per site (Table 9). These dams, if
used, could increase by almost 50% the present US hydroelectric capacity of 65,000 MW.

It might be interesting to point out that from
Table 9 one can estimate what is the amount of power
distributed in small dams as a function of their average
size.

It results that the total power available (P) on
dams of power E is

$$P \sim E^{1/2}$$

i.e. the total power available increases slowly with the
size of the dam indicating the substantial amount of power
dispersed in small dams (Figure 5).

A massive effort to build mini-hydrostations was
made in China[16] and by the end of 1975 over 40,000 of
these stations (with an average capacity of 50 kW) were
in operation. Since China does not have a very large
installed hydropower capacity (~ 10,000 MW in 1975) the
contribution of ministations to the total hydropower is
appreciable.

Table 9.   Number of Existing Dams Sorted by Height and
           Storage Characteristics in the United States

Maximum height (feet)	Maximum Storage (acre-foot)			
	0-99	100-999	1,000-9999	>10.000
0-19	12,432	7,009	1,262	398
20-49	12,883	8,332	1,789	433
50-99	429	919	574	519
>100	100	91	140	602

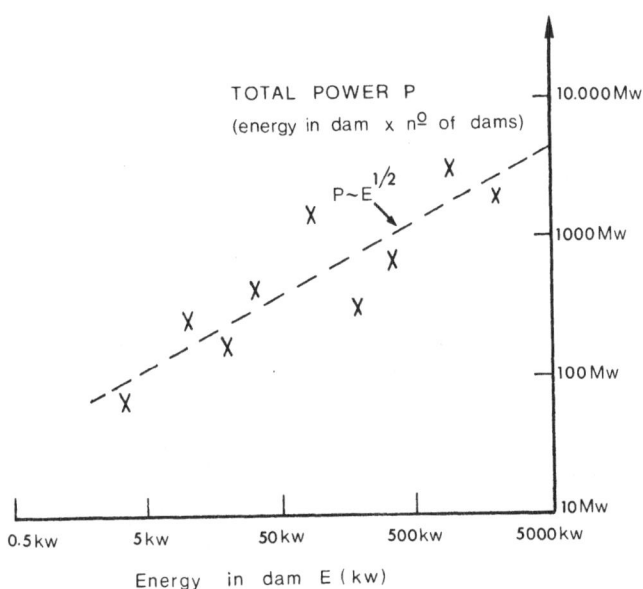

Fig. 5 - Total power available "versus" the average size
of small dams.

A 100 kW minihydroelectric plant, enough to supply
the needs of at least 100 houses, can be installed in a
small waterfall with a flow of 1.5 m^3/sec and a 6 meter
head. If smaller heads are available but the streams are
swift moving, one can still operate small turbines with
good efficiency. Since the available mechanical energy
in a flow of water is gh + 1/2 v^2 per kilogram (h is the
head, v the velocity and g the gravity acceleration,
9.8 m/sec^2), a stream with velocity of 10m/sec is
equivalent to a 6 m mead waterfall.

The price per kilowatt of installed capacity is
competitive with large conventional hydroelectric
stations.[20]

Measurements (or even good estimates) of the hydroelectric potential of small rivers and streams in general
do not exist around the world. Rough estimates can be
made on the basis of precipitation over a given region,
and the region's average altitude above sea level. The
product of these two numbers is a crude measure of the
total hydroelectric power available.

Using this relationship the total hydropower potential of a country like Brazil can be estimated from knowledge of the potential in a region like Europe, where the

potential has been more carefully measured. The area of
Brazil is 8,500,000 km^2, the annual precipitation of
water 2000 mm (which corresponds to 15×10^{12} m^3 of water)
and the average altitude of the country 400 m. The
product of these two numbers is 800,000 while the corresponding product for Europe is 240,000. Since the
hydroelectric potential of Europe is known to be 158,000
MW, the potential in Brazil should be on the order of
500,000 MW, of which 1/3 is concentrated in well known
large rivers where waterfalls do exist or where conventional large hydroelectric plants can be built. The remaining 2/3 should then be distributed in thousands of streams
where minihydrostations could conceivably be installed.

It should be stressed here that in the usual tabula-
tions of hydroelectric potentials[21], small streams are
ignored; it is difficult to find out where the line is
drawn in different countries but probably most resources
below 5,000 kW per site are not included in these
estimates.

3. Direct Solar Collectors

Flat plate collectors are the simplest of all methods
for collecting the sun's energy and heating up water for
domestic (or industrial uses).

The solar rays go through two transparent glass
plates and are absorbed in a darkened metal surface to
which are attached pipes conducting water which is therefore heated up. (Figure 6) The double layer of glass
plates with a layer of air between them (which is transparent to the incoming radiation) acts as a thermal
insulator trapping therefore the infrared radiation inside
the collector; this is the well known "green-house effect"
used all over the world in cold climates to grow vegetables
or flowers in the winter.

The efficiency of conversion of the incident radiation
into heat (carried away by the water) is approximately 50%
and it is possible to produce hot water in the range
40-90°C.

A typical insolation in tropical regions is 4,000
kcal/m^2/day ($\sim$ 0.5 cal/m^2/min). If all this heat could
be captured usefully it could replace the consumption of
approximately 0.5 liter of oil*.

---

* 1 liter of oil = 9×10^6 cal.

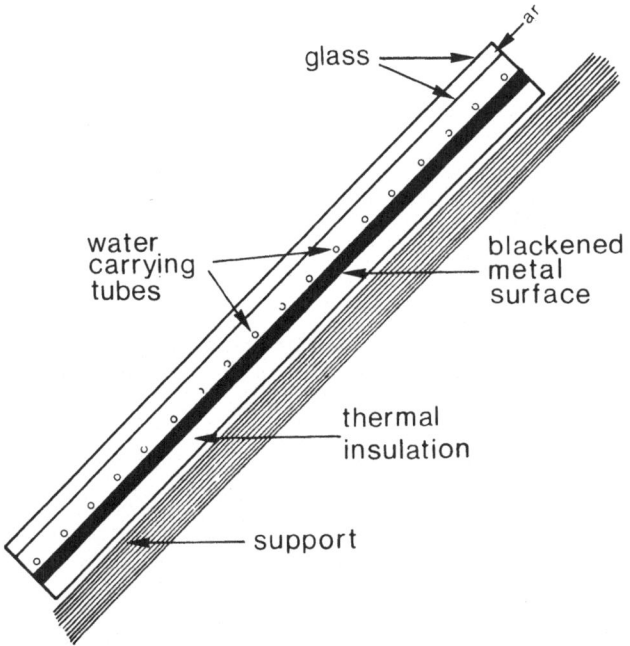

Fig. 6 - Flat plate collector

        In practice a collector of 4 m^2 can supply 300 liters
of water at 40-60°C which is needed for domestic uses for
a typical family.

        Systems in use have a reservoir for the hot water in
which water is driven by the thermo-syphon principle.
(Figure 7)

        Hundreds of thousands of residential solar collectors
are in use in many countries around the world, mainly
Israel, Japan and the United States.

        The technology is simple enough to be used in many
LDC's based on local manufacturing facilities.

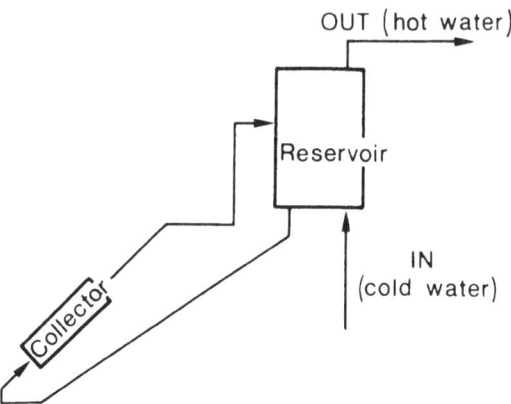

Fig. 7 - System for storing and distributing hot water
         using the thermo-syphon principle.

4.   Photovoltaics

      The basic principles of the conversion of solar
energy into electricity are well known and the characteristics of a typical solar cell[22] are shown in Figure 8.

      Typical maximum efficiencies are of the order of 10%
and typical insolation in a tropical country such as
Brazil[23] can be seen in Figure 9.

      The strong variations of solar insolation from day
to day and the fact that the sun shines only during the
day end up by establishing severe limitations in the use
of photovoltaic cells except in satellites.

      In practical applications on the earth's surface the
average yearly solar intensity can be considered as
$100-200$ W/m^2 which is not very high.   Therefore large
surfaces covered by photocells are needed with corresponding high costs (15,000 dollars per peak kilowatt).

      Most of the present applications of photovoltaic
cells (except on satellites) are being made in special
isolated situations which justify the high costs such as
isolated telecomunications relays, educational TV in
remote sites, power for medical dispensaries and in some
cases water pumping for animal and human consumption.

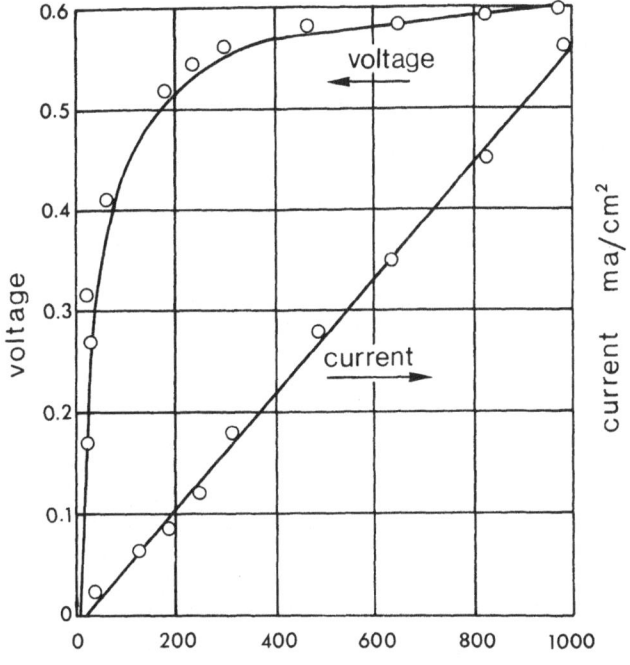

Fig. 8 - Electric characteristics of a typical
         photovoltaic cell

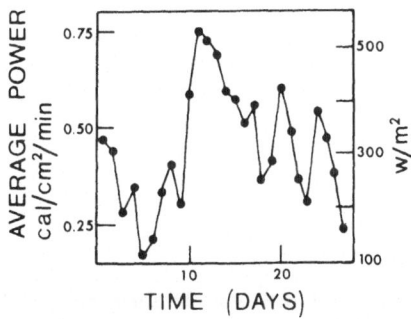

Fig. 9 - Typical insolation in São Paulo, Brazil

Undoubtely photocells would find broad commercial
applications if costs were reduced to US $100 - US $300
per peak kilowatt.

The US Government expects this to happen in the
middle of this decade as the result of a combination of
the following actions:[24]

1. Market Expansion of Existing Technology: Market stimulation through government purchase of a significant fraction of early annual cell production. A total United States government purchase of approximately 11 MW through 1983 is planned. Costs for silicon solar arrays are expected to drop to US $1000 per peak kW by 1984.

2. Develop Large Area Silicon Sheet Technology: One approach to cost reduction is the development of high capacity, low unit cost production techniques for silicon cells. The acceleration of this activity, involving silicon sheet technology, will provide for: (a) an early reduction in production costs of solar cell grade silicon from $65 per kilogram (kg) to $10 per kg; (b) increased efficiency of the solar cell production fabrication (over 75 percent of the silicon material is now wasted); (c) improvement in the ratio of cell to array area (packing factor); (d) development of suitable encapsulation materials to increase array lifetime; and (e) automated production of silicon solar cell arrays. It is expected that this process will reduce the silicon-based solar cell array costs to $500 or less per peak kilowatt.

3. Conduct Thin Film and Novel Materials Research: Use of thin film deposition techniques utilizing silicon, cadmium sulphide, gallium arsenide and other materials may allow the production of solar arrays costing $100 - $300 per peak kilowatt.

4. Develop Concentrators and High Intensity Solar Cells: The use of concentrators, combined with cells from low-cost silicon arrays costing $500 per peak kW, should lead to combined collector/cell costs of $250 per peak kilowatt by 1986.

5. Wind

Although apparently very erratic, the average value of the wind velocity in a given region of the Earth is reasonably constant; in general the monthly average velocity does not show deviations greater than 10 or 15% of the annual average.

The use of wind was very important in past centuries as is well known mainly in the western coastal regions of Europe[25].

In more recent times the farm wind rotor made of metal blades was very important for water pumping in many

regions around the world and the intermitent generation
of electricity.

Recent modern designs such as the Darrieus and Savo-
nius as well as better aerodynamical designs of old models
are in intense investigation in many laboratories around
the world.

One can derive a theoretical expression[26] for the
maximum power (P_{max}) of rotors as being

$$P_{max} = (\frac{8}{27}) \pi r^2 \rho v_i^3$$

r - radius of rotor
ρ - density of the air
v_i - incident velocity of the wind

Table 10 shows P_{max} for a variety of wind velocities
and radius of rotors.

As can be seen from this Table a wind of 10 mph can
supply approximately ~ 30 watts/m^2.

Wind energy has the inherent character of supplying
mechanical (or electrical) power for intermittent chores;
this means that storage might be indispensable for some
purposes such as the use of wind power for lighting or
refrigeration. The economics of wind power seems to be
competitive in many cases, in isolated locations.

There is much discussion of interlinking many wind
power electric generators such as to smooth out the inter-
mitency of the winds without the recourse of large storage
capacities but no definite plans for such ventures have
been established.

Table 10. Theoretical power of several rotors

Wind speed (mph)	Power (kW)			
	r = 3,3 m	r = 8,3 m	r = 16,5 m	r = 33 m
10	1	6	25	100
20	8	50	200	800
30	27	169	675	2700
40	64	400	1600	6400
50	125	781	3125	12000
60	216	1350	5400	21600

III.   LARGE SCALE CENTRALIZED SOLUTIONS

The counterpart of the idea of using decentralized
energy sources to supply the needs of rural areas is the
generation of electricity in large hydroelectric dams or
nuclear reactors i.e. in a centralized way to feed large
cities where consumption is concentrated.

1.   Hydroelectric Power

The world hydroelectric resources are shown in Fig.
10 and can be seen to be very large (2,200,000 MW of
generating capacity at 50% capacity factor). Most of
this potential is located in LDC's where the operating
capacity is presently less than 10% of the total[27].

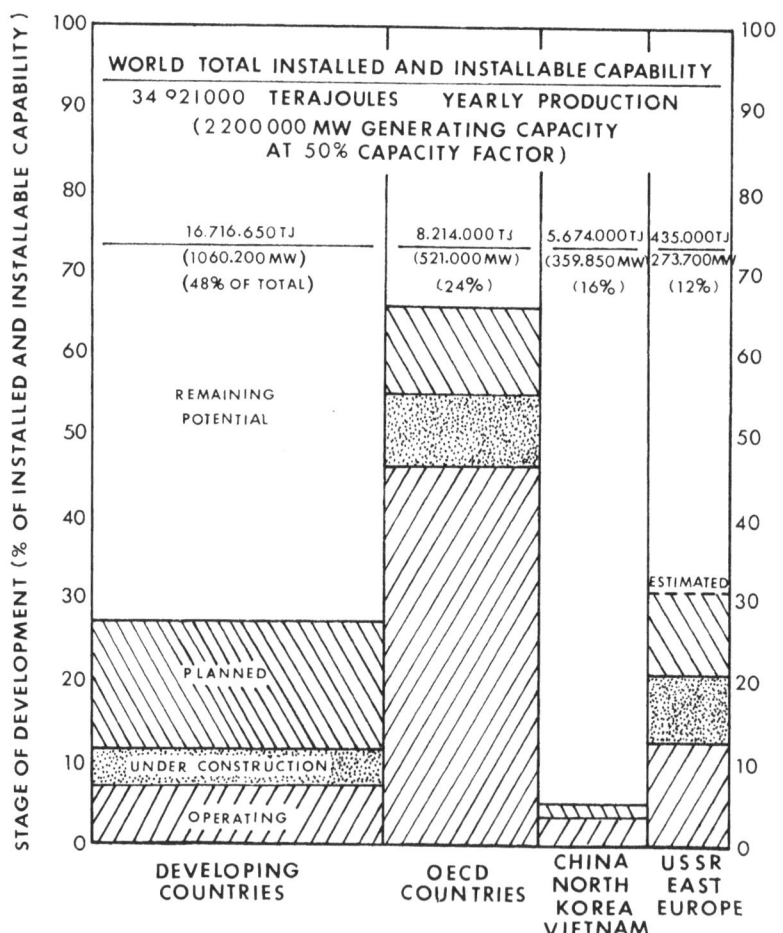

Fig. 10

The annual fossil fuel equivalent energy potential from the 1,800 GW of hydroelectric capacity located in LDC's (counting 1 unit of electricity as being equivalent to 3 units of fossil fuel energy) is 115 Quads (with an average capacity factor of 63% which is the number for 1975). Considerably less electricity will be needed by LDC's before the year 2000.

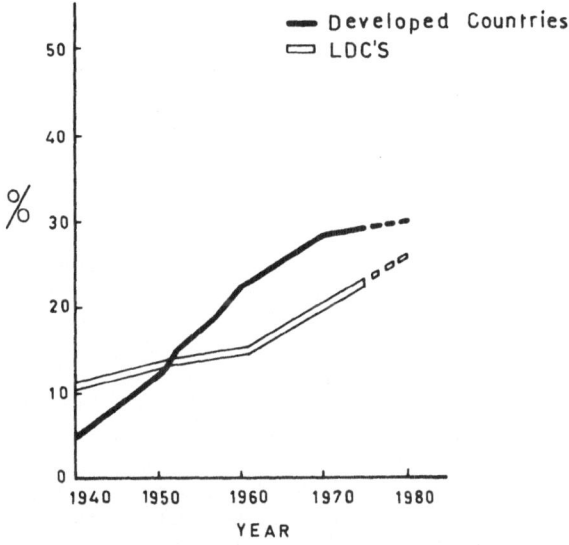

PERCENTAGE OF ELECTRICAL ENERGY
IN TOTAL ENERGY CONSUMPTION

Fig. 11

Because of the inherent need for chemical fuels (for transportation and some industrial applications) there is a practical limit to the degree of electrification a country can reach as seen in Figure 11.

In 1972 electricity accounted for 28% of the energy
consumption in the United States and 20% of the LDC's;
it has been estimated[28] that these numbers will rise to
35 and 25% respectively by the year 2000. The degree of
electrification of LDC's will probably be limited in the
next 25 years mainly by the rate at which plants can be
built and by capital available.

The world hydropotential is actually larger than the
figures given in Figure 10; not included in the numbers
given are waterfalls or dams (existing or to be built)
of less than a 100 feet head and a hydro potential smaller
than 5 MW as discussed in the previous chapter.

One of the most serious problems involved in a larger
utilization of these potentials is that the convenient
locations for hydropower are sometimes far away from
consuming centers.

The suggestion has been made that "captive" hydropower generating stations located near the large hydroelectric sites could be used to produce energy intensive
products such as aluminum. Thus the hydropower could be
transported to demand centers from remote regions in the
form of high value products.

The technological problems posed by the long distance
transmission of the electricity from these remote areas
were not considered completely solved some years ago.

The new technology to reduce losses in the lines
using high voltage direct current (HVDC) transmission
is nowadays the most attractive solution for lines longer
than 1000 km. Table 11 shows the main facilities using
this new technology in the world today[29]; some of them
have been operating for years including the Pacific
Intertie in the Western U.S.. One should notice in this
table the Cabora-Bassa line which will supply South Africa
(Pretoria and Johannesburg) with power from Mozambique.
Upon completion this line will carry 3,600 MW.

2. Nuclear Power

In principle a good case could be made of the role
of nuclear energy in developing countries. Even with
the most stringent efforts to achieve energy conservation the consumption in this part of the world is likely
to increase from the present 61 Quads to at least 160
Quads by the end of the century[30].

Table 11.   High Voltage Direct Current (HVDC) Systems

	Transmission Distance (km)	Rated Voltage (kV)	Rated Power (MW)	Commis- sioning
In Service				
Volgograd-Donbass (USSR)	470	+400	720	1962-65
New Zeland	570	+250	600	1965
Pacific Intertie (USA)	1362	+400	1440	1970
Nelson River, Bipol 1 (Canada)	895	+300	1080	1973
Inga-Shaba (Zaire)	1700	+500	560	1976
In Construction				
Cabora Bassa (Mozambique -South Africa)	1414	+533	1920	1979
Nelson River, Bipol 2 (Canada)	895	+500	1800	1981-82
Under Active Consideration				
Gull Island (Canada)	750/1080	+400	1600	1985
Nelson River, Bipol 3 (Canada)	900	+500	1800	1983-85
Ekibastus Centre (USSR)	2400	+750	6000	?

The uneven distribution of oil, gas and coal reserves around the world does not particularly benefit many LDC's (with the exception of OPEC countries) and by year 2000 will put some of these countries in a position of increasing dependence on energy imports.

Against this background the role of nuclear power in LDC's could be visualized as two-fold:
   a. it could offer a substitute for oil and gas (and possibly coal) which would otherwise be required for electricity production and a means of avoiding an overwhelming dependence on imports.

   b. it is a technologically mature solution.

The word substitution in item a above is used in the sense of being economically competitive and after the 1973 crisis it seemed to many people that nuclear energy would really "take off" both in developed and less developed countries.

However what was witnessed in the last few years was a clear decrease of the expected role of nuclear power: the projections of the International Atomic Energy Agency (IAEA) for the nuclear capacity installed in LDC's in the year 2000 made in 1977[31] were about half of the estimates made in early 1974[32]. As we will show below more realistic projections are even lower reducing drastically the role to be played by nuclear energy in LDC's.

The usual reasons given for the downward revisions of nuclear power in LDC's are the following[31]:

1. the diminished growth in electrical demand that has occurred in many countries during the last several years.

2. the extremely high cost of nuclear plant construction, which has placed financial burdens on countries with existing nuclear programmes.

3. the present lack of commercially available small and medium power reactors, which many of the smaller states would need in order to expand their electric power systems and

4. the growing awareness of LDC's that more attention should be paid to exploitation of indigenous energy sources.

A point seldom discussed in the IAEA (and other papers on the subject) is the question of technology transfer and the perception of the leaderships of LDC's of this problem. The appropriation of modern technology (and particularly of the nuclear technology) is seen as key to development in most of the developing world for the following reasons:

A. On the one hand, due to imitative tendencies, the rising middle class in LDC's favours strongly the adoption of the consuming patterns of more developed countries. In small countries this is done through imports (and trade) but in the larger ones the local production of many goods is necessary. National policies furthering import substitution are in effect in most of the more prosperous LDC's.

B. On the other hand some military, and more generally
the nationalistic civil elements of society, do not want
to remain as secondary "clients" of the larger industrial
nations (United States and Soviet Union, in general) and
want to achieve national independence on all fronts due
to economic, political and strategical reasons. The more
advanced they are, the more uneasy is the relationship
between the "client" and "patron" state as the governments of LDC's become more assertive. This can be clearly seen in the case of the Non Proliferation Treaty which
is considered discriminatory and has not been signed by
the more advanced LDC's.

On the surface these arguments would seem to play
in favour of a larger role of nuclear energy in LDC's;
in practice the opposite has happened because technological transfer of nuclear technology has been rather unsuccessful in most cases.

Table 12 shows the present nuclear and conventional
electrical capacities of 20 LDC's and the projected
nuclear power forecast for year 1990, made in 1975.

As can be seen in this Table, 7 countries (Argentina,
Brazil, India, Iran, Korea, Mexico and Taiwan) account
for nearly 70% of the total projected nuclear capacity.
The main reason for this is that all nuclear reactors
manufactured at present have rather large capacities
($\sim$ 600 Mw) and a sizable electric grid is needed to
accomodate them; this is the case of the 7 countries
listed above.

The 4 countries with the largest nuclear programs
(as of 1975) were India, Brazil, Mexico and Iran.

Although the downward revision of the Brazilian
nuclear program has been the most discussed in recent
litterature it is quite obvious that the nuclear programs
of Iran, India, Mexico are also being phased out.

In Iran even before the downfall of the Shah the
program was under heavy criticism due to a lack of infrastructure and economics. Under iranian conditions the
use of gas for the production of electricity made much
more sense than electricity.

In India a much larger emphasis on coal has reduced
the importance of nuclear power and in Mexico the recently
found oil deposits have dimmed prospects for nuclear energy.

Table 12.   Nuclear and Conventional Electric Capacities
in LDC's
x 10^3 MW

	Nuclear*	Electrical**			Nuclear Power *** Forecasts
		Hydro	Thermal	Total	1990
India	1.689	9.029	14.020	23.689	31.4
Brazil	3.116	18.411	3.385	21.796	11.4
Mexico	1.308	4.691	8.156	12.847	21.6
Argentina	1.505	1.745	7.771	9.856	8.1
Iran	8.982	0.804	4.326	5.130	10
Taiwan	3.800	8.000	32.000	40.000	10.3
Venezuela	-	2.245	2.931	5.176	4.4
Korea	3.598	0.711	4.629	5.340	9.8
Turkey	0.620	1.873	2.477	4.350	5
Colombia	-	2.420	1.430	3.850	1.2
Pakistan	0.126	0.867	1.232	2.236	4.9
Egypt	-	2.500	1.400	3.900	5
Thailand	0.600	0.910	1.865	2.775	1.8
Peru	-	1.500	1.055	2.555	-
Philippines	0.621	1.138	2.369	3.507	4.8
Hong Kong	-	-	2.919	2.919	-
Chile	-	1.462	1.199	2.661	-
Singapore	-	-	1.390	1.390	0.6
Indonesia	-	0.450	0.810	1.260	-
Bangladesh	-	0.110	0.840	0.950	-
Total	25.965	58.866	96.164	156.187	132.1

*   Power reactors operating, in construction or ordered
    as of june 1978.
** 1976 UN Statistical Yearbook.
*** Source: LDC's Nuclear Power Prospects 1975-1990: Com-
    mercial, Economic and Security Implications. Richard
    J. Barber Associates (1975).
    All units below 600 MW removed.

On the basis of this information we assumed that
the nuclear programs of these 4 countries will simply
stagnate at their present levels, i.e. that no new
projects will be implemented. Taken together with the
data of Table 12 this leads to the point labeled
"stagnation hypothesis" in Figure 12 (less than 40,000
MW of installed capacity in 1990).

In the light of the above discussion one could add
the following arguments to the 4 points listed by Lane
et al[31] to explain the decrease of nuclear power forecasts in LDC's.

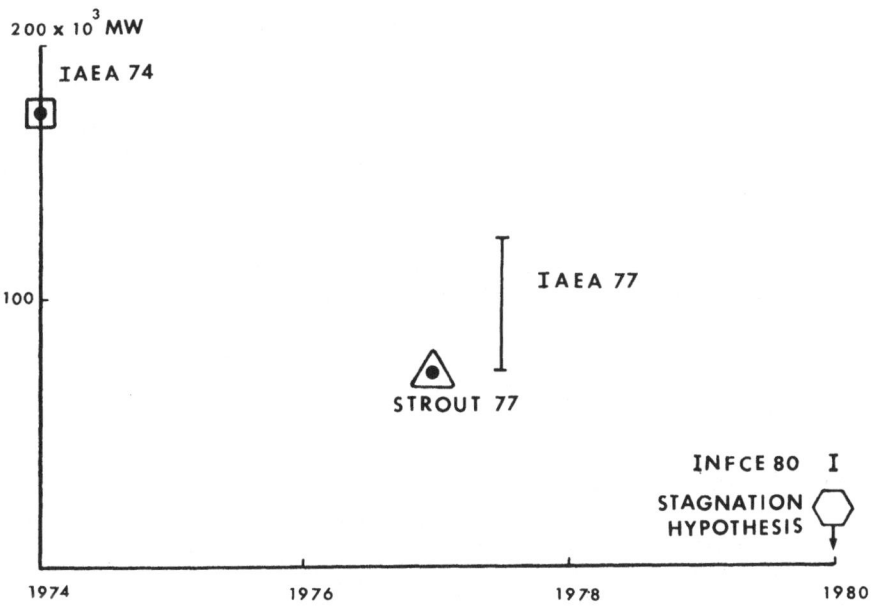

Fig. 12

5. The realization by LDC's that the introduction
   of nuclear energy is not followed by the transfer
   of technology desired by most countries. Nuclear
   power increases political and technological
   dependence rather than the opposite.

6. A small but growing awareness of the risks of
   nuclear energy as far as accidents and offenses
   to the environment are concerned. This awareness
   becomes the more important the wealthier the·
   country (or the particular region where the
   reactors are located). The recent accident at
   the Three Mile Island reactor in Pennsylvania,
   US, dramatized this problem in a particularly

clear way. Since a complete meltdown could occur
in the US with all its experience and trained
personnel, how would such situation be faced in
a LDC much less prepared for it?

7. A general distrust of choices made by technocrats
on the desirability of nuclear energy which they
argue represents the "most modern" in world
technology but which has to be imported into the
LDC's without much regard for indigenous resources
and the current state of technology. The general
feeling exists that the choices of these technocrats are made to satisfy their wishes and the
interests of large foreign corporations rather
than the genuine interests of the country.

With all these factors acting against it, nuclear
energy is unlikely to penetrate LDC's as envisioned in
the IAEA projections and to represent an important
contribution to the energy needs of this part of the
world up to the end of the century.

3. Energy Plantations

The growing of crops is an important example of the
use of the low density of the solar radiation. Although
solar radiation intensity is low and photosynthetic efficiency does not exceed 5% (being generally much lower)
this problem has been solved successfully collecting the
product of photosynthesis over large areas. The harvesting of crops (or harvesting the solar radiation) allows
one to feed the millions of people in a city with the
product of a few thousands of hectares around the city.

Table 13 shows typical productivities of agricultural
products in Brazil; as a rule of thumb 1 ha produces the
food needed per inhabitant.

The same is approximately true for the production
of liquid fuels from energy plantations.

Two main routes are being followed in this direction:
the production of ethanol and methanol which are good
substitutes for gasoline and other liquid carburants as
can be seen in Table 14.

Table 13.   Agricultural Productivities in Brazil

C R O P	Yield (tons/ha)
Rice (non irrigated)	1 - 2
Rice (irrigated)	2 - 4
Black beans	0.8 - 1.5
Soyabeans	1.5 - 2.0
Corn	1.5 - 3.0
Wheat (non irrigated)	0.8 - 2.0
Wheat (irrigated)	1.0 - 2.5
Coffee	1.5 - 2.0

Table 14.   Densities and Energy Content of Carburants

	Density (kg/liter)	Calorific content (kcal/kg)	kcal/liter
Gasoline	0.734	11,100	8,150
Ethanol	0.789	6,400	5,040
Methanol	0.796	4,700	3,740
Wood	0.400	2,524	1,010
Charcoal	0.28-0.44	6,798	1,980-3,000

Although the calorific content of ethanol and methanol are lower than gasoline they are superior fuels in octanes which compensates this to a large extent; on a volume basis a car running on ethanol or methanol uses approximately the same amount of any of these fuels.

For mixtures up to 20% ethanol or methanol in gasoline no modifications in the motors are needed but the use of pure fuels requires extensive modifications. This is being done to a significant extent in Brazil; 1/4 of all cars manufactured in the country (250,000 per year) will run on pure ethanol starting this year and all cars presently run on a mixture of 20% alcohol and 80% gasoline.[33]

The production of ethanol from sugar cane is a well established technology and this is the crop being used extensively; presently 50,000 barrels per day are being produced with plans to reach 200,000 bpd in 1985; since 1 hectare of land produces 3,500 liters of ethanol per

year[*] approximately 4 million hectares of good land will
be needed[**] (5% of the arable land in use in Brazil at
this time).

Additional expansion of the program will be made
using eucalyptus from reforestation projects which allow
the use of marginal lands.

The celulosic chains in wood can be broken by the
action of sulphuric acid (acid hydrolisis process or by
enzymatic methods) leading to sugars that can be converted
into alcohol.

The economics of these processes is such that the
barrel of ethanol costs between 30 and 40 dollars which
is marginally competitive with imported petroleum. Other
advantages are of course energy independence and saving
of foreign exchange.

What are the possibilities of relying in energy
plantations around the world?

Table 15 gives the total and arable land areas of
the world.[34]

Table 15.   Distribution of arable land

	Land area (10^6 ha)			Insolation
	Forest	Crops	Total	watts/m^2
Industrialized nations	1,900	.700	5,500	165
L D C's	2,300	.800	7,400	223
	4,200	1,500	12,900	

[*]An automobile consumes per year approximately 3000
liters of alcohol which means that 1 ha of land can
supply the energy needs of a family.

[**]The production of 1 Quad of energy (10^{15} BTU=252x10^{15}
cal) would require approximately 12 million ha, i.e.
120,000 square km of land.

To produce all the energy needed in LDC's from biomass would require 1/2 of the arable lands or 18% of the forested area. These are very large areas and it is unlikely that they will be used for energy purposes in the foreseeable future; problems to be considered are conflicts with food and traditional forest industries.

There is however a considerable energy potential in crop residues; at least 16 Quads/year are available from this source in LDC's.

In addition to that one should point out that at present only 11% of the world's forest increment in LDC's is being used[34] (2% for industry and 9% for fuelwood). In terms of coal equivalent 3,500 x 10^6 tons of wood are therefore lost every year (Table 16) which are equivalent to more than 100 Quads. This is more than the total present energy requirements of LDC's (61 Quads).

4. Solar Towers

The use of flat plate collectors for the utilization of the sun's energy does not permit the generation of electricity with any appreciable efficiency since the maximum temperature achievable with these collectors is smaller than 100°C, in general.

To improve the thermodynamic efficiency systems have been developed to concentrate the sun's energy incident in a large area in a single point where high temperatures can be achieved.

Table 16.   Incremental Energy in Forests

	Growing stock* (10^9 m^3)	Increment (10^9 m^3)	Total Consumption (10^9 m^3)	Unused** increment	
				(x10^6 TCE)	QUADS
Developed Countries	242	8.8	1.4	3,200	100
L D C's	382	9.0	1.0	3,500	105
Total	624	17.8	2.4	6,700	205

*Estimated to include all wood above ground
**1 TCE = 2.3 m^3 of wood

The most popular one is the "solar tower" concept
in which hundreds of flat plate mirrors mounted on tracking
devices reflect the light incident upon them in a small
spacial region where high temperature steams can be
produced (Figure 13).

The system works only with direct solar rays and not
with the diffuse solar radiation (which is adequate for
flat plate collectors).

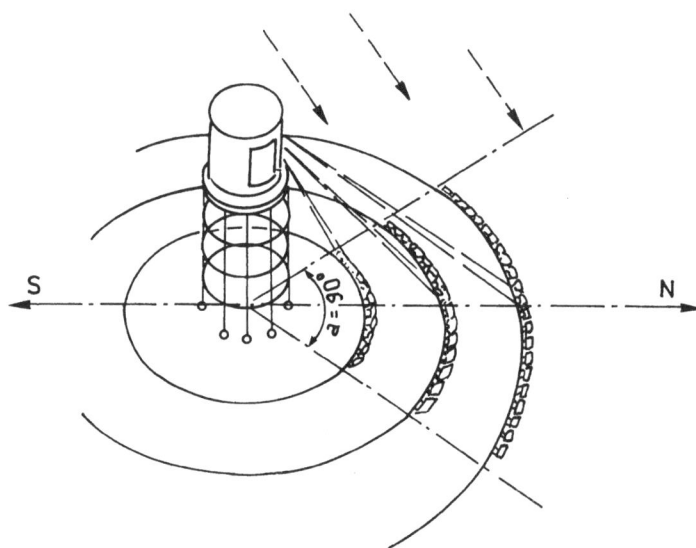

Fig. 13 - Solar Tower

The cost of electrical energy produced in these
systems lies between US$ 1,000 and 3,000 per kW. The need
for storing systems is one of the reasons for the high
cost, but they begin to look attractive in some special
locations.

IV.   ENERGY CONSERVATION IN LDC's

Are energy conservation strategies of industrialized nations relevant to LDC's?

Although the "per capita" level of energy use in LDC's is in general an order of magnitude lower than in industrialized nations, most commercially sold energy in LDC's is consumed by the urban elites, who account for less than 20% of the population.

The energy consuming habits of these elites and the economic activities that sustain them are about as wasteful as those of the industrialized nations. As a consequence, much that can be said about energy conservation for affluent societies is applicable to the uses of commercial energy in LDC's.

All energy conservation experience in the industrial sector of industrialized nations is applicable to LDC's, because of the universal nature of most industrial processes and capital equipment.

The possibilities for fuel savings in industry are relatively large. This is particularly true in the case of heat recovery in industrial processes.

Table 17 shows how primary energy is used in an industrial society[35] (US) indicating that heat represents 48.5% of the total energy consumption.

Table 17. US Primary Energy Consumption

End Use	Percentage of Total
Transportation	25
Miscellaneous Electric	19.5
Feedstock and Other Nonfuel Uses	7
	51.5
Heat	
Water Heating	4
Space Conditioning	19
Industrial Process Heat	24
Cooking and Clothes Drying	1.5
	48.5
TOTAL	100.0

The spectrum of temperatures used in the german industry[36] which is quite similar to all modern industries is shown in Figure 14.

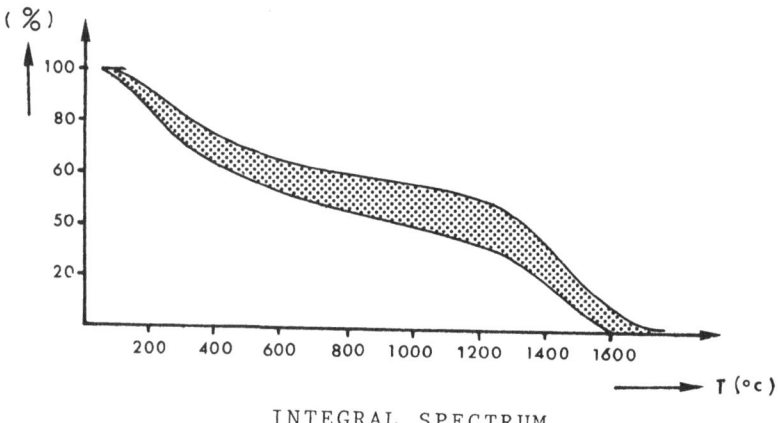

INTEGRAL SPECTRUM

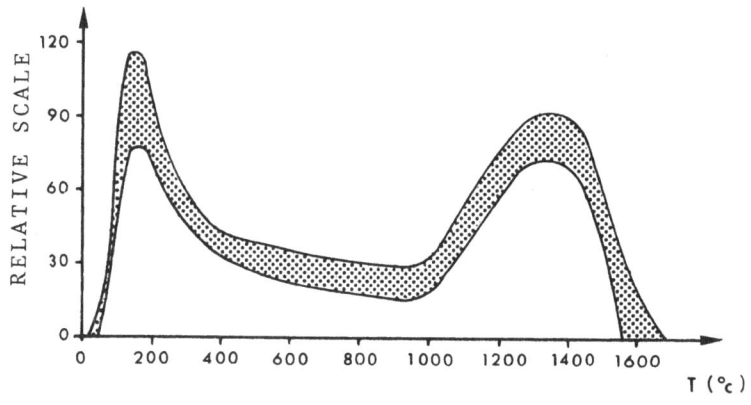

DIFFERENTIAL SPECTRUM

Fig. 14

What is represented in these figures is the temperature at which heat is theoretically required by the
process; as one can see there is a broad peak around
1300°C and another sharper peak around 150°C.

One can estimate that 37% of the energy used in the
US is in the form of hot water, low temperature industrial
process heat, or space conditioning. It is clear therefore
that solar heat captured in simple flat collectors or in
special flat collectors with selective surfaces can represent a large share of all the heat needed in developed
countries.

The problems here are not typical of developed or
LDC's but common to them: any breakthrough in the US,
for possible utilization in its southwestern region could
work even better in tropical countries where the insolation is usually higher.

In some of the LDC's it is quite common for the
government to finance (in general with low interest loans)
housing complexes for workers and middle-income people.
It might be quite reasonable to introduce into the design
of such housing complexes, solar central heating which
will add little to the initial investment and will be
paid off along the years by the resultant economy of fuel
or electricity. To encourage key industries to use solar
heated or preheated water for steam production might also
be very promising to the more widespread use of solar
energy.

In the energy balance of the United States industrial
boilers generating steam for processing and other low
temperature heat uses represents 33% of all industrial
energy consumption or 14% of total US energy requirements.
The steam is produced generally at approximately 200°C
from the combustion of oil, gas or coal which burn with
flame temperatures of ~ 2000°C. This is an exceedingly
wasteful process. It makes sense therefore to first
generate electricity in a heat engine using the high
temperature heat available in combustion and to recover
"waste heat" for low temperature process steam applications.

Several different technologies could be used for the
"cogeneration" of electricity and process steam. In
general the fuel savings ranges from 20 to 30%. It has
been estimated that net fuel savings from cogeneration in
the US could be on the order of 2-3 million barrels of

oil equivalent energy per day by the year 2000.[37]

In addition to that the following conservation
measures in use in industrialized countries could be adapted to LDC's:
- Large petroleum savings are possible through the
  imposition of stiff fuel economy standards for
  new cars, which could lead to fuel savings of 50%
  or more.
- The use of efficient appliances (e.g.,refrigerators)
  and lighting practices can lead to electricity
  savings of 50% or more in modern residential and
  commercial buildings in urban centers of LDC's[38].
- Recovery of useful energy from urban refuse is
  becoming a mature technology in industrialized
  nations. These wastes can be burned directly to
  produce electricity and/or steam for industrial
  process use, or the wastes can be converted to a
  more convenient fuel form for use in a variety of
  applications. For example, pelletized urban waste
  could be a relatively low cost, easy to handle
  fuel which can be substituted for wood or charcoal.
  While 1 kg/day of urban waste is adequate to meet
  only 2% of the average "per capita" energy needs
  in industrialized nations, it would provide 10%
  of average needs in LDC's.

There are however many areas where energy conservation strategies cannot simply be borrowed from industrialized nations because of problems that are unique to LDC's:
- While there is little need for space heating in
  LDC's there is a need for air conditioning.
  Adequate air conditioning can be provided in a
  wide range of circumstances in LDC's without using
  air conditioners by simply adopting building
  designs appropriate for the climates of LDC's.
  Building practices dating back thousands of years,
  supplemented by some modern techniques, can be
  effective in keeping buildings cool without electric
  air conditioners.
- Biomass can substitute for costly oil in providing
  electricity for irrigation and other rural applications through development of low cost, small scale
  external combustion engines based on the Stirling
  cycle.
- In the poorest of the LDC's the problem of cooking
  is of paramount importance.
- The low efficiency of cooking stoves used by some
  2 billion people in rural areas of LDC's is one

of the outstanding energy problems of the world. Introduction of more efficient stoves coupled with efforts to establish woodlots could both stem the trend toward deforestation and improve the quality of life in rural areas of LDC's by making time available for activities other than gathering wood.

More than half of the energy consumed by people in many LDC's goes into cooking. This applies to rural areas and some slum areas around large cities, as can be seeen in Table 18, which compares primary energy consumption in the US and India.

As can be seen in this table the rural population of India spends 63% of their total energy expenditures in this particular activity.

Traditionally cooking is done in primitive stoves using wood as the main fuel. This has had serious consequences in devastating forests and in some sub-Sahara African countries long trips (~ 50 kilometers) have to be taken by families to gather wood for domestic use. It is estimated that 200-300 man-day of work are spent per family in India in the process of collecting wood[39]. Sometimes children are engaged in this work, diverting them from educational activities.

Manure (and crop residues) are sometimes used for cooking, thus consuming one of the important land fertilizers available in poor areas.

The efficiency of existing primitive stoves is low. Figure 15 shows schematically a typical rural stove.

Table 18. Primary Energy Consumption in the United States and India

Activity	United States(%)	India (%)
Transportation	25	3.5
Miscellaneous electric	19.5	4.5(lighting)
Feedstock and agriculture	7	22
Water heating	4	–
Space conditioning	19	–
Industrial processes	24	7
COOKING and clothes drying	1.5	63 (cooking)
Total	100.0	100.0

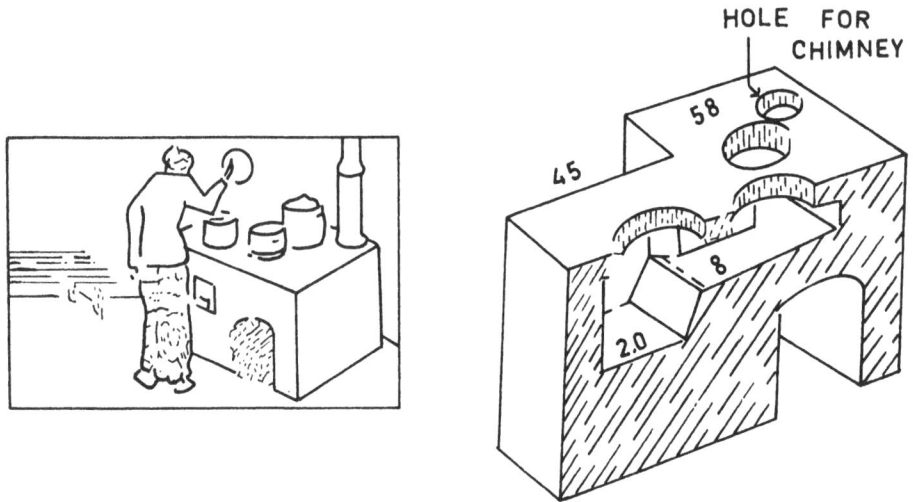

Fig. 15 - Rural Stove

Primitive wood cooking stoves generally have poor performance. One typical problem is lack of control of the air supply to meet combustion needs; another is that the air to fuel ratio cannot be maintained at a constant level everywhere in the burning mass. For example, if air enters at the bottom of the burning mass then the air/fuel ratio decreases as the air moves up and may fall in some places below the level needed to ensure combustion. In this case carbon monoxide is produced; a visual indicator of incomplete combustion is the emission of black smoke made up of fine carbon particles.

The result is that in most stoves used in rural areas one finds a smoky low temperature flame which is both inefficient and unhealthy. The fuel varies in its combustion properties from day to day and even in the course of a day because the humidity content in the fuel can change from batch to batch.

This description of the characteristics of primitive wood cooking stoves indicates clearly why people (even in rural areas) move to the use of bottled gas or kerosene as soon as they can afford it: combustion of these fuels is easy to turn on and off; it is simple to control the intensity of the flame and the fuel burns uniformly.

Typical efficiencies of wood stoves[40] are given in Table 19; in practice they do not exceed 10%.

The overall efficiency of modern gas ranges is 15%; of the total energy input in a typical unit (2.5×10^6 kcal/year) 41% is spent on the pilot lamps, 30% on miscellaneous losses and the remaining is used effectively in cooking; the surface burners are 48% efficient[41].

The overall efficiency of electric ranges is much higher (59%) so the total energy input is accordingly smaller (0.6×10^6 kcal/year at the input to the stove or 2×10^6 kcal of primary energy at the power plant); the reason is that the losses are smaller (21%) and more significantly the surface heaters are 74% efficient due to the close proximity of heater elements and the cooking parts, as can be seen in Table 20.

The need for advanced research, in a problem that looks so pedestrian as a cooking stove, is clear. This is an area where developed countries might contribute to the needed research efforts. Research is needed not only on hardware but on cultural factors as well. Otherwise people would not accept better designs of stoves as it happened when solar cookers were introduced in India.

Table 19.   Wood Fire Efficiencies

Type of fire	Efficiency
Open fire	5 - 10%
Closed fire (one cooking hole, no chimney)	10 - 20%
Closed fire (two or more cooking holes, chimney draft control)	25 - 38%

Table 20.   Comparison of Cooking Methods

	US 1976 Gas range	Electric range	Rural (Wood)
Primary energy input	6,900 kcal/day	1,810-5,700 kcal/day*	20,000 kcal/day
Overall efficiency	15%	18 - 59%	5%
Efficiency of surface burners	48%	73%	-
Reason for low efficiency	Pilot lights	-	Small solid angle
Approximate price of stove	US $294	US $344	Less than US $10**

*Depends on whether primary input is thermal or hydro-
power. Efficiency of thermal plants was taken as 30%.
If transmission losses are also included the actual
range is about 16-47%.
**US $10 was taken as the maximum cost of a rural wood
cooking stove; in general no money is involved in the
construction of such stoves and this is a very rough
estimate.

V.   ENERGY IMPLICATIONS OF SOCIAL CHANGE

It is not obvious that rural life is less energy
intensive than urban life. The fact that a villager in
India consumes 7,100 kcal/day while an average american
consumes 243,000 kcal/day merely illustrates the fact
that most villagers lead a miserable life consuming little
more energy that is needed to stay alive.

Anyone familiar with rural areas in less developed
countries (where approximately half of manking lives)
knows that life in these areas is generally speaking very
difficult, unhealthy and full of hard work and drudgery;
health and sanitation problems, malnutrition, lack of
education and outright poverty are the rule and not the
exception in most of this part of the Third World.

The correct question to ask is "what is the energy
consumption for urban and rural life for comparable levels
of comfort in the same country?"

The answer to this question exists for the United
States[42] and Norway[43] where the total energy consumed was
calculated as a function of income for rural and urban
households (Figure 16);   direct and indirect energy
expenses were taken into account.

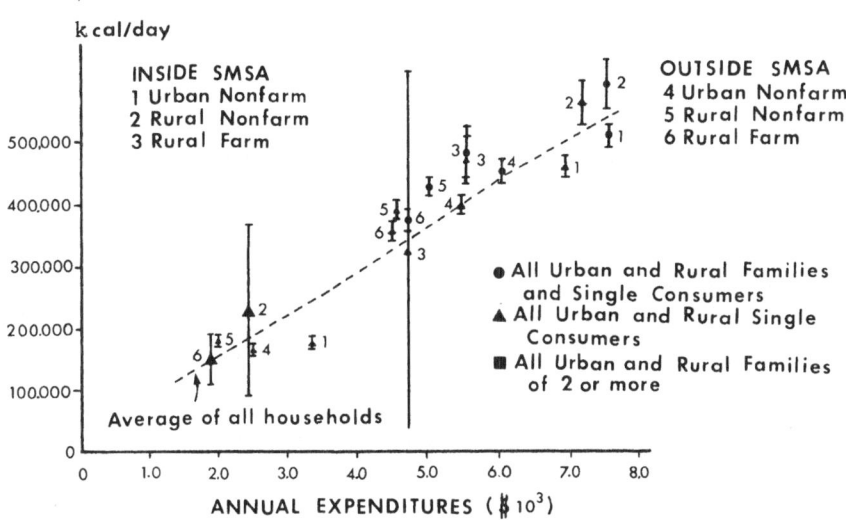

Fig. 16

In both these developed countries there is hardly any difference between the two categories with some indication that rural households are actually more energy intensive than urban. The main reason for that is the larger use of transportation in rural areas and the higher requirements for heating in the winter in isolated, badly thermally insulated farmhouses.

A similar study exists for Brazil, a large less developed country, using results of a household expenses survey conducted in 1974 by the Brazilian Statistics Bureau[44].

Results are shown in Figure 17 for the direct and total energy expenses for rural, urban non-metropolitan (small and medium size cities) and metropolitan areas, in the State of São Paulo one of the most populous states of Brazil.

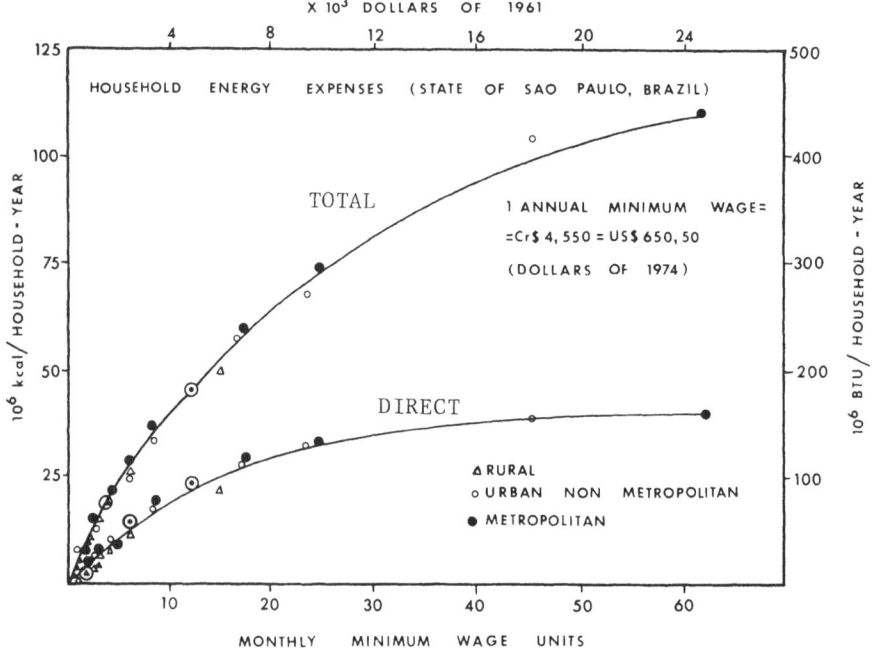

Fig. 17

As one can see, there is hardly any difference in
the total energy expense of households for the same income
(monetary and non-monetary incomes were added since the
latter are very significant in rural areas).

This is in our view a very important and rather
surprising result. Although presently there are few
rural households with large incomes it seems inescapable
that as their income increases they will fall in the
general pattern of consumption in urban areas.

It could be argued that the quality of life in rural
or urban non-metropolitan areas is better than life in
metropolitan areas for the same income. This is probably
true but involves a subjective scale of values. In any
case it seems very unlikely that somewhat comparable
levels of comfort could be obtained in rural and non-
metropolitan areas with less than half the energy needed
in metropolitan areas.

As a consequence if one really wants to improve the living conditions of the poor in most LDC's one should be prepared to expect very high requirements of energy.

The income distribution in LDC's as one well knows is rather unfair with most of the income concentrated in the hands of 10-20% of the population. This can be seen clearly in Figure 18 which shows the income distribution of the population of São Paulo whose energy expenses were analyzed and plotted in Figure 17.

It is obvious that presently, rural and urban non-metropolitan areas have more poor people than metropolitan areas and consequently they consume less energy.

If one multiplies the distribution of population as a function of income as given in Figure 18 by the energy consumption as a function of income one obtains the total energy consumption as a function of income. This is given in Figure 19. The area under this curve is the total energy consumed by the population (approximately 116×10^{15} cal).

It is interesting to notice that if a very severe redistribution of income were to take place, as indicated by the dotted line in Figure 18, (such uniform distribution was not even achieved in the US[*]) the energy consumption of the population would be distributed by the dotted line of Figure 19 which corresponds to a total energy consumption of 135×10^{15} cal. What this indicates is that the total energy requirements of society would not be very different from present ones. Redistribution of income would not lead to an inordinate increase in energy consumption.

This is another way of saying that there is enough energy being consumed nowadays in the world (and inside most countries) but that it is very unevenly distributed as can be seen in Table 21.

---

[*]Distribution of income in the US (dollars of 1972)

	Average income	
Poor	- US$ 2.500	(18%)
between poverty level and US $11.199 - Lower Middle	- US$ 8.000	(42%)
between US $12.000 and 15.999 - Upper Middle	- US$ 14.000	(19%)
above US $16.000 - Well off	- US$ 24.500	(20%)
Medium income	- US$ 11.116	

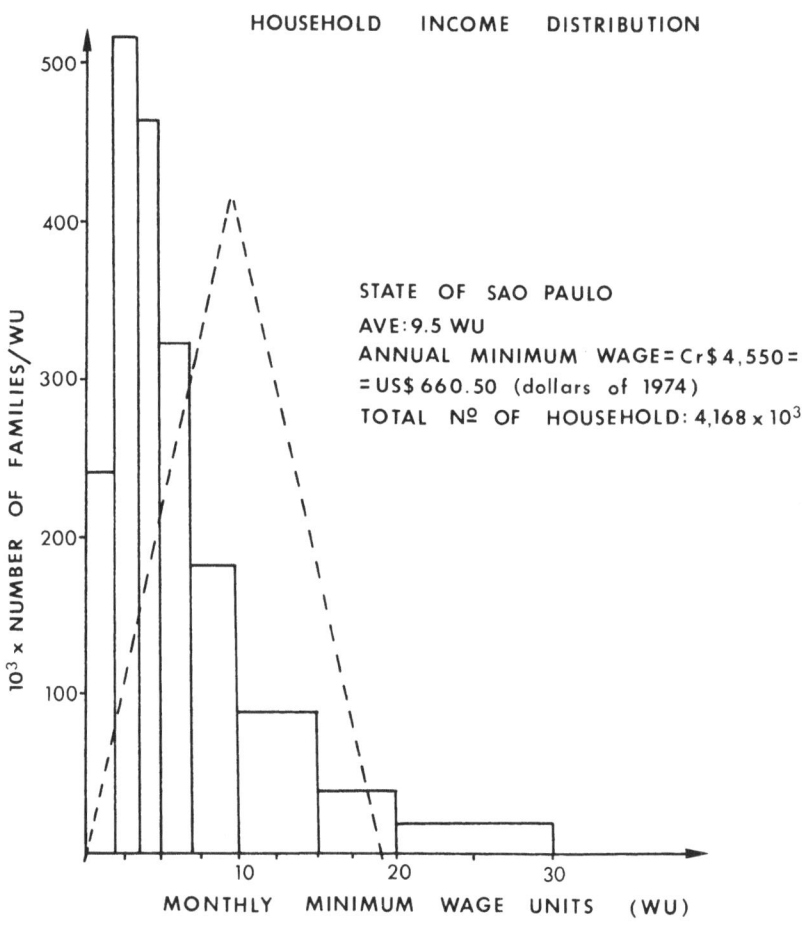

HOUSEHOLD   INCOME   DISTRIBUTION

STATE OF SAO PAULO
AVE: 9.5 WU
ANNUAL MINIMUM WAGE = Cr$ 4,550 =
= US$ 660.50 (dollars of 1974)
TOTAL Nº OF HOUSEHOLD: 4,168 x 10³

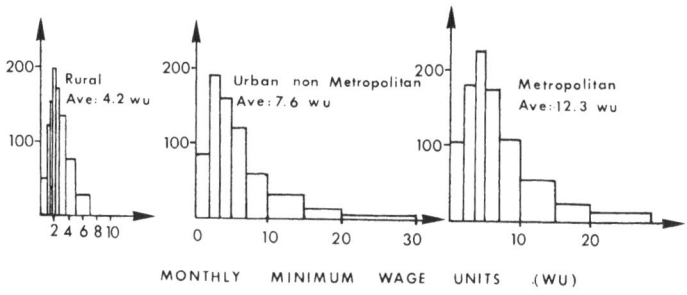

Rural
Ave: 4.2 wu

Urban non Metropolitan
Ave: 7.6 wu

Metropolitan
Ave: 12.3 wu

MONTHLY   MINIMUM   WAGE   UNITS   (WU)

Fig.18

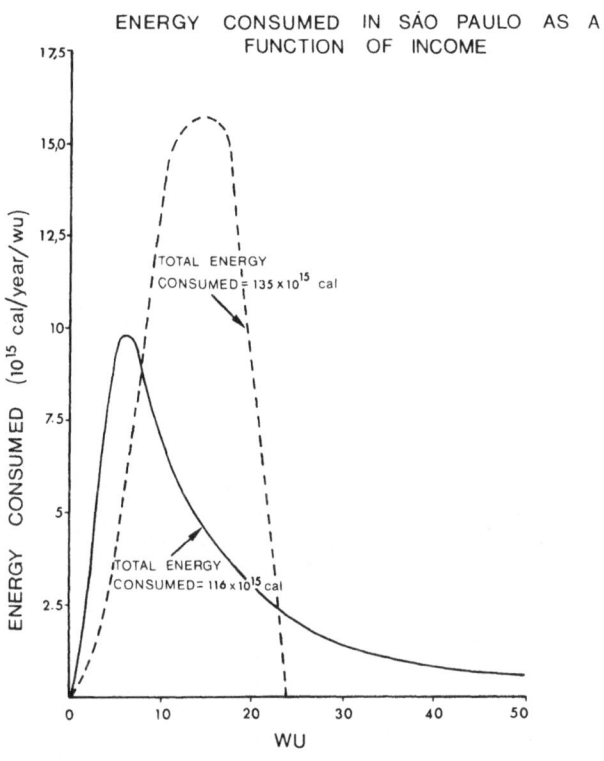

ENERGY CONSUMED IN SÃO PAULO AS A FUNCTION OF INCOME

Fig. 19

Table 21.   Energy Consumption in Selected Countries

	Population (10^9)	Energy consumption "per capita" (kcal/day)
Bangladesh	0.08	2,300
China	0.88	7,100
India	0.6	11,000
Brazil	0.11	22,800
United States	0.214	243,000
Developed countries	1.05	110,000
LDC's	2.95	18,000
World average	4.0	42,500

The average world consumption is 42,500 kcal/day
which is much above the average energy consumption in
LDC's(except in cold climates)which is considered a very
satisfactory level of energy consumption.

Most of the developments in LDC's is ahead of them
and not behind as in the industrialized countries. While
industrial nations are growing so slowly that many energy
consuming activities cannot be fundamentally altered in
a short time, LDC's are changing very rapidly. The face
of most LDC's in 2000 will be very different from today's,
because of high population growth rates, high migration
rates, and potentially high economic growth rates.   The
scale of change taking place in LDC's is suggested by
the fact that by the year 2000 there will be a " second
India"[45] in India with almost another 600 million people.

Such dynamic change means not only that the impact
of introducing more energy efficient technology can be
greater and faster but also that the LDC's have a much greater
opportunity to steer a course of development that is much
less wasteful of energy resources than the course that
has been followed by industrialized nations.   LDC's can
learn how to avoid the mistakes which have left these
nations not only with energy inefficient capital infrastructures, but also with the adverse side effects of
energy wasting technologies, such as polluted air and
congested urban environments.   Among the possibilities
for alternative development are a revitalization of the
railroads as an alternative to roads for freight transport
and urban redesign to minimize the need for transportation.
(Table 22)

Table 22.   Transportation Modes in Brazil*

	kilojoule/ton-km
Truck	2,340
Rail	420
Water	230
Air	83,200
Average	1,840

*The same average for the US (1972) is 890
showing how inefficient freight transportation is in Brazil.

   Public policy initiatives to influence the course
of development along such lines could lead to a saturation of consumer demands at levels far below those in
industrialized nations.   Through better urban design,
for example, the "need" for the automobile could "saturate"
in LDC's at a relatively low level, while living standards
are improved by bringing living, working, marketing, and
community service places closer together and by reducing
the environmental degradation of cities.

   The energy crisis should be a cause for hope rather
than despair for fossil fuel poor LDC's.   The new awareness of energy problems which resulted from the sharp
increase in the world price of oil has focussed attention
on the urgency of shifting away from petroleum based
development strategies and has generated a creative ferment
in the exploration of alternatives.   Just as the " wood
crisis" in 16th century Britain forced a shift to coal
and set into motion a chain of events that culminated
two centuries later in the Industrial Revolution[46], the
present "petroleum crisis" could provide the impetus
needed to shift LDC's to a course of sustainable development.

References

1. T.B.Taylor, "World Energy Alternatives", Princeton
   University (unpublished).
2. "Balanço Energético Nacional", Ministério de Minas e
   Energia, Brasilia (1977).
3. P.D.Henderson, "India: The Energy Sector", Oxford
   University Press (1975).
4. V.Smil, "China's Energy Achievements, Problems,
   Prospects", Praeger Publishers, New York (1976).
5. "Bangladesh Energy Study", Asian Development Bank,
   November(1976).

6. WAES - "Workshop on Alternative Energy Strategies", First, Second and Third Technical Report, the MIT Press, Cambridge (1976).

7. A.K.N.Reddy, "Energy Options for the Third World" - Bulletin of the Atomic Scientists.

8. A.B.Strout, "Energy and Economic Growth in Central America", Ann.Rev.En., Vol. 2, Annual Reviews Inc. (1977).

9. United Nations World Population Conference, Vol. I, "Recent Population Trends and Future Projects" - ST/ESA/Ser A/51 Sales No E/F/S/75 XIII 4, UN, New York.

10. R.Revelle, "Energy Use in Rural India", Sci. 192:973 (1976).

11. R.Tyers, "Energy in Rural Bangladesh", Harvard Center for Population Studies (1976).

12. A.Makhijani and A.Poole, "Energy and Agriculture in The Third World", Ballinger Publishing Co. (1975).

13. R.Loher, "Methane from Human, Animal and Agricultural Wastes", in: "Renewable Energy Resources and Rural Applications in the Developing World", N.L.Brown, ed., AAAS Selected Symposium, Vestview Press, Colorado (1978)

14. P.Meynell, "Methane - Planning a Digester", Schockers Books, New York (1978).

15. A.van Buren, "A Chinese Biogas Manual", Intermediate Technology Publications, London (1979).

16. V.Smil, "China's Energy Achievements, Problems, Projects", Praeger Publishers, New York (1976).

17. V.Smil, "Intermediate Energy Technology in China", Bull. of the At.Scientists, February (1977).

18. P.D.Henderson, "India: the Energy Sector", Oxford University Press, London (1975).

19. U.S. Army Corp of Engineers - "Estimate of National Hydroelectric Power Potential at Existing Dams", July (1977).

20. C. de Lucca, E.Koelle, J.L.A.Junqueira, "Turbinas para Pequenas Quedas d'Água", in: "Energia no Brasil", J.Goldemberg, ed., Academia de Ciências do Estado de São Paulo, Brazil (1976).

21. United Nations Statistical Yearbook (1974).

22. W.Lapedes, McGraw Hill Encyclopedia of Energy (1975).

23. J.Goldemberg, "Energia no Brasil", Livros Técnicos e Científicos, Brazil (1979).

24. M.B.Prince, "Photovoltaic Technology in Renewable Energy Resources and Rural Applications in the Developing World", N.L.Brown, ed., AAAS Selected Symposium, Vestview Press, Colorado (1978).

25. P.C.Putnam, "Power from the Wind", Van Nostrand, New York (1948).

26. S.Penner and L.Icerman, "Non-nuclear Technologies",
    Energy, vol. 2 (1975).
27. E.L.Armstrong, in: "World Energy Resources 1985-2000",
    IPC Science and Technology Press (1978).
28. I.J.Bloodworth et al. in: "World Energy Resources
    1985-2000", IPC Science and Technology Press (1978).
29. E.Rumpf and G.S.H.Jarrett, "A Survey of the Performance
    of HVDC Systems Throughout the World During 1973-74",
    International Conference on Large High Voltage Electric
    Systems CIGRE, Paris (1976).
30. J.Goldemberg, "Global Options for Short Range
    Alternative Energy Strategies", Energy, vol. 4, 733
    (1979).
31. J.A.Lane, A.J.Covarrubias, B.J.Csik, A.Fattah and
    G.Woite, "Nuclear Power in Developing Countries",
    IAEA - CN 36/500 (1977).
32. International Atomic Energy Agency, "Market Survey
    for Nuclear Power in Developing Countries", 1974
    Edition, Viena (1974).
33. J.Goldemberg, "Brasil: Energy Options and Current
    Outlook", Sci. 200:158 (1978).
34. D.E.Earl, "Forest Energy and Economic Development",
    Clarendon Press, Oxford, England (1978).
35. M.H.Ross and R.H.Williams, "The Potential for Fuel
    Conservation", Technology Review, Feb. (1971).
36. Federal Ministry of Research and Technology, "Einsatz
    moglichkerten neuer Energie Systeme", vol. 1, p. 4
    West Germany (1975).
37. R.H.Williams, "Industrial Cogeneration", Ann.Rev.En.,
    vol. 3 (1978).
38. E.Hirst and J.Corney, "Residential Energy Use to the
    Year 2000: Conservation and Economics", Oak Ridge
    National Laboratory Report CON-13 sept. (1977).
39. "Bangladesh Energy Study", Asian Development Bank,
    Nov. (1976).
40. B.D.Ahuja and O.N.Gupta, "A Study of the Efficiency
    of Chulahs", Technical and Research Reports 17 UDC
    643.332, National Buildings Organization, Government
    of India, New Delhi.
41. R.H.Hoskins, "Energy Use and Cost Analysis for Gas
    and Electric Gas Ranges", Oak Ridge National
    Laboratory (1977) - unpublished.
42. R.Herendeen and J.Tanaka, "Energy Cost of Living",
    Energy, vol. 1, 165 (1976).
43. R.Herendeen, "Total Energy Cost of Household
    Consumption in Norway, 1973", Energy, vol. 3, 615
    (1978).

44. G.M. Gil Graça, V.R.Vanin and J.Goldemberg, "Energy
    Consumption Patterns in Brazil: Metropolitan, Urban
    and Rural Areas" (1980) - unpublished.
45. K.S.Kirith, "Second India Studies - ENERGY", The Mc
    Millan Company of India Ltd., New Delhi (1976).
46. J.U.Nef, "An Early Energy Crisis and its Consequences",
    Sci.Am., p. 140, Nov. (1977).